Mustapha Zergoun
1962 East 6th Avenue
Vancouver, B.C. V5N 1P7

D1736911

*Practice Management
for Design Professionals*

Practice Management for Design Professionals
A Practical Guide to Avoiding Liability and Enhancing Profitability

John Philip Bachner

A WILEY-INTERSCIENCE PUBLICATION
John Wiley & Sons, Inc.

NEW YORK / CHICHESTER / BRISBANE / TORONTO / SINGAPORE

In recognition of the importance of preserving what has
been written, it is a policy of John Wiley & Sons, Inc. to
have books of enduring value published in the United
States printed on acid-free paper, and we exert our best
efforts to that end.

Copyright © 1991 by John Wiley & Sons, Inc.

All rights reserved. Published simultaneously in Canada.

Reproduction or translation of any part of this work
beyond that permitted by Section 107 or 108 of the
1976 United States Copyright Act without the permission
of the copyright owner is unlawful. Requests for
permission or further information should be addressed to
the Permissions Department, John Wiley & Sons, Inc.

Library of Congress Cataloging in Publication Data:
Bachner, John Philip.
    Practice management for design professionals : a practical guide
to avoiding liability and enhancing profitability / by John Philip
Bachner.
      p.   cm.
  "A Wiley-Interscience publication."
  Includes index.
  ISBN 0-471-52205-8
    1. Architectural practice—United States—Management.
2. Engineering—United States—Management. 3. Insurance,
Architects' liability—United States. 4. Insurance, Engineers'
liability—United States. 5. Architects—Malpractice—United
States. 6. Engineers—Malpractice—United States. I. Title.
NA1996.B28   1991   90-47916
720'.68—dc20   CIP

Printed in the United States of America

10 9 8 7 6 5 4 3 2 1

For Brenda and Peter

Foreword

Consulting engineers practicing in the geosciences have shown how professionalism is essential to deriving personal satisfaction from one's career and maintaining the status of one's profession.

Geotechnical engineers' "test" began in the mid-1960s. By all accounts, their liability problems at the time were more severe than those of any other professionals, including physicians. The claims being lodged against them were so frequent and large that the cost of professional liability insurance became practically unaffordable. By 1968, conventional coverage became unavailable.

Responding to this situation, principals of ten prominent firms met with a professional liability loss prevention consultant in Chicago. There they agreed to form a group called Associated Soil and Foundation Engineers (ASFE). The organization's principal purpose would be to identify the causes of members' liability problems and to develop means to improve the situation.

Formally established in 1969, ASFE took to its task quickly, mindful that it probably would not have a second chance to achieve its purpose. Based on case histories of actual losses, research soon revealed the problem: Geotechnical engineers had become victims of their own professional inadequacies. But these were not technical inadequacies; rather, they were practice deficiencies—breakdowns in areas such as client relations, human resources management, financial planning, and marketing. Changes in societal and client expectations had created new needs, but geotechnical engineers had not altered their procedures in order to accommodate them. In that respect, geotechnical engineers were not appreciably different from other independent design professionals, except they confronted far more uncertainty in their work, which exacerbated risk and made new approaches mandatory.

Over the years, ASFE has developed an array of programs and materials to help its members overcome practice deficiencies. The efficacy of these efforts has been demonstrated by the progress made. By 1980, members of ASFE—then known as the Association of Soil and Foundation Engineers—were reported to have become the *least* liability-prone members of the design professions. In 1985, at a time when the cost of virtually all insurance had skyrocketed, a survey revealed that the cost of consulting geotechnical engineers' insurance actually declined. In 1987, a survey indicated that geotechnical engineers' professional liability insurance costs were the lowest of all design professionals'.

ASFE characterizes much of what it does as loss prevention. "Enhancement of professionalism" is also accurate. The group's programs and materials address almost every aspect of design professional practice, from procedures for checking the accuracy of technical output, to those created to improve client and project selection, workscope development, personnel training, and dispute resolution, among many others. Fifty years ago, design professionals did not consciously bother themselves about such issues. Today they are significant concerns and, as geotechnical engineers have demonstrated, they must be addressed in a professional manner if a professional practice is to thrive.

One of the most important of all ASFE programs was the original Institute of Professional Practice (IPP). Developed in the early 1970s, its goal was acquainting geotechnical engineers with the full range of professional practice concerns, so they could render their services more effectively—more "in tune" with the times. The program proved to be extremely effective, but ASFE was a much smaller group at the time, and the program eventually created a financial drain it could no longer afford. As a consequence, ASFE sold the Institute to Risk Analysis & Research Corporation, then a division of Design Professionals Insurance Company, forerunner of today's DPIC Companies.

DPIC continued to offer the IPP through the early 1980s. By then, several of its elements had become out of date, however, and the program was withdrawn. At almost the same time, professional liability insurance rates began a precipitous climb. ASFE urged DPIC to revise the IPP and offer it once again. DPIC responded by returning the program to ASFE as a gift, with the understanding that ASFE would update the curriculum and expand its applicability to all design professionals in private practice, not just those involved in the geosciences.

The new program is similar to the original in many respects, but it differs in many ways, too. The Institute of Professional Practice title was changed to Institute *for* Professional Practice, and instead of identifying a specific program, it is an educational foundation whose principal concern is the advancement of architects' and engineers' knowledge of professional practice issues. The initial course developed by the new Institute is Introduction to Professional Practice. This book was developed as its text. Program enrollees are required to read this text, respond to several open-book examinations, perform a research assignment, and participate in an intensive two-and-one-half-day seminar. The program is open to all design professionals and is endorsed by a number of organizations, in addition to ASFE.

ASFE's Loss Prevention Education Committee, chaired by IPP1 Alumnus Kenneth E. Darnell, deserves a great deal of credit for the new program. Along with other ASFE members, leaders of allied organizations, and other interested individuals, they donated innumerable hours to review and comment. While this text stands as a symbol of their accomplishment, their dedication can be honored appropriately only through a continuation of what they began, to keep the Introduction to Professional Practice course continuously attuned to the changing needs and expectations of design professionals and those they serve.

MICHAEL E. MAHONEY

President, 1990-1991
ASFE/The Association of Engineering Firms
Practicing in the Geosciences

Preface

Engineers and architects design America: the buildings in which we live, work, and play; our transportation systems; our public works; the entire built environment. All stem from the talents of design professionals. Thus, it is regrettable that many who obtain architectural and engineering services seemingly take for granted the professionals who provide them. And it is even more unfortunate that so many who provide architectural and engineering services seem to take their own professions for granted.

If the benefit America derives from its design professionals is to continue, it must treat its professionals with respect. To merit this regard, architects, and engineers must treat their professions with respect. They can do this, in part, by upholding fundamental concepts that were articulated by founders of the professions some 450 years ago. Today's circumstances are far different, of course; the world we live in is much more complex. But these complexities do not stand in the way of professionalism. They merely create challenges that long-held professional tenets are ideally suited to pursue.

One of the most significant differences between today's design professionals and those of the past may be their attitude toward money. At one time, financial gain was not that significant an issue. Architects and engineers who epitomized professionalism enjoyed financial security as the natural by-product of their service. Today, critics allege that design professionals are more enamored of wealth than professionalism, and that the pursuit of riches has had a corrosive effect on the quality of their performance and their professional beliefs.

Are the critics correct? In some instances, yes. Undoubtedly, there are those design professionals who, if they did not initially see architecture or engineering as a path to riches, now pursue their practices in that purblind manner. But no one is in a position to prove that those who hold such attitudes are in the majority or that they have ever comprised anything more than a scant minority. Besides, as many of today's design professionals will aver, wealth can be gained far more easily and far more quickly in a variety of other pursuits.

Most persons who enter the design professions today, as centuries ago, are attracted by the personal challenge that the work presents, the opportunity for creative expression and innovative problem solving, and the ability to make important contributions to society. Nonetheless, financial concerns have become important, and in many instances they lead design professionals down dangerous paths, but not so much to pursue riches as to secure economic survival. In essence, it is design professionals' lack of business acumen, not greed, that creates so many of their problems.

ASFE/The Association of Engineering Firms Practicing in the Geosciences established the Institute for Professional Practice in 1987. A principal purpose of the Institute is advancing the management capabilities of design professionals who are completing their formal educations and those who are, or soon will be, receiving managerial responsibilities. The Institute was created to fill a void. In school, architects, and engineers are imbued with ever-more technical knowledge in order to deal with the growing technical complexities that affect their professions. To achieve this end, it has been necessary to sacrifice instruction in what it means to be a professional and in the modern business techniques needed to create an environment that can nurture professionalism. Is it any wonder, then, that so many design professionals seem to stress dollar issues? Many do not have the business knowledge necessary to make financial success a by-product of professional performance. Others, nurtured in an era of high-tech commercialism, have not yet learned what it means to be a professional in every sense of the word.

This text has been developed to provide introductions to some of the most important issues confronting contemporary design professionals. Although much of its content seemingly comprises "how-to" information, the astute reader will recognize that instructional content is heavily influenced by professional philosophy, in part to indicate how traditional tenets can be applied to satisfy contemporary needs.

Throughout this text, there is reliance on the adjective *professional*, referring to both attitudes and skills. Professional attitude comprises the pattern of behavior that society expects from those to whom it has granted professional authority, including the right of self-regulation. Professionals have been granted this authority because their attitude is expected to transcend others'. They are regarded as persons who should put their clients' needs above their own and who can be trusted to perform well in an arena where they generally are immune to their clients' oversight. This immunity is created because professionals' clients—lay men and women—lack the professional skills of those they engage: the high degree of education, training, and experience society expects its professionals to possess before offering their services to the public.

Professional attitudes and professional skills both are necessary to be professional. One cannot exist without the other. And this is precisely who so many contemporary engineers and architects face growing professional problems. They do not possess the full range of skills needed to manage their practices in a professional manner, making it virtually impossible for them to achieve their professional potential or to fulfill society's expectations. The desire to perform well is not enough. In addition to being design professionals, engineers and architects must also be professional business managers. A fine report or highly detailed plans and specifications—even the most professional service—is not well received when it is completed late or its cost exceeds the budget. Careful scheduling is required to perform in a timely manner. Effective human resources management is needed to maintain the efforts of those assigned to implement the schedule. Carefully orchestrated contract formation processes are necessary to communicate the mutual concerns of those associated with a project. Astute business planning is required to secure the types of projects that will help professionals realize their personal goals and ambitions. And the fees derived from these projects must be adequate to support the full range of activities that it is incumbent upon professional organizations to pursue, just as prompt collection of those fees must be made a priority if the profits necessary to maintain a professional organization are to be realized.

Few design professionals possess all of the skills necessary to operate their businesses professionally, making it extremely difficult to provide professional services in a professional manner. This deficiency makes far too many architects and engineers susceptible to the call of those who, for some reason, encourage engineers and architects to minimize their fees. Clients who pursue this tack fail to recognize what design professionals do; that now, as in the past, high quality is always the least expensive in the long term; and that the desire to produce high-quality work and pressure to keep fees low are discordant objectives. When budgets do not permit excellence, excellence is rarely forthcoming. Thoughtful, time-consuming analysis must be supplanted by rules of thumb. What has worked before becomes a substitute for innovation. On the individual project level, costs of construction, operation, and maintenance can become needlessly high. On the societal level, we tether our future to the past. It is wrong for clients to behave in this way, but clients, by and large, are not professional. They cannot be expected to put society's needs above their own. Engineers and architects are expected to preserve society's interests, but all too few possess the professional management skills necessary to do so. And, unless more develop these skills, the professions of architecture and engineering will be at risk, in danger of becoming little more than technical vocations applied to meet clients' wishes rather than their, and society's, needs.

Independent surveys of design professionals' profitability consistently demonstrate that today, as centuries ago, those firms that are most financially successful are also those that place maximum emphasis on client satisfaction. They do not strive for client satisfaction in order to gain wealth. They do so because the achievement of client satisfaction is essential for the attainment of professional satisfaction. And professional satisfaction, when all is said and done, is the real goal of almost all design professionals. But it can be achieved only through the application of professional management skills—skills that historically have made financial security a by-product of the desire to serve well.

This text should under no circumstances be regarded as a "be-all" and "end-all." It provides a brief survey of the various management skills engineers and architects need to achieve their professional ambitions. More information is needed about each subject if professional skills are to be implemented in a professional manner. Read other books. Participate in continuing education. Remember: It is not unprofessional to lack management skills. It is unprofessional to recognize a shortcoming but then do nothing to correct it. How you respond is up to you. So, too, is the future of your profession.

JOHN PHILIP BACHNER

Silver Spring, Maryland
January 1991

Acknowledgments

Development of this text would have been impossible without the oversight and guidance provided by many individuals, all of whom have professional responsibilities that made their free time scarce, and all of whom have the professional attitude that impelled them to give of their time so freely. Special thanks are extended to the following people: Alfred C. Ackenheil, Harl P. Aldrich, Jr., John A. Baker, John H. Brewer, Gary Brierley, J. Richard Cheeks, Elio D'Appolonia, the Hon. James B. Deutsch, Robert L. Donovan, James F. Duffield, Charles DuBose, Douglas G. Gifford, John P. Gnaedinger, Peter B. Hawes, Will M. Heiser, George E. Hervert, Edward B. Howell, Walter Lum, Glenn Mann, James M. McWee, Sandra L. Nelson, Steven E. Pazar, Barbara A. Phillips, John Ramage, Dick Reynolds, Gardner M. Reynolds, Edward E. Rinne, Lawrence H. Roth, the late Robert D. Sayre, William L. Shannon, Lester H. Smith, John G. Thomas, Bent L. Thomsen, Eugene B. Waggoner, Joseph S. Ward, G. Fred Wickman, Arnold L. Windman, and William S. Zoino.

Thanks are extended as well to those who developed the original ASFE Institute of Professional Practice text in 1975, from which many concepts have been taken in the formation of this work. These individuals include Edward B. Howell, Richard P. Howell, Charles H. T. Springer, and Arnold Olitt. Recognition also is due William H. McTigue and Donald E. Clark, who, as chairpersons of the original ASFE Institute of Professional Practice Committee, shepherded the Institute program from concept to reality. Special thanks are extended to the DPIC Companies, which, as owner of the Institute of Professional Practice name and the original materials, donated these to ASFE for the latter's use in advancing professional concepts.

The original chairman of ASFE's Loss Prevention Education Committee,

Kenneth E. Darnell, merits particular praise, serving for five years as the driving force behind development of the new ASFE Introduction to Professional Practice program. Those serving with him, John A. Hribar and Douglas G. Gifford, in particular, merit recognition for their many contributions, too.

Terra Insurance Company (A Risk Retention Group) also merits praise for its many years of substantial financial support, beginning with development of the original Institute program and including the most current version, as well as innumerable other programs and materials that have been incorporated into this text.

Contents

1. **Professions, Professionals, Professionalism**   1

    The Rise and Fall of the Guild System / 1
        The Emergence of the Guilds / 2
        The Decline of the Guilds / 2

    The Rise of the Professions / 3

    The Professions Then and Now / 3
        Systematic Body of Theory / 4
        Professional Authority / 5
        Sanction of the Community / 6
        Regulative Code of Ethics / 7
        The Professional Culture / 9

    The Professions in Transition / 10
        Loss of Professional Distinction / 11
        A Changed National Mood / 12
        Professional Self-Protection / 14
        Fewer Personal Professional-Client Relationships / 14
        More Errors Being Made / 15
        Design Professional Retention Practices and Fees / 16
        Failure to Discuss the Downside / 17
        Lack of Written Performance Standards / 18
        New Interpretations of Civil Law / 19
        Profit Motive / 19
        Insurance / 20

Media Exposure / 20
Lack of Effective Business Practices / 21

Toward a New Professionalism / 21

2. Professional Engagement 25

QBS / 25
    A Typical QBS Application / 26
    Other QBS Applications / 27

Fee-Bidding / 28
    Elements of Fee-Bidding Procedures / 28

The Benefits of QBS / 32
    Fee Considerations / 33
    QBS Promotes Economy / 34
    QBS Gives Leverage to the Client / 34
    QBS Reduces Overhead and Encourages More Competition / 35
    QBS Is Well Suited for the Inexperienced / 35

The Need for Design Professional Activism / 36
    Activism Through Nonprofessional Response / 36
    Informing Others About QBS / 37

3. Executing A Professional Commission 40

The Design Professional's Role: An Overview / 40

Proposal-Phase Services / 42
    Identifying Projects for Pursuit / 42
    Assembling the Design Team / 43
    Proposal Formation / 46
    Formation of Design Concepts / 47
    Interviews / 49
    Workscope Development and Agreement / 50

Design-, Construction-, and Postconstruction-Phase Services / 54
    Predesign Planning and Conference / 54
    Development of Plans and Specifications / 56
    Mutual Review / 56
    Contractor Selection / 57
    Prebid Conference / 58
    Preconstruction Conference / 59
    Shop Drawing Review / 60
    Field Observation / 61
    Postconstruction-Phase Activities / 62

Client Education / 63

4. Professionals and the Law   64

Our Legal System in General / 64
    Statutory Law / 64
    Civil Law / 65

Negligence / 68
    Proving the Claim / 69
    Defending the Claim / 75

Strict Liability / 77
    Buildings as Products / 77
    Plans and Specifications as Products / 77
    Components as Products / 78
    Other Aspects of Strict Liability / 78

Warranty / 78
    Express Warranty / 79
    Implied Warranty / 80

Deceit / 81
    Proving the Claim / 81
    Defending the Claim / 83

Defamation / 84
    Proving the Claim / 85
    Defending the Claim / 87

UnFair Competition / 88
    Commercial Disparagement / 88
    Interference with Contractual Relations / 90
    Interference with Prospective Economic Advantage / 91

5. Contracts for Professional Services   92

Contracts: An Overview / 92
    Categorizing Contracts / 93
    Binding a Contract / 94

Typical Contract Formats / 95
    Conventional Proposals / 95
    Negotiated Terms and Conditions / 96
    Special Contracts for Major Projects / 96
    Model Contracts / 97
    Multiple Contracts / 97
    Client-Developed Contracts / 98

Other Types of Written Agreements / 100
Oral Agreements / 100

The Benefits of Forming and Having a Written Agreement / 101
Mutual Understanding / 101
Establishing Your Own Rules / 101
Sizing Up / 102
Identifying and Allocating Risk / 102

Dealing with Risk / 102

Issues to Consider in Contract Formation / 104
Assumption of Liability / 104
Professional Liability Insurance / 105
Disparate Bargaining Power / 105
Indemnifications / 106
Definitions / 111
Word Selection / 111
Exculpatory Wording / 111
The Role of Attorneys / 112

Negotiating a Contract / 112

Typical Clauses / 113
Certification / 115
Consequential Damages / 118
Construction Cost Estimates / 119
Construction Monitoring / 121
Curing a Breach / 125
Discovery of Unanticipated Hazardous Materials / 126
Excluded Services / 127
Freedom to Report / 128
Indemnification / 130
Jobsite Safety / 132
Limitation of Liability / 135
Maintenance of Service / 140
Ownership of Instruments of Service / 141
Record Documents / 144
Right to Reject and/or Stop Work / 146

6. Professional Liability Insurance   148

The Role of Professional Liability Insurance / 148
A Source of Full Recovery / 149
A Substitute for Professionalism / 149
A Marketing Prerequisite / 149

A Cause of Barratry / 150
A Project Fundamental / 150
A Lever for Reasonableness / 151

The Professional Liability Insurance Industry / 152
The Principals / 152
The Cyclical Nature of Capacity / 155

Aspects of Coverage / 157
Establishing Overall Protection / 157
Claims-Made and Related Provisions / 158
Endorsements and Riders / 160
Project Insurance / 160

Self-Insurance / 161

Insurance Company Formation / 162

Establishing Premiums / 163
Policy Limits and Deductible / 163
Gross Fees / 165
Disciplines Practiced / 166
Geographic Area / 166
Project Types / 166
Subcontracting / 167
Claims Experience / 167
Management / 167
Premium Credits / 167
Overall Impact / 167

When a Claim is Filed / 168

7. Professional Risk Management   170

Dealing With Risk in General / 171
Risk Transfer / 171
Risk Retention / 172

Select Clients and Projects with Care / 173

Insist on an Adequate Fee / 175

Provide Quality Control / 176

Apply Realistic Assignment and Scheduling Procedures / 176

React Quickly to Symptoms of Problems / 177
Everyone Has a Role / 177
Documentation / 178
Feedback Network / 178

Typical Symptoms / 179
Reacting to Problems and Symptoms / 184
Establishing a Risk Management Program / 185

8. Dispute Resolution   186

Civil Litigation: An Overview / 187
    Pleadings Stage / 188
    Pretrial Stage / 189
    Trial Stage / 191
    Posttrial Stage / 194

Alternative Dispute Resolution: An Overview / 194
    Characterizing ADR / 194
    ADR Benefits / 196
    Attaining an ADR Agreement / 197
    Reliance on Attorneys / 197
    The Role of the Insurer / 198

Specific ADR Options / 198
    Informal Spontaneous Negotiation / 199
    Voluntary Prehearing Negotiation / 200
    Mandatory Pretrial Negotiation / 201
    Mandatory Binding Arbitration / 201
    Voluntary Binding Arbitration / 203
    Specialized Binding Arbitration / 203
    Expedited Binding Arbitration / 203
    Voluntary Nonbinding Arbitration / 204
    Mandatory Nonbinding Arbitration / 204
    Court-Appointed Masters / 205
    Settlement Masters / 205
    Early Neutral Evaluation / 206
    Michigan Mediation / 207
    Mini-Trial / 208
    Summary Jury Trial / 208
    Private Litigation / 209
    Mediation / 210
    Mediation/Arbitration / 211
    Mediation-Then-Arbitration / 212
    Med/Arb2 / 212
    Rent-a-Judge / 212
    Resolution Through Experts / 213
    ADR by Covenant / 213

9. Basic Economics of Professional Practice   214

The Pursuit of Profit / 214
    The Purpose of Profit / 215
    Profit's Impact on Fees / 216
    Profit from a National Perspective / 218

Accounting / 218
    Classifying Accounts / 218
    Accounting Methods / 233

Establishing Overhead / 234
    Understanding the Variables / 235
    Applying the Variables / 235
    Lowering Overhead / 239
    Complying with Client Computational Methods / 243

Establishing Fees / 244
    Fee Establishment Techniques / 244
    The Role of Experience / 244
    Contingency Allowances / 245
    Other Concerns / 246
    Identifying Profit Objectives / 246
    Value-Based Pricing / 247
    The Need for Adequate Fees / 249

Financial Reporting / 249
    Income Statement / 250
    Statement of Financial Position (Balance Sheet) / 250
    Cash Flow Report / 250
    Accounts Receivable Report / 254
    Forecast of Assets and Liabilities / 254

Project Management Reports / 254
    Project Planning / 256
    Project Budget and Reports / 257

Other Reports and Concerns / 262

10. Professional Human Resources Management   263

Hiring / 264
    Preparing a Job Description / 264
    Identifying Personality Requirements / 265
    Obtaining Candidates / 266
    Culling / 270

Interviews / 270
    Reference Checking / 272

Orientation and Training / 273
    Program Planning / 273
    Loss Prevention Issues / 273
    Mentors / 274

Personnel Policies / 274
    Manual Content / 275

Person-to-Person Relationships / 278
    Supervisor-to-Employee Communications / 278
    Development of Career Ladders / 279

Team Building / 280

Ownership Opportunities / 282

Termination / 283

Posttermination / 283
The Law / 284

11. Professional Services Marketing and Business Planning   286

Marketing and Business Planning: An Overview / 287

Market Segmentation / 289
Internal Marketing Research / 291
    Project Analysis / 291
    Annual Marketing Unit Analysis / 293
    Trend Analysis / 296
External Marketing Research / 300
    Sources of Assistance / 300
    Doing the Homework / 302
    Creating the Initial Projections / 302
    Third-Party Analysis / 304

Finalizing Forecasts / 305
    Comparing Trends and Forecasts / 305

Goal Setting / 307
    Identifying Individual Goals / 307
    Establishment of the Firm's Professional Goals / 309
    Evaluating MUs in Terms of Professional Goals / 309

Selecting Future Options / 309
    Options in General / 310
    Evaluating the Options / 310

CONTENTS xxv

Putting It All Together / 311
    Developing the Plan / 312
    Finalizing the Plan / 314
    Contingency Planning / 315

Keeping the Plan Current / 315

12. Professional Business Development   316

Marketing Communications / 317
    Image Development / 317
    Image Management / 318
    Establishing a Marketing Communications Plan / 319
    Assessing Publics' Importance / 322
    Marketing Communications Techniques / 323
    Sources of Assistance / 329

Direct Selling / 330
    Networking / 331
    Maintaining Client Relations / 332
    Bird-Dogging / 332
    Keeping Track / 334

Keep the Product in Mind / 335

13. Communicating as a Professional   336

A Few Observations about Professional Communication / 337

Organizing / 338
    Gather Information / 338
    Identify Your Audience / 338
    Establish Direction / 338
    Put Thoughts into Writing / 339
    Create the Outline / 340

Preparing the First Draft / 340

Preparing the Second Draft / 342
    Fine Tuning / 342
    Completing the Second Draft / 355

Developing the Third Draft / 355
    Gunning's Fog Index / 356
    The Acid Test / 356

From Written to Oral / 358

Interpersonal Communications / 358

Index   **361**

1
Professions, Professionals, Professionalism

This chapter provides brief background information about the historic evolution of professions. Developments since the mid-1950s receive particular emphasis, because many observers claim the mid-1950s marked the onset of a professional decline in the United States. Whether or not a decline actually has occurred is a matter of debate. Unquestionably, much has changed; professionals no longer enjoy societal respect as a matter of course. Although the reasons for these developments are given, readers are cautioned that the explanations offered are highly judgmental, based more on the author's personal experience than on historical analysis. Assessments of the current situation also are related. Although these, too, are judgmental, there is likely to be more agreement about their accuracy. As a professional, you should know enough about these important issues to form your own theories and opinions, given their impact on your profession and practice. How you and your peers react to today's conditions could have a profound influence on your own and your profession's future.

THE RISE AND FALL OF THE GUILD SYSTEM

Those who fail to heed the lessons of history, we are told, are condemned to repeat mistakes of the past. Design professionals should thus become familiar with the history of the ancient guilds for, in truth, the guilds and the professions were established to accomplish many of the same objectives. The guilds, however, have become extinct, quashed by the public whose sanction was essential to their survival. The professions also rely upon society's sanction and, it can be argued, many of the most serious problems that confront today's professionals result from their inability to meet society's sometimes conflicting needs and expectations.

The Emergence of the Guilds

The guilds first appeared in the Middle Ages as associations of people who shared a common interest. Although there were many types of guilds—charitable, religious, and social—the most important were those of merchants and craftworkers.

Merchant guilds were developed in the eleventh and twelfth centuries, primarily to afford mutual security. Trade experienced a resurgence after the fall of the Roman Empire. By travelling together in caravans, merchants were less likely to be attacked. This comingling also encouraged merchants to discuss their experiences and ideas, ultimately fostering pursuit of additional means for enhancing individual benefit through common effort. They established cooperative purchasing agreements to improve their negotiating position, reduce unit costs, and increase profits. They set quality standards and prices for the goods they sold, to assure uniformity and minimize competitive advantage. Guild members even agreed on the wages they would pay employees, so none could entice another's workers by offering more money.

Merchant guilds were able to enforce their dictates because they were able to gain control of the marketplace. This monopolistic power ultimately made merchant guild members the wealthiest and most influential individuals in their communities. And, to maintain the support of society, they performed "good works" by helping the poor, supporting the church, and building schools.

Goldsmiths, tailors, bakers, weavers, brewers, and other craftworkers also formed protective guilds. Each restricted the number of its members to assure an even distribution of business and profits, so none gained wealth at the expense of another. Those craftworkers who knew their trade well and had their own shops were called *masters*. Those who knew the craft, but not as well as masters, were called *journeymen*. They worked for a daily wage, anticipating the day when they, too, would become masters. *Apprentices* worked under a master to learn the trade, receiving room and board in exchange for two to seven years of service.

The Decline of the Guilds

By the fourteenth century, cracks began to appear in the foundation of the craftworkers' guild system. The masters of many crafts had become nepotistic, passing guild memberships to their sons and barring journeymen from entering their ranks, creating lasting hostility. In other crafts, however, masters became little more than employees of merchants who had become wealthy enough to purchase the means of production. In these latter industries, craft guilds functioned much as labor unions, becoming vehicles for obtaining fair wages and decent working conditions for their members. The disputes that arose led to strikes and, in some cases, civil war.

The extinction of the guilds can in large measure be ascribed to the self-serving attitudes of their members. The desire to accumulate continually more wealth and power became addictive, an end unto itself that blinded them to the needs of others. Without public trust, they lost public support and were powerless to continue—a lesson learned too late.

THE RISE OF THE PROFESSIONS

The professions, as the guilds, began as associations of people engaged in similar pursuits. Unlike merchants and craftworkers, however, professionals were almost all intellectually inclined; meetings were characterized by deep debate over meanings and techniques. But the professions were not formed to promote intellectual stimulation. Their principal purpose was gaining public recognition, to distinguish their practitioners from others claiming, but not having, similar competence. Consider London's Royal College of Physicians. Established in 1518, its charter states that the organization was formed "to curb the audacity of those wicked men who shall profess medicine more for the sake of their avarice than from the assurance of any good conscience."

From the sixteenth through eighteenth centuries, the professions of divinity, medicine, and law predominated. Others were developing, however, and many of them emerged in nineteenth-century England. The Institution of Civil Engineers (so named to be distinct from military engineers) was founded in 1818. The Royal Institute of British Architects was formed in 1834, followed ten years later by the Royal College of Veterinary Surgeons. The Institution of Mechanical Engineers was established in 1847; the Surveyors Institute, in 1868; the National Union of Teachers, in 1870; and the Institute of Chartered Accountants, in 1880. Americans were in the process of doing much the same, with formation of the American Society of Civil Engineers occurring in 1852 and The American Institute of Architects, in 1857.

THE PROFESSIONS THEN AND NOW

No doubt many nineteenth-century clergymen, physicians, and lawyers looked upon the proliferation of professions with the same type of disdain with which many of today's "established" professionals regard some of those who only recently have sought professional status. In reality, however, the formation of professions is nothing more than the continuation of a process begun centuries before. Then, most people worked with their hands, while professionals worked with their intellect. In contemporary America, the service sector of the economy is growing rapidly, and the same factors that led to engineers and architects organizing themselves into professions now are leading other "white-collar workers" to follow suit.

Are the newest professions really professions, in the classical sense of the word, or are they merely nonprofessional occupations that have assumed professional trappings? A landmark paper on this subject was published in the July 1957 issue of *Social Work*. Its author, Ernest Greenwood, cited five basic attributes of professions:

- systematic theory,
- authority,
- community sanction,
- ethical codes, and
- a culture.

It is worthwhile to review Greenwood's work, not only for its guidance, but also—and perhaps more importantly—as an historical artifact that describes in detail what the professions were like at the time Greenwood wrote about them some three decades ago. As will be seen, there has probably been more change in the professions in the past thirty years than occurred in the prior one hundred.

Systematic Body of Theory

Is it superior skill that distinguishes a profession from a nonprofessional occupation? Not really, because many nonprofessional pursuits require extraordinary skill. The real difference lies in the nature of the skill. As Greenwood explained:

> The skills that characterize a profession flow from and are supported by a fund of knowledge that has been organized into an internally consistent system, called a *body of theory*. A profession's underlying body of theory is a system of abstract propositions that describes in general terms the classes of phenomena comprising the profession's focus of interest. Theory serves as a base in terms of which the professional rationalizes his operations in concrete situations. Acquisition of the professional skill requires a prior or simultaneous mastery of the theory underlying that skill. Preparation for a profession, therefore, involves considerable preoccupation with systematic theory, a feature virtually absent in the training of the nonprofessional .... Because understanding of theory is so important to professional skill, preparation for a profession must be an intellectual as well as practical experience.

Rationality As Greenwood also observed, a profession is not tradition-bound to its body of theory. Instead, a profession possesses an attitude called *rationality*, that is, a willingness to replace any aspect of theory with a newer formulation demonstrated to be more valid. This "spirit of rationality" tends to create an intellectually stimulating aspect to professional involvement. In technological professions, such as engineering and architecture, it also creates a division of labor, that is, those who practice versus those who research. As noted by Greenwood,

> ...if properly integrated, the division of labor produces an accelerated expansion of the body of theory and a sprouting of theoretical branches around which specialties nucleate. The net effect of such developments is to lengthen the preparation deemed desirable for entry into the profession. This accounts for the rise of graduate professional training on top of a basic college education.

Body of Theory Today For the most part, Greenwood's observations about the professions still are valid, but much has changed in the area of nonprofessional pursuits. Continually, more of them emphasize theory to help people deal more effectively with the growing technical complexity of their work. This complexity is also giving rise to specialization, a phenomenon that affects the professions as well. Just as design professionals specialize in certain types of projects or services and physicians specialize in certain parts of the body or certain diseases, so do automobile mechanics specialize in transmissions, motors, or brakes. Similarly, in the field of hairdressing, some practitioners specialize in coloration while others specialize in cutting. Theory must be understood to perform these functions well; numerous educational programs and institutions have been established to teach it.

Many nonprofessional occupations also embrace a spirit of rationality, although for some it is spurred more by competitive pressures—the desire to catch the wave of a fad—than by intellectual curiosity. Nonetheless, it exists, and, as a consequence, systematic body of theory is no longer the exclusive property of professionals.

Professional Authority

According to Greenwood, "Extensive education in the systematic theory of [a] discipline imparts to the professional a type of knowledge that highlights the layman's comparative ignorance. This fact is the basis for the professional's authority..." Greenwood goes on to point out the difference between obtaining a professional service and a nonprofessional service—a difference epitomized by the fact that professionals serve clients while nonprofessionals serve customers. What exactly is the difference? As Greenwood explains:

> A customer determines what services and/or commodities he wants, and he shops around until he finds them. His freedom of decision rests upon the premise that he has the capacity to appraise his own needs and to judge the potential of the service or of the community to satisfy them.... In a professional relationship, however, the professional dictates what is good or evil for the client, who has no choice but to accede to professional judgment. Here the premise is

that, because he lacks the requisite theoretical background, the client cannot diagnose his own needs or discriminate among the range of possibilities for meeting them.

Much has changed since Greenwood developed this concept. We now live with extraordinarily complex technology whose selection, installation, and repair often require advanced expertise. For example, when a television set breaks down, we call on a trained technician to repair it. While a client-professional relationship may not exist, something closely akin to it does, because we have no choice but to trust the technician's ability to diagnose the problem and effect a repair. This trust is heightened by a factor Greenwood labels *functional specificity.* As he explains it:

> The client derives a sense of security from the professional's assumption of authority. The authoritative air of the professional is a principal source of the client's faith that the relationship he is about to enter contains the potential for meeting his needs. The professional's authority is not limitless; its function is confined to those specific spheres within which the professional has been educated.... The professional must not use his position of authority to exploit the client for purposes of personal gratification.

Functional specificity still exists, of course, but it is no longer the exclusive property of the professions. Nor is a client's or customer's fear of exploitation associated exclusively with nonprofessionals. In fact, rather than deriving a sense of security from dealing with someone who assumes authority, many people derive a sense of mistrust. Is a specific service being recommended or directed because it is in the client's or customer's best interests or merely because it will "pad the bill"? To come to grips with this situation, many nonprofessional service organizations have trained their personnel to explain different levels of service to the customer and the different benefits associated with each. Professionals who are in the habit of assuming authority often fail to engage in such "consumer education," creating an image some regard as arrogant. In fact, professionals are no longer trusted merely because they are professionals. Trust usually must be earned on a case-by-case, relationship-by-relationship basis.

Sanction of the Community

Most professions receive formal sanctions from the community, that is, certain powers and privileges that are enforced through the community's police powers. One of these is the right to use a certain title. For example, all states require an individual to be licensed or registered in order to use the title "engineer" or "architect." To hold oneself out as an engineer or an architect without being duly licensed is a violation of law.

In most instances, determining whether or not a legally sanctioned professional requirement has been broken is a function invested in a state-sponsored board comprising members of the profession involved. These individuals are also responsible for determining whether or not an examination is required for licensure and, if so, what it will consist of. They also prescribe the necessary educational attainments and apprenticeship service. The profession also has the right to accredit institutions, that is, to determine whether or not the education received "counts" by virtue of the curriculum offered, the manner in which it is taught, and those who teach it.

Receiving the sanction of the community is somewhat equivalent to being granted a monopoly by the government. Only engineers may practice engineering; only architects may practice architecture. This explains, in part, why nonprofessional occupations strive to gain professional recognition, or at least assume a professional mystique. Their efforts often are pursued under the banner of consumer protection, which is precisely the method used more than four centuries ago, when the Royal College of Physicians was founded "to curb the audacity of those wicked men who shall profess medicine more for the sake of their avarice than from the assurance of any good conscience." As a result, we have certified auto technicians, certified association executives, certified property managers, and so on, with each of these titles being protected by copyright law. Many of the pursuits also require licensure, and the requirements usually are determined—or are at least significantly influenced—by organizations of practitioners.

While the professions may benefit from more extensive sanctions than those granted to nonprofessional occupations, the reasons for having community sanctions remain the same, obscuring even more differences between professional and nonprofessional sanctions.

Regulative Code of Ethics

Codes of ethics are so uniquely professional that the term "ethical" is in many respects a synonym for "professional." But codes and the mores they imply have been particularly affected by the passage of time, as underscored by Greenwood's observations:

> While the specifics of their ethical codes vary among professionals, the essentials are uniform. These may be described in terms of client-professional and colleague-colleague relations. Toward the client, the professional must assume an emotional neutrality. He must provide service to whomever requests it, irrespective of the requesting client's age, income, kinship, politics, race, religion, sex and social status. A nonprofessional may withhold his services on such grounds without, or with minor, censure; a professional cannot.... In contrast to the nonprofessional, the professional is motivated less by self-interest and more by the impulse to perform maximally. The behavior corollaries of this service orientation are many. For one, the professional must, under all circumstances, give maximum-caliber service. The nonprofessional can dilute the quality of his commodity or service to fit the size of the client's fee; not so the professional.

While some architects and engineers may maintain client-professional relationships such as Greenwood described, most do not. There is no longer an ethical compulsion to provide service on demand; groups such as ASFE/The Association of Engineering Firms Practicing in the Geosciences (ASFE) actually counsel their members to select clients with care, as a matter of survival. Somewhat ironically, nonprofessionals tend to be less selective, because they are more affected by the profit motive. And, as for the necessity to render "maximum-caliber" service, both professionals and nonprofessionals offer varying levels of effort, adjusting the extent and quality of their service to the fee that the client is willing to pay. While design professionals generally are obliged to comply with a variety of base-line codes and standards to protect health and safety, they are at liberty to select the means and methods used to achieve such results. They can perform a careful analysis of alternatives in order to specify the most cost-effective construction, or they can simply rely on rules of thumb that reduce their level of effort, often at the expense of construction economy.

Colleague-to-Colleague Relations Although contemporary design professionals might be more motivated than nonprofessionals to "perform maximally," they seldom can insist that their clients authorize a maximum effort, due in particular to competitive pressures applied by their colleagues. In fact, this aspect of professional behavior has undergone a particularly profound transformation, considering Greenwood's description of colleague-to-colleague relations as they existed in the mid-1950s:

> [Ethics] demand behavior that is cooperative, equalitarian and supportive. . . . [Out] of place is the blatant competition for clients which is the norm in so many nonprofessional pursuits. This is not to gainsay the existence of intraprofessional competition; but it is a highly regulated competition, diluted with cooperative ingredients which impart to it its characteristically restrained quality. . . . [Professional colleagues] must support each other vis-à-vis clientele and community. The professional must refrain from acts which jeopardize the authority of colleagues and must sustain those whose authority is threatened.

Change is particularly evident among those attorneys, physicians, and other professionals who draw clientele from the general public. Although their sometimes crass electronic and print media advertising draws their peers' disapprobation, due to the harm it can inflict on "professional image," it is sanctioned by law and cannot be barred. Design professionals' promotional practices are far more restrained by comparison, but they still are far different from what they were three decades ago, when most attained assignments through referrals from peers. The more commercial client attraction methods used today make referrals less consequential, significantly diminishing the importance of responsive colleague relations, active involvement in professional societies, and strict maintenance of ethical ideals. Price competition is one of the client attraction methods used most commonly, due to inadequate understanding of conventional marketing and promotional techniques.

Lowering fees in order to attract clients often results in lower quality service. At one time, clients would have eschewed architects or engineers who competed even partially on the basis of fee. That no longer is the case, radically altering the professional fabric.

Formal Enforcement of Ethics   The manner in which professional conduct is formally enforced also has changed. For many years, competitive norms were governed by the ethical codes established by organizations such as the American Society of Civil Engineers (ASCE), The American Institute of Architects (AIA), and the National Society of Professional Engineers (NSPE). In the mid-1970s, however, certain elements of these ethical rules were weakened or eliminated as a consequence of successful challenges by the U.S. Department of Justice, which saw the rules as restraint of trade. The affected organizations have done little or nothing to reassert their ethical concerns, given the potential costs and consequences. But professionals are far from powerless in this regard, by virtue of the boards of registration that exist in every state. Almost all of these boards are professionally controlled and, as public entities, can develop and enforce regulations. Few of the boards are activist, however, so "ethical behavior" in most states lacks legal definition. That being the case, ethical rules are seldom enforced and, when they are, can be subject to legal challenge.

The Professional Culture

As described by Greenwood, "the professional culture" consisted of *social values, norms,* and *symbols* that, together, comprised the professions' most distinctive aspects.

Social Values   A profession's social values comprise services vital to the community; so vital that regulation is required to prevent unqualified persons from performing them. This regulation creates a de facto professional monopoly, especially so because professionals define the qualifications necessary to practice a profession, accredit the institutions that provide professional education, and have the power to discipline those who violate professional rules. Social values, then, are a two-way street: The public must rely on professionals to maintain certain quality standards; professionals must respond in a manner that merits the public trust that justifies self-regulation.

Nonprofessional occupations are not given as much authority as professions. As already noted, however, the area of professional versus nonprofessional community sanctions is somewhat vague, given the various accreditation programs that nonprofessionals have developed. And, while society may trust a profession more than a nonprofessional occupation, that trust is hardly binding at the individual level.

Norms   The professional norms to which Greenwood referred comprised guides to behavior in interpersonal situations. As he explained:

Every profession develops an elaborate system of these role definitions. There is a range of appropriate behavior for seeking admittance into the profession, for gaining entry into its formal and informal groups, and for progressing within the occupation's hierarchy. There are appropriate modes for securing appointments, of conducting referrals, and of handling consultation. There are proper ways of acquiring clients, of receiving and dismissing them, of questioning and treating them, of accepting and rejecting them. . . . In short, there is a behavior norm covering every standard interpersonal situation likely to recur in professional life.

Do professional behavior norms still exist? Yes and no. Yes, in the sense that many professionals have an opinion of what is, and is not, professional behavior, and will attempt to inculcate their beliefs in their associates and employees. No, in the sense that professionals of the same discipline may not agree on what is, and is not, professional behavior, nor are there any formal or informal rules in this respect, except those that define the "outer limits." To the extent that these outer limits do not constitute a restraint of trade, they are enforceable. Nonetheless, professional associations—and even some boards of registration—are loath to respond to all but the most flagrant abuses, given recalcitrant practitioners' penchant for filing suit against the organization and/or those of its members attempting to enforce disciplinary sanctions. As a consequence, defining behavior norms now is a function left largely to the courts, based on a judge's and/or jury's determination of what is reasonable and what is not. The professions have not lost their power in this area of self-regulation: They have learned that, for the time being at least, it is more practical to leave certain ethical concerns to civil rather than professional tribunals. As a consequence, professionals have yielded some of their authority, further clouding distinctions between professions and nonprofessional occupations.

Symbols Greenwood's "symbols" include such things as a profession's "insignias, emblems and distinctive dress; its history, folklore, and argot; its heroes and its villains; and its stereotypes of the professional, the client and the layman."

The concept of lay professionals embracing certain insignias, emblems, and distinctive dress is somewhat alien to us today, and, frankly, it is difficult for many to imagine them as commonplace in 1957. The professions do have their individual history and folklore, however, as well as argot (i.e., "buzzwords") and—as with any group—stereotypes of insiders and outsiders. Many of these latter symbols do not affect society's perception of the professions, however; nor do they affect perceptions of nonprofessional occupations, which, for the most part, also have history, folklore, argot, and stereotypes.

THE PROFESSIONS IN TRANSITION

Many older practitioners are wont to regard the 1950s as a watershed for the professions in America; the last of "the good old days." Since that time, some say, the professions have entered a steep decline, precipitated in large measure by imposition of professional practice standards that are impossible to maintain and a civil justice system run amok. Others have a different outlook. They contend that the professions have not declined at all; that, if anything, they actually have improved. The problem is, they say, that the professions have not improved enough. They also are more subject to public scrutiny, and the public in many cases does not like what it has seen.

Who is right? Who is wrong? No one really knows; the truth probably lies somewhere in between. One thing is certain, however: Times have changed; society's perception of the professions has dimmed.

As it so happens, the design professions are taking a lead role in attempting to restore public confidence in professional competence and integrity. They are doing this through development of standards of practice, promulgation of peer review, and institution of other programs designed to enhance professional norms. Engineers and architects are responsible for taking such attitudes forward, to help keep their professions responsive to society's needs, so they may continue to earn—and benefit from—public trust. Make no mistake about it: This responsibility is not optional. To any given practitioner, one could say, "What you do and how you do it; what you say and how you say it, all affect the public's perception of the profession you represent." The professions are dependent on positive public sentiment for their survival; they are dependent on the concern of professionals for creation of positive public sentiment.

The following discussion offers explanations for the general loss of professional prestige over the past thirty years. As you have been forewarned, the discussion is based on the ruminations of the author, as an individual who has lived through this period. In that this loss of professional prestige is symbolized by the extent to which professionals are being sued—a problem that was almost nonexistent three decades ago—much of the discussion focuses on that issue. It also addresses some of the measures being implemented to affect needed change. These measures underscore the fact that the professions are working diligently to adjust themselves to contemporary needs and expectations. How successful they will be depends on *you*, at least in part. Your share of your profession is in your hands. It is your *duty* to do the best you possibly can with it, just as your professional forbears did in contributing their share, developing the profession as an instrument to meet society's needs and the personal needs of individual practitioners, yourself included. As Greenwood intimated, but did not say directly, duty to the profession has always been a professional hallmark. Denying this duty is nothing less than denying professionalism.

Loss of Professional Distinction

As the review of Greenwood's paper makes clear, a number of professional hallmarks have become extinct or significantly less predominant. At the same time, many nonprofessional occupations have assumed professional trappings. The convergance of these two trends blurs differences between professionals and nonprofessionals, some contend, resulting in professionals' loss of public esteem.

Although the facts cited to support such a position may be accurate, the conclusions drawn are specious. Engineering and architecture—not to mention law, medicine, and other "learned professions"—still are distinct from nonprofessional occupations. Nonprofessional occupations' march toward professionalism merely represents a commercial realization that customers, as clients, appreciate professional behavior. Stated cynically, professionalism is profitable or, at least, a good marketing tool.

A Changed National Mood

American perceptions have undergone a profound change in the last three decades, marked in particular by less trusting attitudes toward the nation's leadership community. One of the most significant developments of the era occurred when Earl Warren was appointed Chief Justice of the United States Supreme Court. Under his leadership, the Court became a strong and much publicized vehicle for change, welcoming what many regarded as frivolous suits, often filed by society's least powerful. Court rulings made it clear that the rule of law would prevail; that the rights of individuals would not be trampled upon, even in societal pursuit of worthwhile goals.

Did attitudes of the Warren Court encourage change, or did they merely reflect new American attitudes? The answer is elusive, but, unquestionably, liberalism was on the rise. The Reverend Martin Luther King, Jr., led a coordinated challenge to a status quo that had kept black Americans disenfranchised for three centuries. Due, in particular, to the growing impact of television, Dr. King's methods and successes etched themselves indelibly on the American conscience. Soon others applied similar tactics to achieve different ends. And the time seemed right; America appeared ready for change. In 1961, John F. Kennedy became the first president born in the twentieth century. Some of Kennedy's most fervent support came from America's youth, who could readily identify with his oratory and style. Other than the civil rights movement, however, they had little on which to focus their energies and idealism. That situation, too, was destined for change, as the nation stepped up its military commitments in Vietnam. President Lyndon B. Johnson reassured the people that America's youth would not be sent to fight a battle "Vietnamese boys" should be fighting. But an American vessel was attacked in the Tonkin Gulf, the American public was told, and that seemed reason enough to respond with direct and deadly force.

The Vietnam War began with traditional jingoistic fervor and an assumption that a quick victory was at hand. It wasn't, however, and false promises of "light at the end of the tunnel" made the situation worse. Television was "over there" for the first time, too, effecting yet another major change. Just as television brought home the fighting to America's living rooms each evening, so, too, did it create a vast audience for the antiwar protests conducted in cities and towns nationwide. Thousands of American lives were being sacrificed for unclear reasons. "Don't trust anyone over thirty" became a frequent refrain among the nation's youth, reacting to the perceived callousness of "the establishment."

Loss of trust in the nation's leadership was not confined to youth or to war-related issues. The vice-president of the United States was forced to resign after it was revealed that he had taken bribes and kickbacks from architects and engineers. The president resigned soon after, rather than face impeachment proceedings after lying to the American people. Problems of pollution became an issue, as more of America's waters became undrinkable and more of its air became unbreathable. People were being maimed and killed in industrial accidents. Unsafe cars were being delivered to automobile showrooms. Unsafe toys were being placed under Christmas trees. Unsafe medicines were being put on the shelves. Unsafe chemicals were being added to our food. Oil and gas reserves were being squandered. Urban problems were making cities unlivable. Inflation was destroying our workers' buying power. Or so it all seemed. Day after day, night after night, America's attention was being focused on problems stemming from conscious decisions of national leaders who attempted to downplay their mistakes—or try to cover them up altogether. In just a few short years after Dwight Eisenhower had warned America about its "military-industrial complex," the nation had been jettisoned from an era of complacency to an era of confrontation, with "the people" adopting adversarial attitudes toward those whom they had been taught to trust. Calls for patience fell on deaf ears. After all, this was the nation that had found a cure for polio and put a man on the moon. America could do anything it puts its mind to, it seemed, providing the establishment got out of the way.

Throughout this trying period, one American institution served as a constant beacon: the law. Continually more people turned to the courts to right perceived wrongs, and continually more jurists—adopting activist robes—granted them relief. In the process, however, the American judicial system created a major new body of civil law that, by almost guaranteeing compensation, encouraged confrontation over conciliation; litigation over negotiation.

And, for some, the courts became not so much a source of justice as a source of profit.

American attitudes in the early 1990s are vastly different from those in the late 1950s. People seem far less willing to give anyone in authority the benefit of the doubt. Nonetheless, the constant barrage of problems that has marked this period did not shake America's faith in its capabilities; if anything, it revealed how strong that faith really is. And therein lies a problem. Americans by and large expect perfection. When it is not delivered, they are quick to react, and, frequently, their position is vindicated by the courts. The old commercial dictum that "the customer is always right" has more meaning today than ever before. Nonetheless, for many professionals who still cling to traditional methods, the notion that "the client is always right" is a bitter pill to swallow.

Professional Self-Protection

The professions were not at all immune to the investigative journalism and consumerist research that laid bare so many of America's ills. Reports showed that professionals—physicians in particular—were extremely reluctant to take action against their unscrupulous and incompetent peers. With rare exceptions, neither associations nor licensing boards were willing to strip practitioners of their livelihoods. Instead, problems were often swept under the rug by encouraging malefactors to practice in other jurisdictions. For many years, it seemed, professionals had been using their community sanctions not so much for self-regulation as for self-protection.

A great deal of professional prestige was lost as a result of these revelations; professionals, too, were shown to have feet of clay. As a consequence, people were encouraged to take action through the courts instead of review boards. Responding to this situation, the U.S. Department of Justice began an effort to eliminate many of the special privileges that had been granted to professional associations. Virtually any ethical rule affecting a professional's business dealings was challenged, with most being toppled as restraints of trade. The Justice Department took the same approach to state boards of registration that had adopted similar rules, but, in those few cases in which rules had been ratified by a state's legislature, it declined to prosecute. The situation might have been altered tremendously had regulatory bodies been actively policing the professions they were in charge of regulating. But such was not the case, and—insofar as engineering and architecture are concerned—it remains the case today. State boards generally do not take strong action even against those who willfully violate rules. This may be attributable in part to board members' exposure to lawsuits, a problem that some states are remedying through grants of partial immunity. But stronger action by state boards probably will not significantly reduce the number of lawsuits filed against professionals, given the financial rewards lawsuits proffer. Nonetheless, aggressive policing of the professions would certainly help restore public confidence in the professions and the concept of professional self-regulation.

Fewer Personal Professional-Client Relationships

No doubt the number of lawsuits being filed against professionals would be far fewer had traditional person-to-person professional-client relationships been maintained. Today, however, professional service delivery systems make such relationships the exception rather than the rule. Professional service providers have in many instances become organizations of specialists, a trend accelerated by the ever more rapid acquisition of technical knowledge. The trend is not new, of course. It is the same trend that created distinctions between physicians and surgeons, solicitors and barristers, engineers and architects. Although it has helped enhance the quality of service provided, it has almost eliminated cases in which a single individual is relied on to care for another individual's or family's medical, legal, or other professional needs. At one time these traditional relationships lasted for a lifetime, marked by utter faith in the professional's pronouncements and genuinely warm person-to-person feelings. If the client was slow to pay a bill, it could be forgiven. If the professional committed a relatively minor error or omission, it could be forgiven. Far different attitudes prevail when a person barely knows a professional and when the professional seems far more concerned with the problem than with the person who has the problem. Add to this equation the expectation of a perfect result, minimal communication, and general distrust, and it is not difficult to imagine why professionals are so often subject to claims.

The problem is even more pronounced in the design professions. There, most relationships have become organization-to-organization rather than person-to-person, making forgiveness a difficult commodity to come by. In the private sector, projects are developed more by corporations, limited partnerships, and syndicates than by individuals. When problems arise, others' investments must be protected. Established professional-client goodwill can help ease a problem by encouraging a negotiated resolution, but insurers or corporate attorneys may prefer different methods. Nor is establishing goodwill a simple pursuit. The duration of many relationships is limited to the duration of a project, a situation that often results from use of fee-based selection criteria. And, in many instances, no real client-professional relationships are established, as when a prime design professional engages "interprofessionals" who never even meet the owner.

The nature of professional-client relationships can be even more barren when the clients are government agencies. Many "spread the work around," and those who select the design professionals seldom serve as the owner's representative during the design process. When problems occur, they are assigned to attorneys for resolution; forgiveness is against public policy.

In both the private and public sectors, then, most design professional-client relationships are now business-oriented; decisions are governed more by numbers than by instincts. If a lawsuit is judged to be the most effective means for improving the bottom line, little prevents one from being pursued. This puts many design professionals at a distinct disadvantage, because their clients may be far more astute businesspeople than they.

More Errors Being Made

The nation's demand for professional services has been growing constantly, and providers have been scrambling to keep up. At the same time, the fields in which many practice have become steadily more complex. In some cases, responsibilities are assigned to individuals who are only marginally qualified to handle them. In others, professionals and their subordinates are working long hours at a fever pitch. Given such situations, and considering the vast number of professional engagements being pursued everyday, it is not surprising that more errors are being made. Design professionals are particularly vulnerable in this regard because of the enormity of their projects, the interrelationships between systems and components, and the multiplicity of disciplines and individuals involved—situations that make errors more likely and that can greatly magnify the consequences of even minor mistakes. But these problems are hardly insolvable. They can be prevented, in part, through establishment of carefully crafted interdisciplinary teams that are given the time and encouragement to communicate with one another and to engage in mutual plan review. Regrettably, "competitive pressures," that is, the need to keep fees as low as possible, often make such an approach unwise, forcing reliance on other techniques. Computer-aided design and drafting (CADD) systems are one of these, but they are far from fail-safe. Many firms still do not rely on them, and those that do often discover that their systems are not compatible with those of other firms engaged for the same project.

An effective approach for easing time pressures would be to restrict the amount of work a firm accepts. Most design professionals concentrate in just a few markets, however, and activity in them often is cyclical. Many firms thus have adopted a "make hay while the sun shines" marketing philosophy that can at times yield more engagements than they can comfortably handle. In those instances, time pressures and overwork erode quality while also causing delays that fray clients' attitudes, nurturing adversarial outcomes.

Design Professional Retention Practices and Fees

More so than other professionals, engineers and architects are being encouraged to abandon traditional procurement practices, creating a variety of problems that are discussed more fully in the following chapter. The basic problem arises from consideration of fee in selecting firms. Traditionally, fee is considered only after the most qualified firm is selected and the firm and its client meet to establish mutually the scope of services best suited to the project and people involved. By contrast, when fee is used as a selection criterion, design professionals must work with a unilaterally developed workscope—one of their own creation or one that the client has prepared—in order to determine what the fee should be. Such workscopes overrely on a wide range of assumptions and cannot adequately consider the preferences of the design professional or the client, leading to misunderstandings, failed expectations, and disputes.

For many years, most established associations of design professionals ethically proscribed their members' submission of competitively priced proposals. In *National Society of Professional Engineers v. United States of America*, the Supreme Court in 1978 upheld a District Court ruling that the ethical ban comprised a per se violation of the Sherman Anti-Trust Act.

Since all firms of the same discipline draw upon the same pool of design resources, and generally offer equivalent facilities and capabilities, procedures employed to minimize fees often erode quality of service. When the fee is set optimistically to begin with, financial problems tend to become most acute as a project is winding down. That is precisely the time when final quality control procedures, such as shop drawing review, are implemented. In other words, inadequate fees cause many firms to skimp on quality control when quality control is most needed.

Traditional procurement methods can help prevent these problems, in large part because they encourage effective communication between design professionals and their clients. For this reason, the federal government and most state governments rely on traditional methods by statute. For more than a dozen years, Maryland was the only state in the nation that required design professionals to bid on virtually every project. Even though the state employed what could be considered a quality-oriented approach, inherent communication problems resulted in the range of offers varying by 100 percent or more on most projects, as well as innumerable design fee change-orders once a firm began work. The state abandoned its bid-based procurement statute in 1986, resorting to a traditional process for all its major work.

Reliance on fee-based procurement has reached alarming proportions in the private sector, in large part because relatively few design professionals are attempting to reverse the trend. Indeed, many prime design professionals themselves ask prospective design team members to submit bids as a condition for engagement, a practice many architects and engineers believe to be noxious and short-sighted.

To the extent that any professional agrees to accept a fee that is inadequate to support high quality, high quality is not likely to be forthcoming. Given the complexity of much of the work involved and the assumptions that inadequate communication make necessary, fee-oriented procurement only serves to make serious problems worse. The potential for errors and omissions is increased. The potential for misunderstanding is increased. Opportunities for establishing effective professional-client dialogue and relationships are virtually eliminated.

Failure to Discuss the Downside

The air of professional authority Greenwood spoke of can be colloquially summed up as "doctor knows best." But part of "doctor's" knowledge includes awareness of what can go wrong. As a consequence of precedential court decisions, physicians now routinely inform patients of what can go wrong and the statistical probability of each potential outcome. This procedure helps counteract expectations of a perfect result and, when all is said and done, gives patients information they have every right to know.

ASFE was the first design professional organization to encourage its members to discuss the downside of a project: what could go wrong and why. The approach was adopted as one of the procedures used to gain client understanding and acceptance of a concept called "limitation of liability." The concept is basic: Given the potential for error that exists when a geotechnical engineer is attempting to describe subsurface conditions based on sampling techniques, it is unfair to impose unlimited liability on that engineer.

Critics said owners would never accept the proposition of limited liability, but they did, and in surprisingly large numbers.

ASFE has more recently encouraged its members to discuss the downside of not performing certain services and to include appropriate reference to client decisions in their contracts.

Many design professionals still do not address the downside in their discussions with clients. They do not point out the problems that may occur when a given service is deemphasized or ignored; very few make reference to these contractually. Their reluctance in this regard is not derived from a "doctor knows best" attitude. Instead, it seemingly stems from a desire to not upset a client or to not scare one away. Until such time as court decisions and/or professional standards require such discussions, many clients—especially those engaging a design professional for the first time—will continue to expect perfection and will continue to experience the frustrations that often lead to disputes.

Lack of Written Performance Standards

Professionals are considered negligent when they violate the standard of practice and, as a result, someone is injured or damaged. Standards of practice always exist, but they seldom are written. Instead, one must perform research to determine how a professional's peers performed a given task at the time and location in question. It would be extremely beneficial for professionals to develop such standards and to commit them to writing before the fact, in order to help prevent reliance on techniques that are later shown to be violations. In the design professions, written standards would also discourage firms from lowering fees by eliminating or deemphasizing services commonly provided to maintain base-line quality.

Absence of written standards has created a fertile field for professionals who derive much or all of their income from testifying as experts. Many of these individuals are upstanding professionals who do their utmost to fulfill their purpose, that is, to explain technical complexities in nontechnical terms to a lay trier of fact—a judge and/or jury. Regrettably, many others perform as "hired guns"; individuals who serve principally as advocates for their clients' cause, rather than objective professionals. In fact, many professionals have been found liable for breaching a standard of practice that actually did not exist, simply because opposing experts who were not required to authenticate their opinions gained the trust of the trier of fact.

Efforts are at this time under way to help remedy the current situation. Despite some opposition, the American Society of Civil Engineers (ASCE) continues work on a practice guidance manual designed to enhance the quality of construction through more effective engineering practices. And ASFE has helped by developing a set of recommended practices for design professionals who serve as experts. About twenty-five prominent national organizations have adopted these recommendations, giving them almost the force of a standard; adoption by state boards of registration would make them law. Among other things, the recommendations suggest that experts perform that level of research necessary to determine the actual cause of an event, as opposed to setting about solely to prove a client's contentions. They also suggest that the applicable standard of practice should be established through bona fide research.

New Interpretations of Civil Law

Thirty years ago, professionals owed a duty of care only to those with whom they had signed a contract. If errors were made, the civil courts, when used, determined who was at fault and the appropriate amount of damages. Today, as a consequence of numerous decisions, professionals owe a duty of care to *any party* who could foreseeably be injured by their negligence, even when there was lack of privity (no contract) between the parties.

In some areas, third-party claims for purely economic losses are permitted, meaning contractors and subcontractors can seek compensation when an allegedly negligent act results in a monetary reversal. The courts have also established a concept known as "joint and several liability." This holds that, in order to make a damaged party "whole," *any* party who shares in the blame—even at the 1 percent level—could be liable for the entire award, based not on the party's degree of blame but rather on its financial resources.

Design professionals have been particularly affected by these new interpretations of law. They are frequently named defendants and, because they are personally liable for their professional acts, cannot use corporate dissolution to evade responsibility. Compounding these problems, many statutes of repose enacted to afford design professionals reasonable protection have been reinterpreted in such a way as to make them almost valueless. For example, ordinances designed to prevent suits from being filed against design professionals more than ten years after a project's completion have been construed to mean that a plaintiff may file a claim up to ten years after a defect is discovered, even if the defect is discovered twenty (or more) years after the project has been completed.

Some contend these new interpretations of civil law help assure justice; others believe they create injustice. In any event, they have significantly increased design professionals' exposure to professional liability claims. Although not all claims are won, all must be defended, usually at considerable expense.

Profit Motive

Profit is not a dirty word, nor is seeking profit a sin. For professionals, however, profit is not—or should not be—the objective of their work. Each professional should be committed to an ideal, with a realization that striving to attain that ideal in a professional manner will in and of itself create profits. Professionals are not trained as businesspeople, however, and many—perhaps most—lack the management skills and judgment necessary to maximize profit without minimizing service. Some professionals are far more motivated by money than a desire to serve, and a few use their professional positions to gain unjust enrichment, by gouging clients, failing to perform certain services, serving as biased experts, or asking fellow professionals to bid. These and other practices are not motivated by the desire to earn a legitimate profit; they are motivated by greed. When this greed impels a professional beyond the limits of the law, headlines result, making it that much more difficult for the professions to maintain public trust.

Realistically, of course, the problem of lawsuits is created far more by others' greed than by professionals' greed.

Insurance

The phenomenon of professionals being sued created the need for professional liability insurance, a product designed to protect professionals as well as their clients and, later, others who foreseeably could be injured. Unfortunately, professional liability insurance has created two particularly harmful by-products: deep pockets and slack performance.

"Deep pockets" is an expression that connotes extensive resources. These become available just by purchasing insurance. Deep pockets make professionals much more appealing targets for litigation, because "victory" can result in a large reward. And, because the money comes from large, impersonal insurance companies, rather than from the professionals themselves, there is little reluctance to file suit initially or to levy huge damages.

"Slack performance" refers to the attitudes of those professionals who take a somewhat casual approach to quality control from the mistaken belief that insurance provides all the protection they need should a problem arise. Slack performance can also be attributed to many clients who regard an insurance policy as a reason to use the firm that offers the lowest fee and/or to rely on a flimsy workscope.

Today, professional liability insurance is a marketing necessity for many design professionals. Some pay as much as 10 percent of their gross revenues (or more) to maintain it. Unless changes are made, this insurance will foster more problems than it prevents.

Media Exposure

Few in American are ignorant of the fact that lawsuits against professionals can result in huge, multimillion-dollar damage awards. Although these have become more common in recent years, they still are newsworthy and often are covered in both print and electronic news media. When hearing or reading about them, even some well-intentioned people cannot help but ask themselves if a relatively minor problem of some type is not perhaps worth just a few thousand dollars. Perhaps even more compelling are the advertisements of personal injury attorneys, which, appearing in both broadcast and print media, encourage people to seek legal assistance in the event of an accident or injury of any type. Regrettably, many of the attorneys who use this method of advertising interpret advocacy on behalf of their clients as sanctioning activity that is little more than legalized extortion. Until countervailing measures are put into force, such corrosive behavior will continue.

Lack of Effective Business Practices

Professionals in private practice are in business. To succeed, they must deal effectively with a variety of common business issues, ranging from human resources management to marketing. These are the issues that lead to the most problems, at least for design professionals. Case histories reveal that most of their claims are caused, or at least aggravated, by poor business practices, such as unfortunate client or project selection, improper assignment of personnel, inadequate scheduling, or relaxed quality control.

It is not surprising that so many claims and lawsuits stem from weak business management skills. Design professionals learn about management principally through on-the-job training, participation in an occasional seminar, and other relatively informal means. ASFE recognized this deficiency many years ago and established the original Institute of Professional Practice (IPP), designed to expose participants to a variety of business-related issues. ASFE recognized that, for professionals to succeed in private practice, *all* aspects of their business activities should be conducted in as professional a manner as their technical activities. This implies that design professionals should either master the techniques of a given pursuit (e.g., financial management or contract preparation) or rely on another party who has done so. Any area that is overlooked becomes the weak link in a chain. Any weak link puts a firm—even a career—at risk.

TOWARD A NEW PROFESSIONALISM

The professions were established to benefit society. In order for society to achieve this benefit, it was required that professions be granted certain powers in the form of community sanctions. The extent of these powers was allowed to grow, because society for many years did not choose to perform, or was discouraged from performing, oversight. When it finally did, society realized that the professions were not performing as they should; society was not being served as well as it wanted and deserved to be.

The professions have much to do in order to regain society's trust and respect. The design professions are taking a lead role, creating a new professionalism. What is it? Who are the "new professionals"? The answer cannot be found by referring to a dictionary. It is obtained by "reading" society, to define what it means by "professional."

Clearly, "professional" implies more than being a member of a distinguished or distinguishable profession, and it also means more than simply working with one's mind, as opposed to physical labor. It also means more than doing for pay what others do for sheer enjoyment, as in the world of sports.

The most significantly meaning of "professional," for our purposes, is that applied in a laudatory manner irrespective of the particular type of work involved. You have probably heard it said many times that a certain person is "a real pro." It does not require an advanced degree to earn the praise; it can apply even to those who did not receive a formal education. What does it mean? Consider the following scenario.

Something is wrong with your car, and you take it to a repair shop. The mechanic who looks at it identifies several possible explanations and remedies, and the costs associated with each. Next, you receive an explanation of the approach that would be taken were it the mechanic's own car, explaining the pros and cons of new versus remanufactured parts, given the amount of time you expected to keep the car. Throughout this explanation, you sense that the mechanic is knowledgeable and candid; you convey your trust. "Do what you think is best," you say. Later, when you pick up your car—ready when promised—you learn the problem has been corrected. While it was not as simple as it could have been, it also was not as complex and costly. The mechanic meets with you before you depart to point out other conditions you should be aware of, including a weak battery and worn brake linings. The mechanic also lets you know that a few wires were tightened at no additional charge because "it took no time at all."

As it so happens, auto mechanics are among the least trusted of all the nation's service providers. Nonetheless, your experience with this particular mechanic convinces you that the individual is "a real pro."

For many years, members of the professions gained the trust of their clients simply because of their professional status. That no longer is the case. If trust is to be gained, it must be earned on an individual basis. The "real pros," then, are people who can be trusted because they are knowledgeable, capable, and honest in their dealings with their clients or customers. Although they are in a position to take advantage of their clients or customers by virtue of their superior knowledge, they do not. In fact, being a "real pro" means basically to abide by the "Golden Rule": Do unto others as you would have them do unto you.

Where did the expression "a real pro" come from? Probably through comparison of a nonprofessional (i.e., someone engaged in a nonprofessional occupation) with a professional or—more accurately—the image of a professional. In other words, auto mechanics are "real pros" when they treat their customers much as physicians treat—or *should* treat—their patients. But the positive stereotypes of physicians, attorneys, and other professionals are no longer what they once were. As such, if you wish to enjoy "professional" status, simply being a member of a profession is not enough; you must earn professional status on your own, through your conduct. At one time, much of this conduct would have been dictated by the profession, as Greenwood discussed. But the professions no longer have that power. However, if enough professionals seek professional status by striving to be "real pros," the profession itself will follow.

One of the key aspects of professional conduct is effective client communication, not only to develop the workscope, but also to demonstrate character and put people into the equation. If our apocryphal auto mechanic had just been asked, "How much?" without being given an opportunity for explanation, an accurate answer could not have been forthcoming, except by guess, because the true dimensions of the problem could not have been ascertained until after work had commenced. The same is true for design professionals, but all too many are nonetheless willing to forgo communication and submit a fee that may, or may not, be appropriate. Bidding obviates trust; a bargain is struck, and the provider must do whatever it is that has been agreed to for the price submitted. Any variance will likely lead to a dispute. Under these circumstances, it is virtually impossible to perform as a "real pro," nor is there any significant motivation to do so. By asking for a price, clients are saying they do not trust their design professionals to fulfill expressed needs as best they can for a reasonable fee. By submitting bids, design professionals indicate exactly what they will do irrespective of alternative, and potentially superior, approaches that may be available. Clearly, though, no "loose wires" will be fixed, unless agreed to "up front."

Positive attitude and effective communication alone are not enough. Competence also is required and, to deliver competence, quality control is necessary. In the world of auto mechanics, this may involve a road test. In the design professions, it implies double-checking calculations, subjecting plans and specifications to internal and possibly external peer review, examining shop drawings, monitoring construction, and a number of other tasks designed to help ensure that all i's are dotted and all t's are crossed. If, for some reason, the client prefers to relax quality control in order to reduce the fee, that is the client's prerogative but, by contract, any problems that develop as a result should be the client's responsibility, assuming the professional is even willing to work with such a client.

Other aspects of concern also enter into the overall picture, and these relate to business management. The "real pro" auto mechanic was able to have your car ready when promised because a schedule was posted and adequate time was allowed. If others worked on the vehicle, they were trained for their work and effectively supervised. You are not particularly aware of these and related functions being conducted, of course, but, without their proper performance, the car would not have been ready and you would have been frustrated. Clients of design professionals are no different. It is absolutely essential that design professionals control every aspect of their practice in such a way that it supports their overall service goals. In fact, lack of attention to these concerns may be the single most significant cause of design professionals' liability woes. While they were adjusting to the new technical complexities affecting their specific disciplines, many simply ignored the myriad other developments affecting everything else they did, all of which are necessary for the delivery of a professional service.

What, then, is "the new professionalism"? Basically, it is recognition of the many different concerns that are part of, and necessary to, professional performance as that performance has been articulated by contemporary society, and a dedication to performing all associated tasks in a professional manner through education, training, objectivity, honesty, and integrity. It is recognition of everything that needs to be done and doing it all in light of the golden rule. It is, in truth, the same type of attitude that led to creation of the professions a half-millenium ago; an attitude that can be applied by professionals to modern times.

2

Professional Engagement

Design professionals' service to their clients traditionally begins through a meeting or a series of meetings at which the two parties identify the design-through-construction-phase services the engineer or architect will provide to meet mutual needs and preferences. This process, called "mutual workscope development," promotes effective communication between the two parties and establishes a basis for mutual trust. In recent years, however, continually more clients—and not just a few design professionals—have been resorting to fee-based design professional engagement that eliminates the traditional initiation of service and encourages a diminution of quality in order to establish a lower, more "competitive" fee. This chapter examine the various techniques commonly used to engage design professionals and discusses the benefits associated with the traditional approach, that is, qualifications-based selection, or QBS. Because the traditional approach is so supportive of professionalism, discussion also includes guidance on measures that design professionals can use to make others aware of its benefits.

QBS

QBS is an acronym for qualifications-based selection, the term used to connote the traditional method of design professional engagement. QBS has become traditional because it is so compatible with the highly judgmental, creative work that engineers and architects customarily perform. Its applicability rests on the concept that the firm most qualified to provide service is also most qualified to work with clients to identify their expectations and to delineate which services are necessary to achieve those expectations.

Many specific QBS approaches are used. They are characterized by three specific attributes.

1. The client identifies the one firm considered to be the best qualified for a project.
2. The client and representatives of the selected firm confer to discuss the project and establish the design professional's workscope.
3. The design professional's fee is determined after the workscope has been mutually developed.

A Typical QBS Application

Many public agencies and private sector clients employ the following QBS procedures.

Identification of Interested Firms and Solicitations of Statements of Qualifications Public sector clients usually implement this step by advising firms of a project. Notices are sent to firms that have registered with the agency; advertisements requesting expressions of interest usually are placed in local, and sometimes national, publications. A typical notice or advertisement indicates the nature of the project and general services required, permitting interested firms to better focus their submissions. In the private sector, clients commonly obtain referrals from colleagues, among other sources, and contact firms directly.

In responding to requests for expressions of interest, firms usually submit a general brochure, a listing of projects similar to the one under consideration, names of clients, and resumes of key personnel.

Qualifications Review Clients review interested firms' submissions to identify those that appear to be the most qualified. Then they contact those who have used the firms' services in the recent past to ask about the quality of each firm's performance; for example, "Was the design cost-effective?" "Were deadlines met?" "Did the firm implement quality control procedures?" It would also be appropriate to make general inquiries about each firm's pricing practices; for example, "Did the firm charge a reasonable fee?" "Did it submit its bills in a timely manner with appropriate documentation?"

Interviews Clients often contact at least three of the most qualified firms to arrange interviews, preferably at the firms' offices so more can be learned about each. Interviews permit clients to obtain in-depth information about a firm's experience with similar projects, concepts that may be applicable for the project under consideration, and—perhaps most important—the people who will be assigned to the work.

Ranking and Notification Subsequent to interviews, clients rank at least three firms in preferential order and notify the top-ranked firm of its selection. A meeting then is arranged for mutual workscope development.

Mutual Workscope Development and Firm Retention During meetings held for purposes of mutual workscope development, clients and representatives of the selected firm confer to gain understanding of each other's needs and preferences, to better delineate what the design professional work should entail. After the client's project goals and objectives are expressed, design professionals pose questions to gain more specificity, based on extensive experience with similar projects. The design professionals then discuss the various steps in the design and construction processes, the risks inherent in each, and the various services applicable.

Depending on the size of the project and the client's sophistication, mutual workscope development may require several hours or several days. Once the process is complete, the design professional finalizes a scope of services. Then, and only then, is the design professional in a position to establish the fee for the work. Both workscope and fee are reviewed with the client. If the client finds them reasonable, the firm is retained. If the fee exceeds the budget, however, the client meets with the design professional to determine which services can be modified or eliminated to reduce the fee. In all cases, the client is, or should be, apprised of the risks that such changes may impose.

With few exceptions, the best-qualified firm is retained. If agreement cannot be reached, however, the client formally concludes discussions and begins the process again, with the firm considered to be the next best qualified. If discussions with that firm also end in a stalemate, the client is probably working with an inadequate budget.

Other QBS Applications

Abbreviated QBS techniques often are applied for smaller projects. Some public agencies resort only to their experience records to identify the three most qualified firms and proceed from there. Others may identify the one firm they most want to deal with, in some cases considering a need to spread the work among all qualified firms in the area. On the other end of the spectrum, for major projects, an agency may convene a selection panel to identify the most qualified firms and, in some cases, also rank them in preferential order.

The federal government, which procures more engineering and architectural services than any other entity in the world, relies *exclusively* on QBS, as per Public Law 92-582. Known as the "Brooks Law" in honor of its sponsor, Representative Jack Brooks (D-TX), it was proposed at the request of several major using agencies: the General Services Administration (GSA) and the U.S. Army Corps of Engineers, in particular. They had been using QBS for many years prior to the Brooks Law's 1972 enactment because it was so effective. Since then, most states and innumerable local jurisdictions have also adopted QBS.

QBS—as is virtually any other procurement technique—is much easier to implement in the private sector. Most experienced owners and prime design professionals already enjoy established relationships with certain firms. They simply contact the firms they want to work with and enter directly into mutual workscope development.

FEE-BIDDING

"Fee-bidding" is a broad term encompassing a number of practices that share a common element: Design professionals are required to submit a proposed fee as part of the selection process. Since fee depends on workscope, fee-bidding requires design professionals to work with a unilateral workscope; something developed by either the client or design professional without collaboration with one another, eliminating the process of mutual workscope development, and the communication, understanding, and trust it so often engenders.

Elements of Fee-Bidding Procedures

Basic elements of commonly used fee-bidding methods are the following.

Workscope Development Many fee-bidding procedures begin with the client developing a unilateral workscope for circulation to interested firms. Some of these workscopes are so general that they do little more than state the purpose of the project, for example, "Prepare a plan for future development of our county's sewage treatment systems through the year 2010." Others are more specific as to client intent but contain little detail about the specific methods the design professional is to employ or the level of effort anticipated. Even when this information is included, client-developed workscopes almost invariably omit important details and/or require use of techniques that are outdated, inadequate, or improper. As a consequence, when those charged with selection review design professionals' submissions, apples are seldom compared with apples. Even the most sophisticated fee-bid procurements can result in offers that range from one another by 300 percent or more, due to differences in the assumptions offerors must make in order to establish the workscope on which the fee is based.

The problems associated with client-developed workscopes are well illustrated by the experience of the state of Maryland. There, in 1974, legislation was enacted requiring engineers and architects to identify fees in their proposals. The law was enacted in response to revelations that former Governor and then Vice-President Spiro T. Agnew had solicited and received bribes from design professionals seeking state work. It was successfully argued that fee-bidding would minimize opportunities for corruption, despite extensive evidence to the contrary.

Maryland's fee-bidding procedure involved development of highly detailed workscopes (called "programs") that were sent to interested firms whose initial submissions indicated the experience and staffing needed by the project. Those interested in pursuing the work were required to submit a technical proposal and a price proposal, each in its own envelope.

In 1984, the Consulting Engineers Council of Metropolitan Washington (CEC/MW) analyzed 181 separate state projects using data furnished by the Maryland Department of General Services. CEC/MW found that, on average, each project experienced 1.03 design fee change orders worth $23,719. Firms that had been engaged for more than one project were receiving 1.41 change orders per project, worth $28,823.

In the early years of the Maryland "experiment," another CEC/MW survey showed, many firms simply absorbed the additional costs imposed by faulty state-developed workscopes due to their rather naive belief that doing so would help them obtain another assignment. While this outlook is valid when QBS is used, it does not apply to fee bidding. Engineers learned this the hard way, but they did learn. A 1981 CEC/MW study found that 75 percent of the survey's respondents refused even to consider accepting work from the state of Maryland. Respondents said they did not want to "play the game," that is, submit fees based strictly on the state-developed workscope, despite its errors and omissions, and then request changes once the engagement was secured. This practice, called "low-balling" or "buying-in," has long been practiced by many construction contractors but is generally considered unprofessional because it is so adversarial in nature. Design professionals cannot act as their clients' trusted advisors if they seek to gain from their clients' mistakes. Nonetheless, design professionals who are requested to submit bids have few alternatives. If their proposals allow for necessary services that the client may have overlooked, their fees will be higher than others' and they will lose a competitive advantage. Calling errors or omissions to the client's attention will result in the same outcome, since the client will generally inform all interested firms of the need for revisions. In short, among their other failings, client-developed workscopes encourage design professionals to behave unprofessionally, because professional behavior can negate a competitive advantage. In fact, the problems associated with design professional low-balling contributed significantly to Maryland's adoption of QBS for major projects in 1986.

No Workscope: In some cases, the client issues no workscope at all, directing interested design professionals to develop their own. Many design professionals refuse to respond to such requests because of the huge variances that are likely to occur. Most regard it as a costly waste of time to develop technical and fee proposals for clients who apparently do not understand what they are doing.

Identification of Interested Firms Clients who use fee-bidding identify firms by using techniques that are basically the same as those applied in QBS. Occasionally the process is abbreviated by sending requests for quotation (RFQs), including a workscope, to firms on file, or, as has occurred, to all firms listed in a given category of the local Yellow Pages.

Qualifications Review Many clients who use fee-bidding omit qualifications review or invite only prequalified firms to compete. This is seldom wise. While many firms may be generally qualified for a specific assignment, some are far more qualified than others based on the attributes of those who will perform the work.

In the state of Maryland's fee-bidding system, interested firms were informed of relevant criteria, such as staff experience and firm size. Firms' responses were graded against these criteria; only those in the top quartile were sent programs and allowed to submit proposals. Despite these qualifications review safeguards, the system still did not work. In one case, for example, a firm that was exceptionally well qualified for a highly specialized project was considered unqualified because the client representatives who identified the relevant criteria did not understand which factors were important and thus gave no credit to experience that actually was crucial.

If an approach such as Maryland's is to work, those who establish relevant criteria and evaluate responses must themselves be experts. Where will they come from? How will they be retained? And, logically, why shouldn't they be the ones retained to do the work?

Prebidding Conference Some clients hold a prebidding conference in which design professionals ask questions to obtain a clearer understanding of client intent. These conferences are hardly effective substitutes for mutual workscope development, because every question asked and every answer given is available to the competition. Accordingly, engineers and architects are discouraged from posing questions that could reveal innovative approaches or proprietary methods. Such questions are later asked in private, and the answers given are commonly made known to other firms, particularly when the client is a public entity. Of course, firms that hope to profit from client ignorance or errors prefer to have potential misunderstandings remain potential until such time as change orders become necessary.

Technical and/or Price Proposal Evaluation Some clients rely solely on the fee to select a firm and do not ask design professionals to submit technical proposals from the mistaken belief that all firms will probably use the same "standard" procedure or because they do not understand technical proposals and, thus, have no way of comparing them. It is difficult to imagine a worse selection procedure. The submissions of various design professionals are likely to differ radically from one another and from what the client expects.

When firms are required to submit a technical proposal and a price proposal, either a *single-envelope* or a *double-envelope* system is employed.

Single-Envelope System In a single-envelope system, a design professional's proposal typically comprises a statement of qualifications (unless this has been submitted previously), a detailed description of its technical approach, and its fee proposal. Evaluators must determine which of the several offers promises the most value. Even highly qualified technical experts with many years of proven experience find this to be an extraordinarily difficult, time-consuming task, for many reasons. Among them are: Technical proposals are all written differently; there are few, if any, standardized generic terms descriptive of various tasks; and each firm usually proposes a unique approach (even if two offered the same, they would have described it differently). As a consequence, technical merit evaluations are often biased by fee considerations, or fee—the only common denominator—becomes the only factor considered.

Double-Envelope System When a double-envelope system is used, technical proposals and price proposals are submitted in separate envelopes for separate evaluation, usually by different reviewers or teams.

Separate envelopes do not make technical proposals any easier to evaluate. Recognizing this, Maryland employed a "by-the-numbers" ranking system and also insisted that technical proposals be submitted using a format that facilitated their anonymization by departmental staff, to minimize the potential for bias in review. *However*, the state also required submittors to indicate estimated man-hour requirements in their technical proposals. As a result, as the CEC/MW survey indicated, the vast majority of firms receiving the highest technical scores also estimated the lowest or second-lowest time requirements. In other situations, in which estimated time requirements are not included in technical proposals, there is a tendency for the well-written proposals of larger firms to receive almost the same technical evaluations, making price—contained in the second envelope—the true selection factor.

In essence, all available evidence suggests that double-envelope systems are not much better than single-envelope systems, simply because it is an almost impossible task to establish meaningful evaluations of different offers. And it is important to note that the highest-ranked technical proposal is merely that proposal which comes closest to figuring out what the client wants. Thus, the process does not compare favorably to mutual workscope development, in which client and consultant can gain an intimate understanding of mutual goals and objectives.

Double-Envelope/Negotiated Approach Another approach sometimes used can be termed *double-envelope/negotiated*. Instead of awarding the engagement to the one firm that offers the best combination of technical merit and fee, those two or three offering good combinations are identified. The client then meets separately with representatives of each, to refine workscopes, principally to help prevent misunderstandings. Then the client asks each firm to make its "best and final" offer.

On the one hand, the double-envelope/negotiated procedure does help improve workscopes. On the other hand, firms are under tremendous pressure to keep their fees to an absolute minimum, often by reducing or eliminating allowances for contingencies. As in all fee-bidding situations, then, the client is forced to assume a significant risk in order to obtain fee reductions. In most cases, the value of these reductions is little more than the value of the additional time the client must spend in order to implement the approach.

THE BENEFITS OF QBS

Many of QBS's benefits have already been touched upon. Perhaps the most significant is its promotion of design excellence and encouragement of innovation, resulting in structures or systems that perform more effectively and/or are less costly to build, operate, and maintain. Were it not for innovation by design professionals, we still would be living and working in mud huts. With an overemphasis placed on fee, we will be forced to wait many years for what could have been tomorrow. In order to save time and thus keep fees low, design professionals merely propose doing what has been done before.

Another significant benefit of QBS is its promotion of communication between design professionals and their clients. Studies show clearly that more effective communication can lead to many risks being considered before they materialize, to mitigate their impact or to avoid them altogether. Better communication can also reduce the risk of claims that occur because the design professional fails to meet client expectations that were unrealistic to begin with.

Although many believe that QBS is appropriate for any type of service, it is particularly suited to engineering and architecture, because the work involved is so substantially different from virtually all others'. As an example, many physicians, attorneys, and accountants encounter a wide variety of routine challenges that are met through routine responses, permitting any number of reasonably well-prepared practitioners to achieve identical or almost identical results. By contrast, identical results would never occur were different design professionals engaged for the same project. Each firm has a preferred method, and several firms usually are engaged for each project. Their personnel must interact with one another, as well as with those individuals and organizations that become involved later, during the construction phase. The complexities associated even with relatively simple jobs can be enormous, and seldom, if ever, does a project go from start to finish without "glitches" of some type being encountered.

Fee Considerations

Proponents of fee-bidding claim it results in lower fees. While this issue is moot, it is a fact that fee differences are relatively insignificant in the overall "scheme of things." As reported by the U.S. General Accounting Office, design fees account for 1 percent or less of a project's life-cycle costs, but the nature of the design services obtained determines what all the other costs will be. Stated another way, it would be worthwhile to double design professional fees if doing so would yield just a 1-percent reduction in a project's life-cycle expense. But it does not take a 100-percent fee increase to attain the quality needed to derive life-cycle savings far in excess of 1 percent. Very often the cost difference between the highest quality appropriate for a project and minimally acceptable quality is only 10 percent or so. The additional fee is used not only to strengthen quality control, but also to perform close analysis to determine which specific approach is most cost-effective in light of the client's cost evaluation criteria—for example, steel, poured concrete, or post-tensioned; three-lamp luminaries or four; unitary systems or centralized. When such analyses are not performed, most design professionals will rely on techniques that they know will work and that will minimize their man-hour requirements, even though they may not be the most cost-effective for the client. This results in structures or systems that are far more costly to build or operate and maintain.

Focusing on Fees Given the relationship between the cost of design and the costs that arise as a consequence of design, why does the cost of design become such a major concern to so many people? After all, comparable firms active in the same markets tend to charge similar fees. They draw upon the same pool of human resources and usually must offer the same level and extent of services, equipment, and facilities in order to gain consideration from available clientele. Besides this, there is no indication whatsoever that design professionals are wealthier than other professionals or that their firms are more profitable than other businesses. If anything, design professionals make less money than other professionals, and, historically, their firms' profitability has been low, usually about 5 percent or less.

Two factors tend to cause disproportionate focus on design fees. First, these fees tend to be large, often exceeding several hundred thousand dollars in aggregate. Such amounts would still seem small, given the multimillion dollar projects they are spent for, were it not for the second factor, that is, that design usually is considered independent of the work for which it is being obtained. In the public sector, for example, the people who retain engineers and architects often are "procurement specialists." It is their job to obtain whatever is needed at the lowest possible cost. They either fail to recognize that the quality and extent of a design professional's work are influenced by fee, or they just do not care. For them, lower fee means money saved—period.

Larger private sector organizations operate in much the same compartmentalized, bureaucratic manner as government does. Some smaller ones are faced with different problems, however, because they operate on a combination of their own and borrowed funds. Design typically comprises part of the soft costs that are paid "up front," whereas the cost of construction is financed with borrowed money.

QBS Promotes Economy

Claims that fee-bidding is more economical than QBS are ironic, because QBS can lead, not only to lower construction and life-cycle costs, but also to lower design fees. This occurs because the fee negotiated through QBS is more likely to be the fee actually paid. After all, QBS places such a premium on quality performance, that engineers and architects usually prefer to absorb small time overruns as "client satisfaction insurance" that can help them secure the next appropriate commission from the same client. By contrast, when fee-bidding is employed, the fee actually paid is likely to exceed the fee quoted to win the commission, due to the design fee change orders often submitted during the course of the work, to fund tasks omitted from the workscope. Firms have no strong incentive to absorb extras, because they know that the client will still ask several firms to bid on the next project and that a low quoted fee will still be essential.

In some instances, of course, clients have limited design budgets that must be accommodated. Even in these cases, QBS is the preferred alternative, because it causes clients and design professionals to work together in deciding on workscope cutbacks. This gives clients far more control over their own risks and permits design professionals to assign personnel who can produce solid results within the time and budgetary constraints imposed. The end result is a scope of services that minimizes design fees, maintains sensitivity to client and design professional risk preferences, and reduces opportunities for misunderstandings.

QBS Gives Leverage to the Client

As many clients already know, QBS provides a more advantageous bargaining position than fee-bidding. Once the workscope is agreed to, design professionals being selected through QBS realize that the only thing standing between them and the contract—plus a better "shot" at future work—is the fee. Many design professionals state unequivocally that the fees they would accept for implementing a given workscope established through QBS are less than those they would charge were fee-bidding used. Part of the reason is the leverage mentioned; another element is the professional satisfaction that is derived when QBS is employed. When the gamesmanship associated with low-balling is not a factor, everything is focused on performing as agreed to, fulfilling the client's informed expectations. Still and all, design professionals cannot go overboard in this regard. They require a fee sufficient to support the level of quality appropriate to the work, particularly as that work winds down, when the last—and often most important—quality control procedures are applied. No one is well served by an inadequate fee, especially so when it results in a less than complete professional service.

QBS Reduces Overhead and Encourages More Competition

QBS helps firms reduce their costs of business acquisition, because responding to QBS solicitations is so much less costly than responding to requests for price proposals. Materials sent in response to QBS solicitations are relatively standard: brochures, lists of projects and clients, testimonial letters, and so on; in essence, information that addresses a firm's qualifications in general or places emphasis on certain types of projects in particular.

Responding to a request for a priced proposal requires much more cost. For primes, it starts with selection of a team, followed by a close analysis of the request in order to prepare the technical proposal and the price proposal. The proposals must be carefully worded and reviewed, since they could become binding as submitted.

It is not unusual for firms to invest $2,500 worth of time in responding to "routine" requests for technical and price proposals. On major projects, primes can spend more than $10,000 worth of time. While QBS participants ultimately must invest a similar level of effort, they do so only after they have been selected as the best qualified to do the work. With fee-bidding, such investments are needed on every project. Thus, over time, clients as a group can help reduce the overhead they ultimately must pay by relying more on QBS. Perhaps more important, however, is the fact that more firms are likely to express interest when QBS is employed because it is so less costly to do so. Besides this, firms that compete principally on the basis of quality are far more likely to respond to a QBS request, just as firms that compete more on the basis of price are more likely to respond to a request for a price proposal.

QBS Is Well Suited for the Inexperienced

Sometimes fee-bidding is used because those engaging a design professional have little or no experience and either do not know any better or simply are fearful of using QBS because they have no concept of what is, and is not, a reasonable fee. In such cases, fee-bidding is probably the worst approach that can be used, since the client cannot effectively evaluate a firm's credentials, experience, or technical proposal. It would be far wiser for such an individual to rely on QBS but to retain the services of others for assistance in implementation. These others could comprise another design professional, a group of several design professionals, or experienced colleagues.

THE NEED FOR DESIGN PROFESSIONAL ACTIVISM

It is easy to understand why those who procure design professional services want to keep fees as low as possible and why they believe fee-bidding is an important tool for that purpose. How many would have a different attitude were they better informed of the problems associated with fee-bidding and the benefits of QBS? Hundreds, if not thousands, based on experience in Wisconsin. There, the state American Consulting Engineers Council (ACEC) and American Institute of Architects (AIA) chapters have pooled their resources to hire a coordinator to speak with local public officials who seek bids. The coordinator's success rate is almost 100 percent. Design professionals can help their own cause substantially, not only by supporting the engagement of coordinators, but also by becoming directly involved. All too many either respond to fee-bid solicitations without comment or simply discard them. Those who do feel it appropriate to speak up often will turn the matter over to an association, rather than risk offending a prospect. As association officials who get involved in such issues will point out, however, it is difficult to mount an effective argument when the client notes, "If my procurement methods are so terrible, why is it that I received 15 responses to my request and not one negative comment?"

Activism Through Nonprofessional Response

The design professions' best ally at this time may very well be those practitioners who respond to fee-bidding with the specific intent of taking advantage of the client. Those who take this step can rationalize their action by saying that clients who want a professional response use QBS, because QBS is the technique used to obtain virtually all professional services. Why respond professionally to a nonprofessional request (i.e., fee-bidding) when a professional response will almost assuredly result in someone else obtaining the work? Such arguments have merit. In the early years of Maryland's fee-bidding system, many design professionals who obtained engagements performed top-quality work for ridiculously low fees. They did so under the erroneous assumption that the financial loss they experienced was tantamount to an investment in obtaining future work. The state could hardly be faulted for bragging about the fine results it was obtaining. After all, it was getting far more than it was paying for. It did not take too long for firms to learn their lessons, however, leaving the competition principally to those with different attitudes. For them, low-balling was nothing more than an effective business practice for application in a price-competitive environment but, their superior knowledge clearly gave them an upper hand, ultimately causing the state to opt for QBS.

The positive results that have been garnered from the activities of low-balling design professionals should not be taken as an excuse to forgo professional responsibilities. To allow benignly a situation to get far worse before it gets better hardly represents a professional attitude, especially when professionals have every ability to inform others about QBS and its benefits.

Informing Others About QBS

The low-balling opportunities created by fee-bidding and the increased leverage QBS gives to clients make it clear why fee-bidding can be more profitable than QBS. This being the case, why should engineers and architects want to encourage the use of QBS? There are many answers, and most relate to professionalism.

Among other things, QBS helps establish more professional relationships with clients. If clients customarily establish workscopes and design professionals apply their talents to take advantage of clients, even in part, it will be impossible to establish trust—the glue of all client-professional relationships. Instead of serving society, engineers and architects will prey upon it, ultimately leading to loss of community sanctions and professional standing.

Consider, too, how fee-bidding can retard professional growth. The unilateral workscopes devised by clients are seldom more than standard, "off-the-shelf" approaches that have been used before. Much the same can be said about the unilateral workscopes prepared by design professionals. As a consequence, most new work could become little more than an iteration of prior performance, discouraging innovation and tethering the future to the past.

Overemphasizing the bottom line can also encourage design professionals to do whatever they possibly can to minimize overhead. Equipment, facilities, and activities that ordinarily would be obtained to enhance quality will be forgone; the most talented designers will be encouraged to find work in other fields, where salaries are more commensurate with their competence. The goal of design will be meeting minimum criteria, relying on rules of thumb, and using whatever has worked before.

The issue of professional liability also is important. The exorbitant costs of professional liability insurance experienced in the latter half of the 1980s were due principally to the frequency and size of the claims being filed. Many of these claims could have been prevented through better quality, if only fees permitted it. But, all too often, prime design professionals accept a project for too little and then, by reliance on fee-bidding or "jawboning," encourage others engaged to do likewise. QBS creates a vehicle for obtaining a reasonable fee for a better scope of work, one that is in the client's best interests. It also encourages the communication that can prevent misunderstandings and unrealistic expectations—outcomes that often lead to the frustrations which trigger claims.

In essence, the design professions have a huge stake in working for adoption of QBS. This stake makes it essential for design professionals to educate others, in order to gain recognition for, and acceptance of, QBS. Such efforts can be implemented on an *individual basis, firmwide basis, group basis,* or a combination of the three.

Individual Basis On an individual basis, design professionals should always use QBS when selecting other design professionals. They should be aware of, and subscribe to, all the benefits of QBS and should be eager to encourage its use when appropriate, as when an engineer or architect is asked to submit a fee proposal without opportunity for mutual workscope development. Although a letter can be sent in response, consider doing more. Where possible, call the client to explain why a request for quotation (RFQ) approach is not desirable.

When an RFQ is issued by a prime design professional, contact the prime to determine why. If it is to comply with the wishes of the client, those receiving the RFQ should suggest that they and the prime speak with the owner.

Reactive forms of education should be augmented with "proactive" forms. In this regard, one of the most effective approaches is public speaking. It permits design professionals to face important groups and to answer questions.

Initial efforts can be geared toward groups whose members use the services of engineers and architects, for example, local chapters of the Building Owners and Managers Association International (BOMAI), as well as local groups of specialized owners, such as those associated with shopping centers and malls, other retail establishments, and care facilities. Many members of such groups may also belong to organizations such as Rotary or the Chamber of Commerce, which also should be reached. Groups of property managers, lenders, insurers, and others who influence owners also are important, because they, too, have a stake in the successful outcome of a project.

Firmwide Basis   On a firmwide basis, start by educating *all* employees. This can be done by issuing written material and then by holding discussion sessions. During these sessions, or at other times, it may also be appropriate to identify who among area clients and prospective clients frequently resort to fee-bidding and what can be done to change attitudes.

As an adjunct to use of written material or as an alternative to it, consider conducting in-house seminars—before the start of work, at noon, or after work—to explain to one and all the benefits of QBS and the consequences of alternatives. Chief executive officers (CEOs) themselves should be actively involved in firmwide activities. For example, they may direct that, within the next six months, they are to receive copies of all RFQs for their own personal follow-up and that others in the firm are to give two or three talks before certain types of groups, are to write an article, and so on.

Group Basis   A group-basis approach is the best of all, especially because it comprises both individual and firmwide approaches and thus tends to marshal the resources of the entire design professional community. If there are local chapters of design professional organizations in your area, each should be encouraged to discuss the issue of design professional procurement. Where possible, the groups should adopt a goal of encouraging the use of QBS. Numerous strategies are available for attaining this goal. One would be to form a steering committee, umbrella group, or joint organization to apply available manpower and funds to attain common ends. The group could establish a speaker's bureau to reach numerous other area groups; it could employ a coordinator, as in Wisconsin; it could send delegates to meet with major owners and prime design professionals, state and local legislators, and purchasing officers, and, if appropriate, retain a communications professional to develop magazine and newspaper articles, issue news releases, prepare talks and testimony, and create and place advertising. Where there are no local chapters, a general local group can be formed (e.g., Design Professionals of Anytown), which has as its sole purpose defending the use of QBS and encouraging its greater application. A group such as this—or any similar group—could also encourage involvement of groups of trade allies, such as contractors, attorneys, insurance agents, and others, who can speak convincingly from experience to indicate why QBS is beneficial and why fee-bidding creates problems. Local firms and individuals could also form political action committees (PACs).

One of the basic advantages of groups is their ability to raise money for a cause. Thus, when it comes to supporting the use of QBS in the public sector, a group could bring in nationally respected speakers. As an example, certain former heads of the Corps of Engineers, Naval Facilities Engineering Command, and Postal Service construction branch are outspoken advocates of QBS. Executive staffs of ASFE, ACEC, ASCE, NSPE, AIA, and other national associations can provide suggestions in this regard. Many can also provide materials for distribution.

Unquestionably, the lack of attention given to QBS in the recent past has resulted in the current situation and has made the task of maintaining the traditional approach more difficult. But clearly, design professionals have far too much at stake to simply give up. QBS can be restored to its former position of being the near-exclusive method for selecting design professionals, for the benefit of all.

3

Executing a Professional Commission

The extent of service an engineer or an architect is to provide in the execution of a professional commission is determined during a project's proposal phase. In recent years, all too many design professional workscopes have been abridged, reflecting pressures imposed by time and budgetary constraints, as well as those stemming from engagement methods that encourage a low fee. This chapter examines the major services engineers and architects should provide in order to initiate and fulfill their project responsibilities in a professional manner. Many of these services are designed to give a project consistent leadership and direction from beginning to end, and to enhance communication, coordination, and cooperation among all project participants. Failure to provide the full spectrum of services cannot help but contribute to the number of claims and disputes arising from contemporary construction.

THE DESIGN PROFESSIONAL'S ROLE: AN OVERVIEW

At one time it was common for design professionals to speak of their work in terms of "commissions." Today, the reference is often to "jobs." The difference is more than cosmetic. A commission implies a client-professional relationship in which the professional is granted the authority to do what is necessary to achieve the client's goals. A job implies a piece of business, without traditional professional overtones. In fact, all too many design professionals today approach their commissions as jobs, because they fail to perform or even to offer to perform services commensurate with their roles as project leaders.

Thirty years ago, a commission meant that design professionals were in charge of the overall project, from beginning to end. Work was pursued with the attitude that no detail was too small to escape oversight. The prime design professional served as the guiding light and the authority, focusing the attention of all involved on the work that had to be done in order to achieve the common goal: providing a finished project that would meet the client's expectations. Although some owners may still expect design professionals to perform in this way, they seldom do. The team orientation that used to be the hallmark of effective construction has become something of a rarity. Relatively little effort is made to establish a sense of team effort among design professionals, let alone among design professionals and the other individuals and organizations involved. Lacking a unifying force, those associated with a project tend to be more concerned with their own individual responsibilities—and liabilities—than they are with interrelationships with others or the needs of the project as a whole. Some say that the concept of commission has fallen victim to the technical complexities of modern construction, the need to rely on innumerable specialists to deal with these complexities, and the difficulties associated with project management. It is even alleged that these complexities comprise a principal cause of the claims and disputes that on many projects are virtually inevitable.

While construction most certainly is more complex today than it used to be, design professionals have better tools and techniques to deal with the situation. Thus, it is not unreasonable to conclude that the problems arise not so much from construction's complexities as from failure to manage these complexities well. Design professionals would have to share the blame. They have been willing to reduce their involvement in order to offer lower fees. It is easy to blame clients for accelerating these trends, but the professions of engineering and architecture do not belong to clients; they belong to engineers and architects. While customers may always be right, clients are supposed to lack such infallibility. After all, it is a professional responsibility to refuse an engagement if accommodating a client's preferences will compromise professional standards. But turning down work is far easier said than done. Thus, in recent years, professional standards bearing upon the extent of service required to fulfill a commission have been changed and—regrettably—few attempts have been made to educate clients as to what professional service used to—and still should—comprise. This situation cannot be changed easily, but construction claims data alone suggest change is needed. In this regard, engineers and architects should understand how to initiate a commission in a professional manner and should advise their clients of the full range of services required to execute a commission in a professional manner.

Does professional performance require a larger investment on the client's part? It can be argued rather convincingly that the cost benefits derived from enhanced professional performance create value that is far greater than the fee premium required to obtain it, because the improved communication, coordination, and cooperation created can result in far fewer delays, overruns, claims, and disputes—problems that can add hundreds of thousands of dollars to the cost of a project. Overall, then, enhanced execution of professional commissions is in the best interests of clients, just as it is in the best interests of design professionals and the public, whose health and safety engineers and architects are pledged to preserve.

Unquestionably, the higher fees associated with more extensive service create additional opportunities for profit, but the possibility of being charged with self-interest should not dissuade design professionals from offering services that historically have been provided and that demonstrably are needed today more than ever before. In fact, turning the situation around will profit all parties associated with construction. For design professionals, the most significant profit will be the personal satisfaction derived from the superior results of their professional service.

PROPOSAL-PHASE SERVICES

A construction project requires a team effort. As with team sports, results are largely determined by the degree to which members of the team share a common goal and work with one another to achieve it. When individuals are concerned more about their own preferences than those of the team, the team suffers; the goal becomes that much less attainable.

In construction, the prime design professional should function much as a head coach, establishing procedures designed to fuse an assemblage of individuals into a true team and to take that team to victory. Design professionals should initiate team formation procedures by identifying the types of commissions they will seek, because this is the step that determines who the team members should be. As such, execution of a professional commission starts during the proposal phase of a project. This phase begins with identification of desirable commissions and concludes with an agreement.

The discussion that follows, and the discussion in the next section as well, assumes the context of a conventional project in which the prime design professional remains in charge from beginning to end and construction is coordinated by a general contractor. Many of the specific services described are even more important when the owner retains a project manager and/or decides to rely on "fast-track" or "multiple prime" construction contracting.

Identifying Projects for Pursuit

Execution of a professional commission begins by identifying projects for pursuit. The process used should be a direct outgrowth of a firm's business plan, which, as discussed below, should be structured to attain the professional goals of a firm's principals. In some cases, these goals have not been articulated, making "attainment of prosperity" or "making payroll" goals by default. As a consequence, engineers and architects sometimes trap themselves by accepting projects that they are not fully equipped to execute or that entail risks they are not in a position to control.

By clearly defining the types of projects that will provide the most professional fulfillment for a firm and its personnel, engineers and architects can focus on the attainment of commissions whose challenges elicit a professional response. When work is approached as if it were "just another job," the desire to perform well professionally is muted. Stilling professional response is often the first step to a bitter experience.

In many cases, the nature of a firm's practice locks it into a given market. The clients who people that market may have unfortunate attitudes; that is, they may look for lowest cost, even if it means "spotty" service. In such circumstances, principals and project managers should not only maintain effective attitudes but also should strive to encourage professionalism among their own staffs *and client personnel as well.* Saying "it's just not possible" makes it impossible. As continually more problems result from conventional procedures, continually more opportunity exists to effect improvements, through more professional service. By working for improvement now, even with clients who are particularly resistant to change, change is far more likely to occur.

Design professionals who commonly obtain their commissions through subcontracts with a prime—"interprofessionals"—should embrace the same attitudes as other engineers and architects. They, too, should establish professional goals and business plans, instill the desire to perform in a professional manner, and demonstrate willingness to educate their clients—and their clients' clients—about the proper execution of a professional commission. As prime design professionals, they should refuse engagements that require expertise they do not possess or that require acceptance of risks they are not in a position to control.

A firm that pursues any and all projects it can, without defining its goals and objectives, is not performing in a professional manner. Instead, it is acting as a commercial entity that happens to purvey engineering and architectural services. While this approach may permit short-term profitability, it cannot help but lead to professional bankruptcy in the long term, if not financial bankruptcy as well.

Assembling the Design Team

For building projects, the design team typically consists of the architect as prime design professional, working with independent firms that provide structural, mechanical and electrical, and geotechnical engineering. Additional specialists may be required, retained either by interprofessionals or the prime. When public works are involved, the prime typically is a civil, civil and structural, or environmental engineer, supported either by independent specialists or those maintained on staff.

The prime must formulate a policy with respect to those subconsultants who may be asked to serve on other teams. Will this be permitted? If so, will it be permitted for all subconsultants or just a selected few, such as those who happen to possess a rare expertise or who satisfy other criteria? If some are permitted to serve on more than one team but others are not, what will the prime do to help prevent decay of team morale? Can the situation be accommodated well?

In the event that a subconsultant is permitted to serve on more than one team, how will that subconsultant be treated with respect to the balanace of the proposal development process? Of course, a pledge of confidentiality is essential. But what about participation in the preproposal meeting (discussed below)? How can a prime be assured that an idea developed during its preproposal meeting will not resurface at another prime's preproposal meeting?

The situation can at times be difficult to manage, but it is far from impossible. Primes need to identify the issues that must be considered and then to evaluate the pros and cons of each in light of the project and people involved.

The *timing of team assembly* is important. Interprofessionals report that the task is all too often left to the last minute, requiring team members to execute concepts and workscopes developed by the prime. This is no way to build a team, especially so because some team members may feel it is demeaning. If anyone should treat professionals with respect, fellow professionals should.

All team members should be given an opportunity to create their own workscopes and to comment on the overall scope of service that will be proposed to a client. This is far more than professional courtesy; it is basic to professional liability loss prevention. Specialists are far more knowledgeable about appropriate approaches than others and are far more aware of the new or innovative methods that can be brought to bear. Timely team formation can thus result in more effective proposals, to help obtain the engagement, as well as better communication between team members, more coordination, and more cooperation.

Timing is not the only factor essential to effective team building. *Experience, capabilities,* and *availability* also are important concerns. Accordingly, prime design professionals should contact a prospective client before assembling a team to learn more about project requirements and thus to define better the attributes specialists should possess.

Will a firm be able to assign appropriate personnel at the time their service is required? Sometimes this concern is not addressed. Instead, firms may be retained simply because their experience looks good "on paper," thus enhancing a proposal. In reality, however, a firm's experience may be of little consequence. A firm that has designed thirty hospitals in the past five years is not really in a position to claim hospital design experience if it no longer employs those who headed the hospital design projects. Prime design professionals have important responsibilities in this respect. No matter how appealing a prospective team member's experience may sound, the prime should take steps to ensure that capabilities are genuine and that the firm will be able to assign seasoned "veterans" when they are needed.

Fee considerations should, ideally, have little bearing upon design team selections. Primes should select the best firms available to do the work, providing they are compatible with other team members and their fee structure is reasonable. Should an owner express a strong desire to minimize fees, it would be appropriate to acquaint prospective team members with the situation. Will they be willing to take on a project with such restrictions? If so, fine; if not, it is far better to learn about it sooner rather than later. Primes need to be diplomatic in this regard. Interprofessionals should not be told, in essence, "We know how you prefer to execute a commission, but, in this case, you've got to keep your fee down." Subconsultants can take this as an insult. What they hear the prime saying is, "We want to take advantage of your reputation, but we don't want you to apply the techniques you've used to earn it." Why do primes sometimes commit this blunder? Because they don't regard it as offensive. For many, its implications are a fact of life. For example, when the state of Maryland relied on bidding to retain design professionals, it would often counter certain criticisms of its methods by noting that, on any given project, twenty or thirty "top" firms would express interest. "If our methods are so inferior," Maryland officials would comment, "why is it that so many good firms are anxious to get involved?" The actual reason is basic: Many firms seek all the work they can get in order to maintain their staffs. They have grown used to adjusting the level of their service to "market conditions." To their credit, many of these firms often produce good work despite their clients, often by using commercial tactics, for example, design change orders, in order to enhance the extent and/or quality of their service after the commission is in hand. Nonetheless, *professionals should never ask other professionals to cheapen their service in order to serve on a design team*. Design teams are made up of people. How people approach their work determines how that work will be executed. A professional response by all team members is in the best interests of all parties. Professional response is the usual reaction to a professional inquiry. The manner in which professionals are asked to participate on a team thus forms the foundation of the team's attitudes. Any approach that overemphasizes fee at the outset immediately focuses professional participants' attention on the bottom line instead of on the project itself. Concern shifts from services that best fit the project to those that best fit the fee. Those who believe otherwise; those who believe design professionals always proceed in a professional manner regardless of fee, are naive. Somewhat ironically, they ascribe superhuman virtues to professionals whom they treat in an unprofessional manner. A nonprofessional client can be forgiven for this lack of professional awareness; a prime design professional cannot. If a project is to be pursued effectively, the design professionals involved *all* should function as a team. In order for them to do this, they must have a common objective. That objective should be meeting the client's expectations. When a professional must continually worry about minimizing performance in order to earn a reasonable profit, self-interest naturally begins to predominate.

Proposal Formation

The nature of a design professional's proposal and the techniques used in putting one together are heavily influenced by the engagement practices of the client. When QBS is used, proposal development is a multistep process comprising creation of a statement of qualifications, development of concepts, interview participation, and, ultimately, establishment of a mutually developed workscope and agreement on fee. When fee-bidding is used instead of QBS, the various steps usually must be conducted as one.

Team Participation Needed Irrespective of the procurement method used, it is imperative to emphasize team participation to the greatest practical extent, given that some team members may be serving on more than one team. Even when this is the case, however, primes can make a concerted effort to sow the seeds for communication, coordination, and cooperation—three essentials for the establishment of quality and prevention of claims and losses. If the prime is to make the design professional team a genuine team, something more than just an assemblage of firms, work should begin as soon as possible, preferably during the proposal phase. Interprofessionals should be wary of primes who are unconcerned about team development. Primes should be wary of interprofessionals who are unwilling to act as team players.

Initial Proposal Meeting It is not uncommon for the prime to develop the proposal on its own, after each team member submits the materials needed to create a statement of team qualifications, for example, brochures, lists of pertinent projects and clients, and Standard Forms (SF) 254 and 255, federal government forms (used by many state and local governments as well) that describe firm experience in general (SF254) and that are particularly relevant to the project in question (SF255). Team members seldom voice complaint about such an arrangement, since each saves time. But the value of the time saved may not be nearly as great as the value of the quality lost. This is why proposal development should begin with a meeting of prospective team members whenever possible. Holding a preproposal meeting does more than just simply support team-building concepts. It fosters an exchange of ideas and experiences. In fact, such meetings can yield important information that can help make the proposal more to the point for the specific client involved or that can result in the selection of additional team members who are specialists in areas that might otherwise have gone overlooked.

Meeting Participants The individuals asked to attend the proposal development meeting should be selected with care. Often they consist of firms' principals and/or marketing personnel. While these individuals will have valuable insights to offer, they usually are not the people who will be responsible for the commission, once it is obtained. This is the work of project managers. To help assure their commitment, they should be at the meeting, too, to be involved with the project from its inception. Besides, they often are in a position to offer suggestions that in some cases can make the difference between obtaining the project and being an "also-ran."

Team Composition As an initial element of the meeting, the prime design professional should introduce each participant to the others. If the prime has not already done so in writing, any restrictions on discussion should be made clear, to help ensure that persons serving on more than one team are not put into a compromising, or potentially compromising, situation. When this is not a problem, participants should reveal everything they know about the project and then should discuss their experience with the type of project and client involved, to obtain comments about prior projects that should be referenced in promotional materials. Discussion should also focus on the full range of disciplines that should be considered. Numerous specialists may be required; it would be an error to omit any, such as illumination system design, landscape architecture, or fire prevention engineering, among many others. When there is uncertainty about the need for a given specialty, it usually should be included with an advisory note that it is available if needed, followed by information about the firm or individual that will provide it.

Contractual Issues Once team make-up has been resolved, the matter of contractual relationships and general conditions should be discussed. Overlooking these issues can result in subsequent squabbles that can disrupt the team. To help prevent this problem, the prime should furnish copies of the contract that will be used and ask meeting attendees for review and comment by a specific date. The prime should also prepare and distribute a schedule of subsequent activity, for example, identification of additional consultants, provision of promotional materials, and so on. The schedule should indicate when the draft proposal will be circulated for comment and when it will be finalized.

Second Meeting When QBS is being used, a second meeting may not be needed, unless it becomes necessary to add firms to the team, resolve concerns about the draft proposal, or deal with other important issues. When fee-bidding is involved, subsequent meetings are necessary to prepare technical and price proposals, illustrating why fee-bidding can result in so much more overhead for project acquisition.

Formation of Design Concepts

The manner in which design teams develop their concepts for a project also is affected by the client's engagement technique.

QBS versus Fee-Bidding When QBS is employed, design teams usually refrain from developing or committing to specific design concepts until they make the "short list", that is, until they are informed they are being considered for the project, along with several others. By contrast, fee-bidding procedures require full articulation of design concepts before a proposal is submitted. In either case, the design concepts that emerge are formed unilaterally by the design team. When QBS is applied, however, the design concepts are subject to extensive modification, by virtue of the discussions that comprise mutual workscope development. As a consequence, QBS encourages design team members to broach whatever methods (including alternates) they prefer. The project is well served because the owner can draw upon several concepts, while design professionals are encouraged to apply all their professional capabilities, a situation that can heighten their enthusiasm.

When fee-bidding is used, technical proposals are based on assumptions about client preferences. It usually is prudent in these cases to assume the client is working for a low fee; why else would the client have opted for fee-bidding? In some cases, of course, clients state that they will give far more credit to the technical proposal than to the fee proposal. And they mean it. But, in these cases, several firms often are awarded the maximum possible technical merit and experience "points," making fee the *sole* selection criterion. Experienced firms recognize this and thus strive to keep the proposed fee as low as possible, no matter what "weighting" is assigned to it. As such, whenever fee-bidding is used, fee almost always becomes an overriding issue that can result in significant limits to the services firms propose and their enthusiasm for the project.

Formation of Design Concepts All parties are well served when design concepts are developed in a professional manner. The prime design professional should set the tone by developing an overall concept, explaining why it was selected, and inviting comments. As a professional, the prime should be mindful of Greenwood's spirit of "rationality": Criticisms should be welcomed and adjustments should be made as appropriate, to help assure the full team "buys into" the approach. Then, once the overall concept has been accepted, other team members can articulate their discipline-oriented concepts, also welcoming discussion.

A professional approach to concept development involves design professionals more intimately in the project, helps them understand and appreciate one another's objectives and concerns, and encourages development of well-considered strategies.

Shortcuts Shortcuts are sometimes used at this stage. Typically, they involve use of standard verbiage and reliance on concepts that have worked before. While shortcuts may save time and money in terms of proposal development, they can sap enthusiasm for a project and lead to greater construction expense. Nonetheless, there may be few alternatives, particularly when fee-bidding is involved.

Commit Concepts to Writing Once design concepts have been agreed to, they should be committed to writing for team review prior to interviews.

Interviews

Interviews are a common aspect of QBS procurement, particularly when major projects are involved and/or the client has not previously dealt with a firm. Although the principal purpose of interviews is discussion of design concepts, they also permit owners to learn more about interested design teams and design teams to learn more about their prospective client. The importance of the "eyeball-to-eyeball" contact that such interviews promote cannot be overemphasized. It has a significant bearing upon the client's selection and should have a substantial effect on specifics of the design team's technical proposal, since that proposal, to be effective, must be sensitive to the client's attitudes, in addition to the project's needs.

The interview is held either at the client's office or the prime design professional's office. Unless they have been there before, sophisticated clients usually prefer to visit the prime's office to obtain more information about the firm and its operations. Ideally, the entire design team should be represented at the interview, with each firm being limited to one or two representatives (depending on team size), so the client is not overwhelmed. Individuals selected to represent a firm should be up to the task. If a project manager has the personality and experience to do it, so much the better. If a project manager will not be effective, however, someone more competent should be used.

Particularly in the case of major projects, it is wise to conduct at least one mock interview in preparation for the "real thing." Someone with experience should play the role of the client, asking questions typical of those that clients often pose. All participants should take notes about the responses given. After the mock interview is over, responses should be subject to discussion. The process could then be repeated, perhaps two or three more times, until everyone is satisfied that they are ready.

The format of the interview bears consideration. Should slides be used? A film or videotape? The prime design professional should confer with the client or client's representatives to learn more about the interview before it transpires. How much time will the client be spending? Does the client prefer audiovisual presentations? What types of information does the client want to know most about? At the very least, such inquiries demonstrate strong interest in the commission.

Some prime design professionals limit interview participation to their own personnel, particularly when subconsultants serve on more than one team. In these situations, primes need to recognize that they do not possess the full range of expertise needed to reply effectively to all questions that the client may pose. Primes must be careful not to promise more than can be delivered.

Preparatory mock interviews can be of particular value in these situations or when the client states a preference to meet only with the prime.

Prebid Meetings When fee-bidding is used, interviews—if they are employed at all—generally take the form of "prebid meetings." These are conducted to acquaint interested firms with project goals and objectives, and to address any questions raised. The full design team usually should be represented at such meetings.

In preparation for a prebid meeting, design team members should identify their questions when they meet to discuss design concepts, or, alternatively, they should meet to identify their questions before they commit to a design concept. In developing questions, team members must recognize that any they pose will be heard by their competitors.

Some public clients permit private question-and-answer sessions, whether or not they hold prebid meetings. The substance of these meetings usually must be made known to other interested firms. This is not necessarily the case with private-sector projects, so the client's position should be assessed beforehand. Obviously, when a firm notices errors or omissions in a client-developed workscope, it will enjoy a competitive advantage by keeping that knowledge to itself. As such, the competitive pressures of fee-bidding can create a situation in which the design team's best interests are served by withholding information. This is not what most would regard as professional behavior. But it is legal, and it is a situation created by the client, not the consultant. This is just another reason why astute clients and concerned professionals prefer QBS. It is the only procurement method through which the design team's best interests are served when it demonstrates its desire to serve as the client's trusted professional advisor.

Workscope Development and Agreement

When QBS is used, the owner considers the team best qualified for the project to be best qualified to provide guidance for developing the project's workscope. By participating in the process, the owner becomes part of the team, helping to assure mutuality of goals. When fee-bidding is used, mutuality of purpose cannot be easily established. The workscope itself is seldom more than a "rehash" of methods used before, by the client or design team, and seldom creates the "fit" associated with QBS workscopes.

Developing the QBS Workscope and Contract QBS workscopes commonly are developed in two settings. First, the design team assembles its preferred workscope on a unilateral basis. Second, the preferred workscope is presented to the client for review and discussion. The discussion process results in a mutually agreed to workscope and a solid understanding of each party's goals and objectives, needs and preferences.

In all cases, the initial workscope should comprise all services needed for proper execution of a professional commission. Critics of QBS sometimes charge that certain of these services are frills, offered more to line design professionals' pockets than to meet the client's needs. Such charges evidently are based on inadequate knowledge and the belief that somewhat routine projects do not require the level of quality control applied for major undertakings. In fact, the vast majority of all claims and losses stem from somewhat routine projects, simply because they are taken for granted, lulling participants into complacency. Initial workscopes should thus be comprehensive for *all* projects, and those suggesting services should be prepared to justify the reasons for each. Important services should not be eliminated simply because the client may object to the fee. Some clients may be unaware that such services are available, and, upon learning about them, may be eager to authorize their use, especially if they may be able to prevent certain problems they have experienced in the past. To help assure that the design team's intent is not misconstrued, all services offered should serve the client's best interests. If, after hearing explanations, the client wants to eliminate some, so be it, *but* such a client should be required to underwrite any risks its decision imposes, assuming the team is still willing to take the work forward.

A number of specific approaches are available for developing the workscope. The best emphasize team involvement, as when the prime requests each team member to submit a workscope by a given date, after which the prime assembles a complete workscope in draft. The draft, which indicates the services to be provided by whom and when, is submitted to the team for review. A team meeting then is held where project managers and others discuss the draft and offer suggestions. This type of approach reinforces team concepts and takes advantage of team capabilities. It also familiarizes team members with scheduling requirements, points of interface between different team participants, reporting responsibilities, and the prime's policies regarding matters such as documentation and issuance of memoranda.

Once a preferred approach is agreed to, fall-back positions can be established to identify the nature of changes individual team members would agree to in the event they become necessary. The risks created by service reductions should be analyzed closely, from both the client's and design professionals' viewpoints.

Participants should also discuss their fees, the manner in which fees may be affected through modifications, and the manner in which these modifications may affect other costs. Once discussions are complete, the technical proposal can be finalized and submitted to the client for review prior to workscope finalization.

Workscope finalization takes place at meetings with the client. The client will often be represented by several individuals, each of whom has specific areas of concern. The design team should be represented in full. There is no set routine or pattern to such sessions. In some cases, as when certain federal government procurements are involved, the client's representatives may be knowledgeable of technical issues and may pose highly specific questions. On the other hand, a client may say, in essence, "Do whatever you want" or, more likely, "Do whatever you want as long as it doesn't cost more than...."

Initial workscopes are based on various assumptions, and one way of beginning a meeting is to inform the client about these assumptions. In this regard, each member of the design team should feel free to ask important questions, to learn more about the client's perceptions as they affect that design professional's work, and, as necessary, to modify client attitudes, in order to bring them more in line with reality or conventional practices.

It is at this stage of the project that the client and the design team meet for the first time, to bear down on project specifics. Firm representatives should view this meeting as an opportunity to "flesh out" the client's goals and objectives, to adjust the workscope to accommodate these, and to help clear up any potential or actual misunderstandings or false expectations. Clients who are working with restricted schedules or budgets need to understand the risks any workscope adjustments may create and the needs of design professionals with regard to managing or transferring these risks.

Overall, workscope development may be the single most important element of a commission, because it makes the client part of the team, results in common understandings and expectations, and focuses all parties' attention on the project, reinforcing the common desire to fulfill the commission in a professional manner.

Once a workscope has been agreed to, the design team can establish a fee for its implementation. If the fee exceeds the client's budget, another meeting can be held to discuss workscope changes. With a few rare exceptions, a workscope and fee are agreed to whenever QBS is used. If not, it almost always means the client has unrealistic expectations or is asking design professionals to accept what the wisest would agree are intolerable risks.

Contract Development   After workscope and fee are established, the prime design professional prepares a contract. If the contract's general conditions have not previously been reviewed by the team, they should be reviewed at this point, before they are submitted to the client. Review helps assure that general conditions included in team members' contracts with the prime are appropriately considered in the prime's contract with the owner. It also enhances quality control with respect to such factors as workscope and schedules.

In some cases, the prime's contract will be signed "as is." In other cases, disagreements or questions will make a negotiation session necessary. Negotiation sessions usually are attended by the prime and the owner. In this way, the prime can tentatively accept a client's proposed revisions, subject to the approval of other team members. This creates an opportunity for the prime to review the proposed revisions with other team members, in order to determine appropriate responses. For example, if the client is unwilling to accept certain limitations of liability, the team would be able to suggest the fee premiums required to offset their increased risks and/or the additional services necessary to minimize those risks. Based on discussions such as these, the prime would be able to return to the owner, possibly accompanied by other members of the design team, to discuss the alternatives available to reach agreement. Once these issues are resolved, one final meeting would be necessary, to sign the finalized contract. This last meeting could be attended by the prime's staffer who will issue bills and the client's representative who will pay them or authorize payment. They can review payment terms and conditions at the meeting, and the prime's financial staff can learn when bills should be issued, the format required, to whom they should be sent, when payment can be expected, and whom to call with questions. This is more than a courtesy: A great many disputes can be traced to payment problems. Answering questions at the outset reduces the potential for problems.

Fee-Bidding Approaches A far different approach must be used when the workscope is unilateral, based on a client submission or one of the design team's creation. In either case, a variety of questions will exist, but some will be unanswerable. As a consequence, the workscope includes some assumptions, most of which promote lower fees by offering less service. As such, when fee-bidding is used, the design team sometimes get together not so much to develop a workscope as to determine the lowest level of effort participants can live with. Because such workscopes elevate risk, it is vitally important to have effective communication, coordination, and cooperation. In order to save valuable time, however, design team members may rely on shortcuts; in fact, the design team might not meet at all. These time pressures prevail in fee-bidding more so than QBS because fee-bidding is more speculative: Five or more teams may be "in the running," requiring each to develop workscopes and fees. When QBS is used, the team asked to develop the workscope knows the commission is virtually theirs. Ironically, those projects that most need communication, cooperation, and coordination to overcome inherent problems are precisely those that are least likely to receive such a degree of care.

If it is impractical or impossible for the design team to meet to develop a workscope, then—at the very least—the proposed submission should be subject to comprehensive review before it is submitted; at least one meeting should be held to bring the design team to agreement. Failure to take this step can prove disastrous. The workscope or other elements of a submission may require agreement to terms and conditions some team members dislike or may even find impossible to abide by, for example, insurance. It also is important for the team to review and/or suggest workscope alternatives, as well as the listing of assumptions that should be included. In many cases, assumptions are not identified at all, and this omission can lead to serious problems. Those who adopt a "we'll cross that bridge when we come to it" attitude may discover that, when it comes time to cross the bridge, there is none.

DESIGN-, CONSTRUCTION-, AND POSTCONSTRUCTION-PHASE SERVICES

The various procedures discussed above, when followed, are employed by design professionals for purposes of quality management. Design team members recognize that investing the time required for quality management is a professional necessity; something that is needed to support professional performance should they be accorded the commission. By contrast, most of the quality services performed later generally must be authorized by the client. As it so happens, however, many firms do not reference these services separately in their workscopes, even if they do intend to perform them. This can be a mistake. Quality management services should be clearly identified and described, so the client realizes what they are, why they are suggested, and that they will be pursued in something more than a casual manner.

Due to competitive pressures, certain quality management services may not be included in the basic workscopes associated with fee-bidding engagements. The services involved would be those that exceed the level of care and effort dictated by the current standard of practice. In such cases, the additional quality management could be offered as optional services that the client is strongly encouraged to use. As with communication, cooperation, and coordination, these services are least likely to be used when fee-bidding is involved, and it is in precisely these circumstances that they are most important.

Predesign Planning and Conference

After a design team wins a commission, it should develop or reconfirm the schedule for completion, in accord with its contract, and identify the individuals assigned various roles. Once again, team building is essential, to focus all participants' attention on the project. Each design team member's project manager should meet with appropriate staff to resolve scheduling and assignment issues and to discuss the project. Scheduling and assignment information should be reviewed by the principal in charge, assuming this is appropriate, to make any revisions required. Once finalized, the information should be submitted to the prime design professional for development of the overall schedule and assignments. The prime should then conduct a predesign conference at which the schedule and assignments are subject to mutual review and comment prior to acceptance through consensus.

The predesign conference can be a brief working meeting or something more elaborate. Elaboration is suggested, because it can give the project an aura of importance and thus elevate each participant's sense of responsibility. How often should a more elaborate approach be used? Consider this: Any project a firm accepts, no matter how small, can lead to a multimillion-dollar lawsuit that threatens the very existence of the firm and the careers of all those associated with the project. (*This is not an overstatement.*) As such, from that perspective alone, *every* project is important. That being the case, it always is in a firm's best interests to demonstrate its principals' concern and enthusiasm. Those who react to such approaches with a somewhat skeptical or jaded manner are precisely the types of people whose attitudes can lead to unfortunate outcomes.

Elaboration does not require a great deal of time or expense. For example, the prime could prepare a project notebook for each design team member's principal in charge and/or project manager. The notebook could be embossed or labelled with the project's name. Its contents might include a copy of the contract between the team member and the prime, the team member's schedule and overall project schedule, the team member's personnel assignments and overall assignments, a list of other project managers and their telephone numbers, and so on. Notebooks could be presented at the predesign conference, with the prime reviewing their contents. The review would be enhanced by having slides made of certain common elements, to establish an audiovisual presentation. The cost of the notebooks and slides would probably be less than a hundred dollars.

The conference itself can be made more elaborate, by expanding the attendance list to include the supervising principal of each firm, the project managers, and possibly other members of staff. The owner and owner's representative also could be invited. Pursued on a somewhat grand scale, which would be more appropriate for major projects, the prime design professional would serve as the master of ceremonies, first introducing the owner to explain the history of the site, if applicable, the purpose of the project, time and budgetary issues, and related concerns. The prime's representative then could display (if available) preliminary sketches, possibly through a slide show, identifying basic parameters of initial concepts, followed by an introduction of each of the other design professionals involved, to discuss the approaches they intend to use and any of their unique features. The prime could conclude the presentation with a review of schedules and introductions of each project manager.

While a predesign conference does not have to be elaborate, involving all who will contribute—especially the owner—demonstrates and reinforces the fact that all are essential to the project's overall success. It also reinforces the concept of team, encourages a high level of communication, coordination, and cooperation, and focuses everyone's attention on the project. An elaborate meeting may cost a few hundred dollars, and it could easily be orchestrated by the prime's marketing department. (It might even be appropriate to invite selected members of the news media.) Time requirements likewise would not be onerous, because the meeting could be held early in the morning or after work. The value? If just one project member identifies and helps correct an incipient problem because of the "attitude adjustment" created by the meeting, it will have paid for itself. In truth, however, the benefits might go far beyond that.

Development of Plans and Specifications

Whenever fees are kept to a minimum, the work associated with developing plans and specifications also is minimized. Instead of analyzing several alternatives to identify the one that will yield the best results in light of client-developed criteria, "rules of thumb" will be used. This often results in oversizing of elements, such as beams and columns, transformers, and mechanical equipment. While such procedures reduce design fees, they elevate construction and life-cycle costs. This fact should be explained to clients.

The matter of details also should be discussed. All too often the extent of detail provided is minimized, permitting contractors far more leeway and thus far more opportunity to misinterpret design intent. If less detailed plans are to be developed, it is essential to perform close review of shop drawings, as well as a high level of construction monitoring. Failure to do so can result in errors, omissions, ambiguities, and other problems that lead to claims.

No matter how low clients want fees to go, design professionals should never abandon their duty to protect the public's health and safety. All plans and specifications, and the calculations behind them, should be double-checked before leaving a firm. This is a basic duty of care that design professionals owe to the public, their clients and, ultimately, to themselves and their professions.

Mutual Review

Particularly when time pressures are applied, a variety of quality control-oriented procedures may be sacrificed. The time saved initially often is inconsequential; the time lost later can be monumental. One of the more important quality control functions is mutual plan review. The process should start within a firm, as internal peer review, during which plans developed by one design professional are reviewed by a colleague. Plans then would be sent to other design team members whose work affects or is affected by them for review and comment. This procedure does far more than recognize the interrelationships between professional team members. It also reduces the potential for errors and omissions, or for interference between one discipline's plans and another's. As plans evolve from preliminary to near-final, review can be continued.

Sometime during the course of a project, and perhaps more than once, the prime could call the design team leaders together for an update session, to review progress to date, changes that have been made, schedule revisions, and other issues of concern. Although face-to-face meetings are best, such sessions could be conducted through teleconferencing.

Creation of a team concept can also result in valuable peer pressure, as when one member of the team is delayed and thus holds up others. Instead of having to face the disapprobation of the prime alone, the project managers would be confronted by several concerned individuals. Just the thought of this happening might be an effective preventive.

Note that mutual review can be extended to specifications and other design professional "deliverables" as well. Small problems caught might yield significant savings later on, due to one individual's "special knowledge." As an example, an indoor pool creates certain types of corrosion problems that make proper specification of structural fasteners crucial. If the structural engineer is not familiar with this need, chances are the mechanical engineer is. Nor does the information have to be specialty based. The structural engineer who reviews cooling tower plans may know that cooling towers can spread legionella bacteria to neighboring areas, creating a pollution exposure that may be uninsurable. Upon learning this from the structural engineer, the mechanical engineer may call for modified cooling tower design and/or the use of permanent water quality monitoring equipment.

Mutual plan review can be identified as a specific service in the workscope, work schedule, assignment list, and fee proposal. Some of the tasks will be basic, that is, necessary to achieve the prevailing standard of practice. Other tasks may be suggested to elevate quality and reduce risk. A client who dilutes this aspect of work would impose unnecessary risks on engineers and architects. In that case, the client should bear the risk, through partial indemnification and/or a risk-funding-based fee premium.

If mutual plan review is not listed separately, the client might assume that all firms perform it to the same level indicated in the scope of services. This is not usually the case, nor are clients likely to request the service on their own. By listing the service separately, it is called to a client's attention. This gives the design team an opportunity to explain its purpose and benefits, and to strongly recommend its use.

Contractor Selection

Surveys reveal that negotiated retention of construction contractors is becoming steadily more popular among private-sector owners, and with good reason. The change orders commonly needed during a construction project can make bidding a misleading procedure, since the dollar figure used to select a provider is rarely the amount actually paid. Contractors who ultimately win the work may do so simply because a higher bidder included the cost of something the low bidder neglected to consider or purposely overlooked, knowing the point to be arguable and thus subject to a change order later on.

According to contractors, a root cause of bidding problems is a level of detail in plans and specifications that leaves far too much to the judgment of contractors. Some have even gone so far as to say that inadequate detailing is the direct result of pressures put on design professionals to keep fees as low as possible. And the position seems to have merit. In some cases, public works projects whose cost has exceeded $500 million have experienced change orders of less than 3 percent, in large part because the owner insisted on (and paid for) highly detailed plans and specifications.

Even when plans and specifications are detailed, however, bidding can involve gamesmanship. Those offering a bid act somewhat as the adversaries of those seeking it. For many, the goal is to identify all contestable issues "up front," make assumptions about each that tend to lower the asking price, and then claim errors, omissions, ambiguities, and conflicts or inconsistencies later on. This is not to say that construction contractors cannot be trusted. However, when the financial pressures of bidding are applied, many people—even design professionals—will act against their better judgment or nature. As already noted, design professionals are not immune to this syndrome.

ASFE was the first association of design professionals to go on record in favor of negotiated procurement of construction contractors, due to the significant benefits that can result. Key among these benefits is selection of contractors based upon their capabilities, that is, their ability to do a good job for a reasonable price. They can be involved far earlier in a project, to review plans and specifications in their preliminary stages, to comment on their constructability, and to otherwise offer comments and suggestions. This review helps prevent problems that often result in delays, claims for extras, and disputes. In addition, early involvement also permits contractors' representatives to become better acquainted with design professionals' representatives, building on the concept of teamwork and project, rather than fee, orientation.

Although the price quoted for negotiated work may be higher than that obtained through a bidding procedure, the price actually paid may be the same or less, due to fewer changers and better contractor attitudes. Contractors also realize that the best assurance of obtaining another negotiated project from the same owner is to perform good work and, on occasion, bend rather than break.

If negotiated contracting will be used, design professionals would serve to identify the qualifications of candidate firms and assist the owner in selections. This same service could be provided if bidding is used, as part of a prequalification process that would permit only reputable contractors to become involved. Prequalification is far more effective than disqualification after the fact, especially so because a low-bidding contractor who is not awarded the project, due to prior poor performance, might file suit against the owner and/or design professional who vetoed or recommended against acceptance of the contractor's bid. This potential should be recognized, called to the client's attention, and be contemplated in a contract's general conditions.

Prebid Conference

When bidding is to be used to select contractors, prequalified or not, the design professionals should hold a prebid conference. It may also be appropriate to hold specialty conferences for subcontractors.

Prior to the conference, design professionals should review their plans and specifications "one more time" in an effort to identify any errors, omissions, ambiguities, inconsistencies, or conflicts. Contractors should be told of the following: any modifications to bidding documents; the need to identify any possible problems, such as errors or omissions, and treat them as priced alternatives; the time schedules involved; various penalty provisions; the basis upon which selections will be made; and such other information as pertains to the project.

As with virtually all other meetings in which design professionals participate, the prebid conference should be tape-recorded and transcribed, or copious notes should be taken for development of minutes or a summary, for the record. Not only does creation of a written record enhance communication and thus prevent problems, but also it can be used to help block any spurious claims.

Attendance at the prebid conference should be made a precondition for consideration of a contractor's submission.

Preconstruction Conference

Design professionals should host a preconstruction conference to help avoid misunderstandingss, identify the roles each will play during the construction process, and introduce field personnel. If appropriate, separate preliminary conferences should be held between specialists, for example, between the mechanical engineer and the mechanical contractor.

Note that a preconstruction conference can be made more elaborate, for the same reasons that a predesign conference would be: to elevate the project's importance in the minds of participants, to focus attention on the project, and to establish a sense of team effort. As with the predesign conference, part of the process could be introducing various team members to others attending, in part to help recognize their importance as members of the team and that others are counting on them to perform well. The owner and various design professionals could be on hand to show slides of renderings and to discuss certain unique features of the project.

Some design professionals scoff at the notion that somewhat elaborate preconstruction conferences can alter a contractor's attitudes. Such design professionals may be as jaded as they accuse the contractors of being. In fact, not too many years ago, virtually all parties associated with a substantial project approached it with eagerness and anticipation, as if they were about to embark on an adventure. In some instances, this attitude was heightened by the nature of the project, since it would significantly affect an area's skyline or people's living standards. The built environment is far more crowded today, but, in truth, many of the projects being embarked upon are far more substantial and far more complex than any of those in the past. This aspect of significance has not been extensively utilized in team development, contributing to a self-fulfilling prophecy: Few attempts are made to foster a sense of project pride in the belief that it cannot be fostered. Is that belief genuine, or is it merely the cause cited for not making an effort? Unless an effort is made, potential benefits cannot be realized. And, when all is said and done, human nature is still human nature. The factors that motivated people a century ago still exist today.

Shop Drawing Review

Shop drawing review is a vitally important design professional service. It is the final element of design review, helping to assure that design detailing performed by contractors conforms with the intent of design. "Design intent" is a somewhat vague term, however, and one that design professionals have never standardized. Thus, when inadequate review of shop drawings allegedly contributed to the Kansas City Hyatt failure, where more than one hundred persons were killed, it was left to Judge James Deutsch to define what "design intent" really means. In essence, he said that it is the intent of the design for the design to work. As such, close review of contractors' shop drawings is essential insofar as building design (as opposed to construction methods) is concerned, particularly when the drawings involved require any degree of design judgment.

The need for close review of shop drawings became continually more prevalent from the late 1950s and early 1960s, as the extent of design detail was reduced in response to time and budgetary pressures, placing continually more design responsibility on contractors. Regrettably, the same pressures that resulted in less design detail also resulted in less attention being paid to contractors' submissions. Alarmed by this state of affairs and fearful that design professionals' liabilities were being significantly expanded, at least one professional liability insurer encouraged engineers and architects to use less-than-definitive wording on their shop drawing stamps, by employing phrases such as "REVIEWED. NO EXCEPTIONS TAKEN." The insurer particularly discouraged use of the word "APPROVED," fearing that it would make design professionals liable for any errors or omissions committed by the contractor. Regrettably, such wording tended to obfuscate the issue of responsibility rather than clarify it, and some design professionals relied upon exculpatory wording as a device to further reduce the rigor of their shop drawing review. Not only did this lead to more claims and disputes, but also it put decision-making authority into the hands of judges and juries after the fact, as opposed to the hands of those directly involved and in a position to resolve issues beforehand.

It is suggested that design professionals should include shop drawing review as a separate element in their workscopes, in order to focus attention on the issue and establish adequate time and fee allowances. It is also suggested that procedures be clearly identified and adhered to. It is generally agreed that proper procedures should involve the following.

1. Design professionals should identify the shop drawings required, the schedule contractors will adhere to in preparing initial and follow-up submissions, and the schedule design professionals will adhere to in reviewing initial and follow-up submissions.

2. Contractors should review each shop drawing before submitting it, to establish its acceptability in terms of the means, methods, techniques, sequences, and operations of construction, and safety precautions and programs incidental thereto, all of which are the contractor's responsibility.
3. Design professionals should review each shop drawing and approve or disapprove it as to its conformity with their design intent and compliance with information given in the construction documents.
4. Contractors should call to design professionals' attention any shop drawing that varies from what the design professionals have called for and should recognize that approval may or may not constitute an acceptance of a change, depending on contract provisions.
5. Contractors should pay design professionals for the review of any submission that varies from what design professionals have required.
6. Design professionals should return any shop drawings they have not required.

Shop drawing review procedures should be referenced in general conditions of the construction contract and should be discussed at the preconstruction and/or prebidding conference. Effective shop drawing review stamp wording might comprise the following:

> DESIGN PROFESSIONAL has reviewed this submission SOLELY for its conformance with DESIGN PROFESSIONAL's design intent and information given in construction documents and, solely in those respects, DESIGN PROFESSIONAL APPROVES the submission. Specific conditions related to that approval and responsibilities of the various parties apply to this approval in all cases. Details of these specific conditions are provided in Paragraphs _____ through _____, inclusive, in General Conditions of the Construction Contract.

Field Observation

Field observation is one of the most important of all services that design professionals perform in fulfilling a professional commission. It involves full-time review of construction, site remediation, or other field work based on plans and specifications. It is far different from the "occasional site visits" that are called for in some model contracts and that seldom are effective.

From a somewhat negative point of view, field observation discourages contractors from cheating on the quality they build into a project, by providing the oversight needed to encourage conformance with plans and specifications and to catch shortcomings, intentional or otherwise. It also should be borne in mind that even the most carefully developed plans and specifications are subject to errors and omissions, among other problems. Through effective field observation, these can be caught before molehills grow into mountains. In this regard, however, it is important to recognize that field observation is best performed by or under the direction of those who developed the plans and specifications. A third-party reviewer monitors to check conformity with plans and specifications. A reviewer employed by the designer is in a far better position actually to question plans and specifications.

Field observation is another service that used to be provided as a matter of course. It, too, has been relied on less in recent years, due to its cost and because of the liability imposed. Now, in large measure due to the problems created by lack of observation, many owners see it as a service that saves more than it costs, while design professionals regard it as a service that reduces liability exposure more than it increases it.

When applied to construction, field observation, i.e., construction monitoring or review is not the equivalent of construction superintendence or management. Either of the latter involves a much higher degree of responsibility during the construction phase, including contractor coordination and, in some cases, construction site safety. Construction monitors are on site to review the manner in which plans and specifications are being implemented and are available to help resolve any problems that may occur with respect to approved plans and specifications. Field personnel do not tell contractors how to perform construction; the means and methods of construction are the contractors' responsibilities. Likewise, contracts should make clear that contractors are responsible for the quality of their work; field monitoring is provided to help enhance quality, not guarantee it.

Although field observation seems to place the design professional in a somewhat passive role, the passivity exists more on paper than in fact. Advising a contractor that the owner will reject the work has virtually the same force and effect as a direct rejection of the work.

For the most part, the prime design professional, structural engineer, and geotechnical engineer should provide full-time observation. When particularly complex or widespread activities are being undertaken, more than one reviewer may be needed. Those performing mechanical and electrical construction may or may not require full-time monitoring. In all cases, however, specifications should call for an appropriate level of quality control testing at various stages of the project, in order to help minimize the need for backtracking later on.

Postconstruction-Phase Activities

For many design professionals, the project is over once the client accepts the work. Additional services often should be offered, however, to help minimize problems and to provide what some term a "complete design service." Some of these follow-up activities include provision of all appropriate manufacturers' warranties and guarantees, preferably included in an indexed notebook or series of notebooks, as well as operating instructions and guides. In some instances, it is appropriate for design professionals, such as mechanical and electrical engineers, to prepare special guides or manuals for proper equipment operation and maintenance, as well as to offer training sessions for operating and maintenance of personnel, and occasional reviews during the first year or two of building operation. Other design professionals can perform similar work, not only as a service to their clients, but also to learn more about the actual performance of what they have designed and specified. Architects and interior designers could perform a walk-through six months after initial move-ins and submit a report to the owner; a landscape architect could do much the same. If the project is built on somewhat "tricky" soils, it might be appropriate for the geotechnical engineer to install settlement monitoring devices.

CLIENT EDUCATION

Execution of a professional commission should involve far more than the minimum, and it is up to design professionals to make clients aware of this fact. Complete service is in the best interests of the client because it is cost-effective and helps reduce the likelihood of the unplanned, unbudgeted surprises that today plague the construction industry. It is in the best interests of the many construction contractors who want to perform effectively. And it is in the best interests of design professionals, due to the reduction of risks and because of the far higher level of personal satisfaction design professionals derive from reassuming their former position of prominence in a manner that can make a meaningful difference in the quality of the project, both as a finished structure and as a process. Because that role is not being fulfilled as extensively or as well as it could be today, there is at times an absence of leadership, direction, and thoroughness that permits problems to occur and fester until it is just too late to resolve them quickly or easily.

If engineers and architects are to enjoy the many benefits associated with proper execution of a professional commission, they must first educate their clients about what is involved and why they should call for a heightened level of service. This means discussion, not only during the workscope development stage of a project, but also as a matter of general activity. Promotional materials should be developed and distributed; client groups should be addressed. It is not a task that can be accomplished overnight, but it can be accomplished. It must be, if design professionals are to perform as true professionals and professionally execute the responsibilities that are theirs.

4

Professionals and the Law

All Americans are subject to the rule of law. Tort law—that branch of the judicial system dealing with civil suits—is particularly important for design professionals, because it governs both their professional and commercial dealings. This chapter examines tort law as it relates to negligence, strict liability, warranty, deceit, defamation, and unfair competition. Design professionals need to understand these issues in order to avoid legal exposure. In some instances, avoidance amounts to careful selection of words; in others, adherence to behavior that for many years has been considered "ethical" or "professional." By becoming conversant with the issues involved, engineers and architects can avoid many of the pitfalls their peers have encountered.

OUR LEGAL SYSTEM IN GENERAL

Our legal system is divided into two distinct branches: statutory law and civil law.

Statutory Law

Statutory law comprises rules of behavior enacted by a legislative body, such as Congress, a state legislature, or city or county council. Alleged infractions are prosecuted by "the people" through the prosecutorial system established by the government whose laws have allegedly been broken. A *trier of fact*, that is, a judge and/or jury, determines if the allegations are valid. If so, the trier of fact also determines what the penalty will be: the legally prescribed maximum or something less than that.

Statutory law is often referred to as criminal law. Serious crimes, such as crimes of violence, are called *felonies* and often are punished through incarceration. *Misdemeanors* are crimes of far less consequence; punishment is usually limited to a relatively small fine, such as that occasioned by a parking ticket.

Civil Law

Civil law is that branch of the legal system in which individuals, rather than the state, take action against other individuals. The specific procedures used are discussed in Chapter 8.

Derived principally from English common law, civil law is implemented through an adversarial system in which the aggrieved party (the *plaintiff*) takes action against the party alleged to be at fault (the *defendant*). Note, however, that statutory law and civil law both may be applied for the same infraction. For example, a landowner who signs a complaint to have criminal trespass charges lodged against an individual can also file a civil complaint against the same person in order to recover the cost of repairing the damages caused by the trespass.

Civil law has two main divisions: contract law and tort law.

Contract Law Insofar as contract law is concerned, the issues typically being examined are whether or not the contract has been *breached* (broken), the meaning of certain words or phrases, and the enforceability of various terms and conditions. The legal enforceability of contracts makes them meaningful. If contracts could not be legally enforced, commerce would come to a standstill. Contracts and contract law are discussed in the following chapter.

Tort Law Tort law concerns itself with torts, that is, civil wrongs for which a court of law will grant a remedy. The existence of civil wrongs presupposes *norms of behavior* by which society expects people to abide. Triers of fact determine what these norms of behavior are, whether or not they have been violated, and, if so, what the remedy should be.

Although torts are not established statutorily, many are relatively straightforward, because the issues involved have been dealt with so frequently before. In establishing norms of behavior and related issues, therefore, courts look at prior rulings (*precedents*). This is why several national design professional associations have established legal defense funds. Their purpose is to appeal unfair or unreasonable decisions in an effort to prevent them from imposing unreasonable standards on others. Left unchallenged, such decisions could become the equivalent of new law, not only in the jurisdiction where the issue is resolved, but also in others where the precedent is reviewed. Thus, in recent years, much of the new tort law emanating from civil courts in jurisdictions such as California has been adopted by civil courts throughout the nation, including those on the federal level. This underscores the fact that norms of behavior are dynamic and, for this reason, tort law continually evolves.

It is essential for design professionals to be aware of tort law rulings that exist where they practice, because many have a significant influence on service providers, in general, or on engineers and architects, in particular. As an example, design professionals for many years were protected by the concept that they owed a duty of care only to those who had contracted for their services. If architect Jane B. had a contract with owner Ted A., Jane was liable only to Ted; a third party could not claim negligence—a tort—against Jane because the third party was not a party to the contract. Today, this concept has all but vanished. Now tort law generally holds that licensed design professionals owe a duty of care to *anyone* who foreseeably could be damaged or injured by their professional acts. This concept illustrates why knowledge of tort law is so vital. If a client asks you to reduce your workscope to such an extent that others could be damaged physically or monetarily, you could be exposed to allegations of negligence, even though your client finds no fault with your service. In such instances, you would be well advised to explain the problem to your client and work with a more adequate workscope, obtain an indemnification through which the client would cover your losses, or refuse the commission.

Just as tort law presupposes that norms of behavior exist, so, too, does it presuppose that people have a *duty* to uphold norms of behavior. For this reason, a tort is also defined as a *breach of duty* for which the court will grant a remedy. In establishing "duty," courts generally apply what is called the "reasonable person doctrine." For example, consider the case of an ice cream truck driver who stops in the middle of a block. A child who runs across the street to buy a popsicle is struck by a car. The child's parents file a "claim in tort" against the company that owns the vehicle, seeking to recover *damages*—monetary compensation—for the child's injuries and mental anguish, as well as their own pain and suffering. They allege, through their attorney, that it was unreasonable for the company (through its agent, the driver) to stop the vehicle in the middle of the block, because doing so encouraged the child to run across a busy street.

Was the company unreasonable? A number of factors will be considered in making that determination, especially the degree to which a reasonable person should know right from wrong. A reasonable person in this case would mean a reasonable ice cream vending company, and the plaintiff would try to establish that, by virtue of the company's experience, it should have known children would cross a street to reach the truck, thereby imposing a duty on it to stop at a crosswalk or take some other type of measure to protect children's safety. Assuming that the trier of fact accepts this line of reasoning, the amount of damages awarded would likely depend on a variety of factors. For example, if the company can show that its driver unintentionally violated both company policy and training, chances are the award would be kept low—something less than the amount claimed by the plaintiff. After all, the company was aware of its obligations and made a good faith effort to prevent exactly what occurred. By contrast, if the company had been warned about such problems before but did nothing to prevent them from occurring again, the trier of fact might award the full amount sought and then add punitive damages on top of that, to punish the company for its "wanton and willfull" disregard of prior warnings.

It is particularly important to note that any one incident can give rise to several claims. For example, suppose you need to deliver a set of plans and instruct an employee, Frank A., to drop them off on his way home. While doing so, Frank speeds, has an accident, and injures someone. In reviewing Frank's driving record, the attorney for the injured party learns that your employee has received three speeding tickets within the past year. Frank is sued for negligent driving, and your firm is sued for negligence in assigning tasks, because you failed to check Frank's driving record before asking him to drive on your firm's behalf. Why would opposing counsel sue your firm? To increase the likelihood of recovery. By naming more defendants, chances are at least one—and perhaps more than one—would be found at fault. Besides, your firm's pockets are probably far deeper than Frank's. If the court rules for the plaintiff, Frank may only have to pay, say, $10,000, because that is the limit of his insurance coverage and he has no other assets. But your policy may have a limit of $1 million or more.

Would your firm be found negligent for Frank's accident? That would depend on the trier of fact. But be forewarned. Many of today's civil courts seem principally concerned with making injured parties "whole." In other words, when someone suffers a loss, many courts seem to believe it is their responsibility to find a way to compensate that person, even though that person may have been partly at fault.

The court's desire to compensate injured parties has given rise to a concept called *joint and several liability*. It applies when a trier of fact determines that two or more parties breached their respective duties (i.e., were negligent) and thus caused injury. The doctrine states that each party is responsible for the "whole" injury or amount of damage. Thus, by applying the doctrine of joint and several liability, it could be ruled that a party that was 5 percent at fault must nonetheless pay 100 percent of the damages, simply because that party can afford to do so. While this may be unfair, the courts obviously have decided that it is more unfair to deny compensation to an individual simply because other responsible parties cannot pay their share.

The doctrine of joint and several liability has been a major factor in the growth of civil claims, a trend accelerated by the actions of personal injury lawyers. Most of these attorneys operate on a contingency fee basis, meaning that their being paid is contingent upon their clients winning the case or otherwise obtaining compensation. The amount of their fee is determined as a percentage of what their clients receive (one-third to one-half is not uncommon). This arrangement creates a strong financial incentive to obtain as much money for their clients as possible through whatever means are available. In order to obtain more business, some of these attorneys advertise through both print and electronic media, urging people to seek the "pot of gold" to which their alleged injury or damage may entitle them. Such advertising, along with the publicity accorded major awards, has contributed substantially toward making ours a litigious society, one in which people sue others "at the drop of a hat" in order to right any perceived wrong. The way in which the civil justice system operates has also resulted in many frivolous or meritless suits being filed. These generally are based on claims that will be extremely difficult for plaintiffs to prove but that may yield money nonetheless, in the form of a cash settlement paid by defendants in order to get charges dropped. Defendants often are willing to go along with this form of legalized extortion simply because it is "better business" to pay a claimant $5,000 to drop charges than it is to pay an attorney $10,000 or more to get a case dismissed by the court.

Many of the frivolous suits filed against design professionals in the construction industry are elements of "shotgun suits." These occur when a damaged third party files claims against all or virtually all parties associated with a project as opposed to those parties who most logically would be at fault. Many attorneys believe shotgun suits are prudent, not only because some of those named may be willing to settle out of court, but also because it often is difficult to ascertain who specifically is at fault until research is performed. This research takes the form of *discovery*, something unique to civil procedure. Discovery gives each party to a claim the right to examine other parties' evidence and witnesses before the matter is heard in court. While some lawyers limit the number of persons and entities against whom claims are filed, others do not. While some do not use the legal process as a means to extort money from those who could not conceivably have been at fault, others do. Those who do state that their action is motivated by *advocacy*, that is, their duty to do as much as they possibly can for a client. It also is appropriate to point out that virtually all new law created through judicial interpretation began as a so-called frivolous suit. For example, the first case in which it was ruled that a design professional owed a duty of care to a third party no doubt was considered frivolous when it was filed.

The tort law issues most likely to affect design professionals are *negligence, strict liability, warranty, deceit, defamation,* and *unfair competition*. These are discussed below. In reviewing the commentary, note that techniques for avoiding claims are readily available, and almost all involve performing in a more professional manner, by modifying patterns of behavior, communicating more accurately and effectively, or by recognizing certain risks before they are encountered and using the contract to specify the manner in which they will be handled should they materialize.

NEGLIGENCE

It is essential for design professionals and their clients to understand that an error or omission is not necessarily a negligent act, due to the doctrine of professional negligence. The issues can be demonstrated through the hypothetical case of Arnold R., the principal of a consulting mechanical engineering firm. He was commissioned by Douglas M., an architect, to design the mechanical system of a sixty-unit condominium office building. Douglas told Arnold that the developer wanted the building to be energy-efficient and was willing to pay a premium to help assure effective design. Arnold did the necessary work, and the system he designed was so effective that it was featured in advertisements for the project. Arnold even won several awards for his design.

Six years after the building was completed, Arnold received a call from a local building official. "The people over at the condo office building are having a real bad time with their HVAC system," the official said. "They called me to come take a look, to see if any codes were violated." Arnold was stunned. This was the first he had heard of any problems.

Not long after speaking with the building official, Arnold received a summons and a complaint. It was alleged by the unit owners association that he was negligent in the design of the heating, ventilating, and air-conditioning (HVAC) system and, as a result, that occupants were warm in summer, cold in winter, and often suffered from respiratory and other ailments due to inadequate ventilation. Numerous system modifications had to be made, and payment was being demanded from Arnold. In addition, as a consequence of the uncomfortable conditions, productivity of office workers had suffered severely, and the plaintiffs were also seeking recovery for that loss. To make matters worse, Arnold three days later was served with a class action summons and complaint filed on behalf of ninety-five persons who worked in the building, claiming their illnesses resulted from Arnold's negligent design. Arnold tried to get in touch with Douglas M., who was on vacation. When he returned, he called Arnold immediately. He, too, was being sued. The developer of the building, a corporate entity, was not named because the corporation was dissolved shortly after the last units were sold. The contractor was also out of business.

Proving the Claim

In order for the plaintiffs to win their negligence claims, five conditions must be proved as fact.

1. The defendants were required to abide by a standard of practice.
2. The defendants owed a duty of care to the plaintiffs.
3. The defendants breached that duty of care.
4. There was a causal connection between the breach and the alleged injury.
5. The injury was real.

Condition 1: Standard of Practice In filing claims against professionals, it is not usually necessary to establish that a standard of practice applies, because professionals historically have been bound to abide by standards of practice. However, it is necessary to establish precisely what the applicable standard of practice was at the time it allegedly was breached.

Defining Standard of Practice Stated generically, a standard of practice is:

> the ordinary skill and competence exercised by members of a profession in good standing in the community at the time of the event creating the cause of action.

It is worthwhile to dissect this definition.

Ordinary skill and competence means just what it says: *ordinary*. Professionals are neither required nor expected to provide "the highest level of skill," if only because the highest level, by definition, can be practiced by just a few. It is recognized by the courts, although not necessarily by all clients, that committing an error or an omission is not a per se act of negligence. Professionals are human; they are not expected to be perfect. Instead, as a basic precept of tort law, professionals are required to abide by the standard of practice by which their peers abide.

Members of a profession in good standing explains who a professional's peers are, namely, other qualified professionals: those who are appropriately licensed and provide the same or similar services.

In the community does not necessarily relate to design professionals in the same town or city. At one time it did, but today, due to advances in communication and travel, *community* can be interpreted as global. The key determinant is the nature of the work. In fact, are there local, statewide, or regional differences? If there are, then it would be necessary to establish this fact by identifying how practices in the area in question are different. If there are no differences, however, then the community can be virtually anywhere. If a relatively rarified expertise is involved, then the community may have to be the nation, simply because few others perform the same procedures nearby. In some instances there may be only a few other professionals in the world who do the same thing.

At the time of the event creating the cause of action refers to the fact that one must evaluate a claim of negligence based on the standard of practice prevailing at the time the allegedly negligent act was committed, as opposed to the time at which the claim is filed or the case is heard. In this instance, Arnold committed the allegedly negligent act six years before the claim was filed. Chances are that another year or two—possibly more—will go by before the case is heard in court.

Condition 2: Duty Owed Did Arnold owe a duty of care to those who purchased the units initially? To those who bought units on resale? To workers? Absence of a contractual relationship, or *privity*, not being a defense, the answer is, yes. Arnold knew or certainly should have known that the developer's units would be sold to others and that these others, in turn, might sell them to new owners. Arnold also knew or should have known that people would be working in the spaces involved and that it was necessary for them to be comfortable in order to work in a productive manner. He also knew or should have known that air quality in a building is vitally important, thus requiring an HVAC system capable of eliminating impure air.

Note that duty of care is not so easily established in all cases. For example, assume that the building originally was designed as a rental project, with the developer planning to maintain ownership of it for an indefinite period. After it was designed and nearly completed, however, the developer decided to convert it to office condominium units. Would Arnold in that instance have owed a duty of care to the unit owners? Probably not, because the duty was not *foreseeable* at the time Arnold developed the design. *However*, if the developer had gone out of business, a consumer-oriented judge might feasibly have held that the conversion *was* foreseeable. After all, it was precisely that kind of thinking which resulted in absence of privity being eliminated as a defense.

Condition 3: Breach of the Standard of Practice In order to show that a professional has breached the applicable standard of practice, it first is necessary to establish what the standard of practice was at the time it allegedly was violated. Doing so almost always requires reliance on an expert witness, that is, a fellow professional whom the court formally recognizes as an expert. Experts enjoy special standing in a court of law, in that their opinions, unlike those of other witnesses, are allowed to stand as evidence.

In cases involving professional negligence, it is common for each side to retain an expert. It is the expert's duty to perform the research necessary to establish the true cause of the problem and to explain that cause, among other issues, in language that the trier of fact can understand. Although paid by the plaintiff or defendant, *an expert is supposed to serve the trier of fact*. In other words, the expert is supposed to be neutral. Thus, when experts disagree about given issues, as commonly is the case, the disagreement should stem from differing judgments. Regrettably, all too many architects and engineers are willing to subvert their service as experts by acting as "hired guns," advocates for the side retaining them. Instead of trying to determine what actually went wrong, they will merely try to prove that their client's version of events could be credible, even though the opposition's version may be even more credible. Opposing counsel can defeat or "impeach" the testimony of hired guns by pointing out errors, contradictions, or other weaknesses in their testimony. This can be difficult to do, however, because most attorneys lack an intimate understanding of engineering and architecture. Attorneys can overcome this deficiency by working diligently with their own experts, but few make enough time available for this all-important effort.

The problems associated with expert witness testimony are particularly prevalent when it comes to establishing the applicable standard of practice. In essence, experts are permitted to say that, because a fellow design professional did not perform X, Y, and Z, the standard of practice was breached. Although experts may be expected to perform research to back up such pronouncements, they are not formally required to. It was partly for this reason that ASFE led the effort to develop *Recommended Practices for Design Professionals Engaged as Experts for the Resolution of Construction Industry Disputes*. Adopted by more than twenty-five of the most prominent organizations in the U.S. construction industry, the document has near-standard status, and clearly identifies how experts should act. This document requires those addressing a standard of practice to perform some legitimate research to identify what the applicable standard was at the time the negligent act allegedly was committed. Such research could take the form of reviewing plans filed with a local building licensing department or interviewing other professionals active in the area at the time the allegedly negligent act was committed, among others.

As noted above, disagreements between experts should arise solely from differences in professional judgment, not advocacy. Regardless of the cause of disagreement, however, the trier of fact will have to determine which expert is more credible. Since a jury is more difficult to educate than a judge, plaintiffs in negligence cases often request a jury trial. The jurors' decision as to which expert opinion is correct may depend not so much on the logic presented as the expert's physical appearance, attire, demeanor, and manner of speaking.

Until such time as the courts or professionals are able to require that research must be performed in order to address standard of practice issues, today's practices will continue, which is exactly what happened in this case. Arnold's expert demonstrated that the design was completely acceptable and, in fact, an award winner. By contrast, the expert retained by the plaintiffs, a multicredentialed college professor, averred that the system was just too small to meet the comfort and health requirements of the building's occupants.

Condition 4: Causal Connection Merely showing that a professional may have breached the standard of practice is not necessarily sufficient to win a suit. It must be shown that the breach caused an injury or damage. This requires the plaintiff to show that the negligent act was the *actual cause* or *proximate cause* of the injury or damage.

Actual Cause Actual cause is determined in either of two ways, depending on the jurisdiction: *but for* or *main factor*. Through the "but for" approach, it must be shown that, *but for* the defendant's negligence, the damage would not have occurred. Through the "main factor" approach, it must be demonstrated that the negligent act was a *main factor* in the events resulting in damage. There is a significant difference between the two approaches, as shown by the following case.

BUT FOR: Arnold, in an attempt to minimize energy consumption, designed a central plant that others would have considered undersized. He justified this by making certain assumptions about how the building was sited with respect to insolation (solar radiation received over an area), the type of glazing (glass) to be used, and internal heat gains from people, equipment, and lighting. Ventilation rates were set below those required by standards, but Arnold obtained a code exception by demonstrating to officials that code-mandated ventilation rates were far more than necessary to assure healthful conditions. (Later, in fact, code-mandated ventilation rates were lowered.) Furthermore, a thorough investigation revealed that, unknown to Arnold, the building's orientation was changed from the original plans and a different type of glazing was used; the ductwork had many more bends in it than originally planned because of conflicts between structural and mechanical plans, and very little had been done to maintain the central plant over the years—for example, duct insulation ruined by a water line leak was not replaced.

Under the "but for" rule, it could *not* be said that Arnold's professional acts resulted in damages, because it could not be said, "*But for* the undersizing of the central plant, the heating and cooling system did not perform well." A number of other factors intervened.

MAIN FACTOR: Under the "main factor" approach, Arnold might be exposed, because—it could be argued—undersizing of the central plant was a main factor that resulted in inadequate heating and cooling.

Proximate Cause   *The doctrine of proximate cause* would be applied to determine if Arnold's professional acts were closest (most proximate) to being the sole cause of the problems. In making this determination, the courts generally apply three considerations:

- the number of possible causes,
- the presence of intervening causes, and
- the foreseeability of the consequences.

In this case, the number of possible causes probably would have been limited to undersizing alone had the central plant not performed properly from the start. But that was not what happened: The HVAC system performed adequately for several years and then began to deteriorate. A number of intervening causes arose: a siting change, glazing modification, additional bends in the ductwork, and lack of maintenance. None of these intervening causes, except for the last one, would explain the deterioration in performance.

Were any of the intervening causes foreseeable to Arnold? Just one: the potential for inadequate maintenance. But Arnold recognized this possibility and thus prepared a comprehensive operation and maintenance manual for the system he designed. Arnold kept a copy of the manual and demonstrated that it included the various manufacturers' instruction guides, as well as copies of all applicable warranties and guarantees. The manual also included an introduction that said, in part,

> The mechanical system of this building should be considered somewhat like the motor of a sports car. It has been designed to derive maximum effectiveness from

the smallest possible size, to minimize energy consumption without having to sacrifice comfort. Note, however, that just as with a sports car, effective maintenance is absolutely essential to obtain good performance. Without effective maintenance, performance will deteriorate: More energy will be consumed, and less comfort will be obtained.

All things considered, then, even if the undersizing was considered a breach of the standard of practice, the plaintiffs could not show any causal connection between that breach and the alleged damages; they could not prove that undersizing was either the actual cause of the damages or the proximate cause.

Condition 5: Real Injury The plaintiffs were seeking compensatory damages from Arnold to recover the cost of replacing the central plant and the value of the productivity lost due to uncomfortable conditions and the absenteeism occasioned by the frequent illnesses allegedly caused by the inadequate HVAC system. In determining how much Arnold owed, if anything at all, the jury was instructed to compute the cost of remedying the wrong, that is, the cost of replacing the central plant and of compensating unit owners for the lost productivity of their personnel. Pain and suffering might also have been an issue, had the plaintiffs alleged that lower staff productivity caused them mental anguish. (How much anguish is worth, and how much a jury should award for each sleepless night is something a jury must decide on its own.)

Before awarding anything, the jury had to decide if the injury was real. Arnold argued that the extent of injury was grossly overstated. In fact, the central plant did not require replacement; only repair. And, insofar as the claims of lower productivity were concerned, the plaintiffs had to prove that lower productivity actually occurred and that it was caused by central plant inadequacies. Linking central plant inadequacies to lower productivity was not difficult, due to the testimony of employees. Actually demonstrating that lower productivity occurred was a far more difficult task. The plaintiffs were unable to furnish documentation showing that identical tasks performed by the same people were taking longer due to inadequate heating or cooling. They also were unable to prove that absenteeism was above normal and that any of it was due to illnesses caused by the HVAC system's inability to remove contaminants from the air. The employees filing the class action suit failed to demonstrate real injury for much the same reason. As a result, since the plaintiffs could not demonstrate that the productivity-related injuries were real, they could not collect. And, because they failed to show that anything Arnold did was responsible for problems with the central plant, they failed to prove negligence.

Unjust Enrichment A surprisingly large number of cases are filed when there are no actual damages, simply because the client or some other party knowingly or unknowingly seeks *unjust enrichment*. This situation typically occurs when a design professional makes an error or omission that requires correction through field modification, via a change order. The owner is suddenly looking at an unanticipated bill for, say, $25,000 and thus files suit against the design professional, seeking to recover that sum. However, the cost of installing the omitted item, had it not been omitted, would have been $25,000. As such, the owner really has not been damaged at all. Nonetheless, a suit will be filed, and people will spend countless thousands of dollars and hundreds of hours of time in an attempt to resolve a dispute that never should have existed.

Why do unjust enrichment disputes arise? In the case of Arnold R., it would appear that the expert retained by the plaintiffs stirred up controversy (and potential business) where there should have been none. In fact, a comprehensive review of the building by a qualified mechanical-electrical engineer seeking honest answers would have revealed the true nature of the problems, and everyone would have been better off. Instead, a hired gun was retained, and he sought business by claiming negligence in design, in order to obtain an engagement for expert witness services and for redesign of the central plant.

Defending the Claim

A number of approaches are used to defend a claim of negligence. The defense demonstrated above is known as the *missing elements defense*, in which the defendant seeks to show that one of the elements needed to prove negligence is missing; that is, no standard of practice applied, the plaintiffs were not owed a duty of care, no breach occurred, there was no causal connection between the breach and the injury, or the injury was not real. Other defenses commonly used when negligence is alleged include *contributory negligence, assumption of risk,* and *statutes of limitation or repose*.

Contributory Negligence   In those jurisdictions where the contributory negligence doctrine is still employed, it is only necessary to show that the plaintiff somehow contributed to the problem in order to get the claim dismissed. This is also known as *the totally clean rule*, since plaintiffs must demonstrate that they had absolutely nothing to do with causing the problem. In Arnold's case, it would only have been necessary to show that lack of maintenance contributed to the problem in order for Arnold to walk away, assuming the jury concurred.

Many states have now converted to the *doctrine of comparative negligence*. This holds that each party pays damages proportionate to culpability. Thus, if a plaintiff is 70 percent at fault, the plaintiff would pay 70 percent of the award and may also be made liable for 70 percent of the defendant's costs.

Assumption of Risk   An assumption of risk defense holds that the plaintiff knew or should have known about certain risks and, therefore, the plaintiff must bear the consequences should any of these risks materialize. For example, those who attend a baseball game know they might be hit by a foul ball. If they do get hit, they have no standing to recover, since they assumed the risk.

This points out a very important precept in professional liability loss prevention and your function as a professional: *Let the client know what the risks are; make all warnings explicit and obvious.* This helps prevent the types of problems that can lead to negligence claims, and, should they occur anyway, they tend to make the client liable for them, because the client knew of the risks but accepted them nonetheless.

Statutes of Limitations and Repose In many instances it will be argued that a claim (for negligence or other causes), even if merited, is time barred, because the statute of limitations or statute of repose has run its course. There is a major difference between the two types of statutes.

A *statute of limitations* begins after a defect has been discovered, typically because it has caused damage. For example, John D. lives in a condominium. The roof begins to leak during a rainstorm, damaging John D.'s rugs and some important papers. Investigation reveals what supposedly is a roof design defect. The state has a five-year statute of limitations that affects construction defects. As a result, John D. has five years from the time of the roof defect being discovered to file a suit against the designers, even though their design was completed more than eight years ago.

A *statute of repose* begins after construction is substantially complete. In the case above, John D.'s claim would have been time barred if the state had a five-year statute of repose rather than a five-year statute of limitations, since more than five years had elapsed since substantial completion.

In recent years, many states have attempted to establish statutes of repose offering specific protection to design professionals and construction contractors. Some of these have fallen on appeal, with the justices ruling it unconstitutional to grant such relief to a specific class of persons. As a consequence, many statutes of repose have been converted to statutes of limitations. Those that have withstood court challenges generally include wording that identifies why the legislation was enacted, namely, to offer legitimate protection to a class of individuals who, without such protection, would find it difficult to remain in business and thus continue to provide valuable services that benefit society.

The problems associated with statutes of limitations should be obvious. In essence, they create a lifelong liability for design professionals. For example, a ten-year statute of limitations would give a plaintiff ten years to file a claim for an allegedly negligent act committed thirty years ago!

If a five-year statute of repose existed in Arnold's case, he could have sought dismissal of the suits against him on the ground that they were time barred, and he probably would have won. By contrast, if a five-year statute of limitations was in force, the case would have continued as long as claims were filed within five years of the alleged defect having come to the attention of the unit owners.

Do not assume that statutes of repose offer ironclad protection. Many judges are loath to enforce them. In some cases, they may fall on appeal. In others, judges will look for a way around them. For example, in one case it was ruled that a plaintiff did not file a claim for faulty design soon enough (the deadline was missed by just several weeks). However, an appeals court ruled that there was still time to file a claim based on negligence in construction observation.

STRICT LIABILITY

The doctrine of strict liability generally applies to manufacturers of a product. It is far more stringent than the doctrine of professional liability, since negligence does not have to be proved. It merely must be shown that:

- there was a defect in the product,
- the defect existed at the time the product was transferred to the purchaser or user,
- the defect caused or contributed to an injury, and
- the product failed in normal use.

Even though design professionals usually are held to the doctrine of professional liability, strict liability claims may at times be filed.

Buildings as Products

For the most part, a building is not subject to strict liability claims because it is not (yet) interpreted as being a product in the conventional sense. There are some exceptions, however. A federal court has ruled that, in Pennsylvania, based on certain state court decisions, a single-family house can be considered a product. Reportedly, California courts have ruled the same way. Factory-built units can also be exceptions, be they for housing, small offices, or warehouses. Another exception relates to components of a structure, such as boilers or windows. A fourth may be anything completed on a *turnkey* basis, that is, where owners agree to pay a certain amount for a completed structure that the party with whom they are dealing then has designed and built. In the latter instance, however, design professionals may not be exposed to strict liability as long as they are engaged in a conventional manner. If the design professionals own all or part of a closely held company providing a structure or if they will receive an ownership portion as compensation, the doctrine of strict liability may be imposed, because a design professional who owns all or a portion of a company delivering a product might be considered the manufacturer of that product.

Plans and Specifications as Products

Another aspect of strict liability concerns plans and specifications. These are not products in the conventional sense. Instead, they are *instruments of service* that traditionally remain the property of the design professionals who prepared them. Continually more owners now demand instruments of service, however, and, as a result, they are construed as products by some courts. Accordingly, if there are defects in them—errors or omissions—it may be possible to convince a judge or a jury that the doctrine of strict liability, rather than professional liability, should apply. Avoiding liability in such instances would require design professionals to perform flawlessly, since they would be responsible for any defect, not just those for which negligence is proved.

As in the case of many other risk exposures, the potential for this problem can be contemplated when drafting a contract. It can be made clear that the documents in question are instruments of service, that they traditionally remain the property of the design professional, but "in this instance" are being given to the owner-client.

Components as Products

Another area of exposure relates to components of a structure, since these often are manufactured products. Design professionals generally will not be held strictly liable for the failure of components they specify, although they may have to disprove a claim that their specification of such a product was professionally negligent. Design professionals who design *and fabricate* a component of some type could be held strictly liable should the component fail.

Other Aspects of Strict Liability

The foregoing examples of strict liability all relate to design. More than design may be involved, as when warranty is the issue. For example, if you warrant that your design will produce certain results, and if it fails to do so, you could be held strictly liable for that failure even though the design was not negligent. Breach of warranty and other tort exposures for which you may be held liable are discussed below.

WARRANTY

The doctrine of warranty arises in contract situations. It comprises a promise that things are exactly as they are represented to be. When applied to engineers or architects, it subjects them to the doctrine of strict liability.

Consider the case of Jane R., a civil engineer in independent practice, who was retained by Dr. F. to review a homesite. Part of Jane's report to Dr. F. contained the wording, "There are presently some rather severe drainage problems with this site. In spite of these, you can build on it."

As it so happened, the drainage problems were worse than Jane envisioned. After construction was completed, a landslide caused massive damage to the physician's new house. The doctor sued both Jane and the developer, charging both of them with negligence and breach of warranty.

There are two types of warranties: *express* and *implied*. In this case, the claim alleged breach of express warranty.

Express Warranty

The criteria for express warranty are relatively straightforward. They are the existence of:

- an express warranty comprising language such as, "I warrant that . . ." or "This product is guaranteed . . .," or
- an absolute promise of something, such as the accuracy of a test, or
- a self-imposed obligation to convey a product or property in a particular condition, or
- an indication that individuals or business entities assume responsibility for the accuracy of their statements.

It was the last criterion that was used against Jane R. She held herself out as an expert in her field. She conducted a review of the site and, if her fee was inadequate to provide a thorough review, she failed to indicate the problem beforehand or in her final report. The language used—"you can build on it"—included no qualifications. Was there any reason for Dr. F. to have any doubts about Jane's findings? No. Was anything said to encourage Dr. F. to analyze the site more closely or to have special work done? No. In other words, due to unfortunate word selection, Jane exposed herself to a totally needless action in tort.

What should Jane have done? Very simply, she should have conveyed her findings more accurately. Something along the following lines would have been more representative of the situation:

> Our review of the site indicated some severe drainage problems. These should be reviewed in far more detail. A plan should be developed to help eliminate any problems through recontouring of the land and installation of subdrains, if necessary. A subsurface exploration should be conducted to establish the stability of subsurface materials. Generally speaking, however, it is my opinion that this site can be made acceptable and safe for construction of a home. In any event, however, additional review should be conducted and is strongly recommended.

Unfortunately, the only thing Jane R. said was that the site could be built upon. It was built upon, failed, and Jane had to take some of the responsibility.

Design professionals should not speak in absolute terms except when it is called for. There is a major difference between, "Perform A, then perform B, and C will occur," and "Perform A, then perform B, and C *probably* will occur." Some may argue that "probably will occur" is "weasel wording." This is not the case at all, providing it is accurate. If there is even the slightest chance that C will not occur, it is inaccurate to say it will. If you say it will as a design professional, you are expressly warranting that it will.

Design professionals also are induced into giving express warranties when they are required to sign certifications. In legal jargon, a certification is a promise made in writing, which can be tantamount to a warranty. Typical of the certifications required are those that state that a structure was completed in accordance with approved plans and specifications. In fact, design professionals can certify or warrant such a finding only when they and their associates monitor the work of *everyone on site, at all times*. Failing that, there is no way for design professionals to know that everything was done according to plans and specifications. The liability exposure thus accruing is a major one, since it involves strict liability rather than professional liability. In essence, design professionals could be "hung," not because they committed an error or omission, but rather because something that they said was so wasn't so. In such cases, they would not be covered by professional liability insurance, since no act of professional negligence was committed. (This issue is covered more fully in the following chapter.)

Implied Warranty

An implied warranty of suitability pertains mostly to manufactured products. In essence, the product should not cause injury when used in the intended manner. If it does, then the doctrine of strict liability applies.

Implied warranty may be applicable to design professionals in those jurisdictions where plans, specifications, and reports are considered products. In such instances, a breach of implied warranty claim may hold up if it can be shown that:

- the documents contain a defect,
- the defect existed when the documents were delivered,
- the defect contributed to the structure being insufficient, thereby causing injury to a party, and
- the structure did not undergo any unusual use.

In some states, it is left up to the jury to decide whether or not the design professional is part of the manufacturing process. In the majority, however, it is held that plans, specifications, and reports are not products. None of these states is immune from overturning prior rulings, of course.

On a more positive note, a particularly favorable ruling on this entire matter was obtained in the California case *Allied Properties v. John A. Blume & Associates*. There the court held, "Where the primary objective of a transaction is to obtain services, the doctrines of implied warranty and strict liability do not apply." Of course, not all states agree with California, and jurists in "the Golden State," as those elsewhere, are wont to change their minds. Accordingly, it is prudent (and professional) to have a written contract for all work and a provision in that contract that makes clear that no warranty or guarantee is offered; for example.

> The purpose of this AGREEMENT between DESIGN PROFESSIONAL and CLIENT is the rendering of services by DESIGN PROFESSIONAL on CLIENT's behalf. DESIGN PROFESSIONAL will perform in a manner that upholds the standard of practice maintained by DESIGN PROFESSIONAL's peers performing similar work in the area at the time. No other warranty or guarantee, express or implied, is offered, and none should be inferred.

Despite such wording, it is possible that a jury still may be required to determine whether or not the written statement materialized in fact. For this reason, it is essential for design professionals to conduct themselves properly on site. In this respect, it is imperative that those retained to observe construction or remediation activities do exactly that: observe. While it may be most expedient to tell a contractor's worker what to do or how to do it, taking such action puts the design professional into what feasibly could be construed as a manufacturing process. On site, design professionals and their representatives are performing a service to their client. They are there to observe what the contractor is doing. If the contractor is performing poorly, design professionals should report that fact to the client. It is *not* their responsibility to correct the contractor. Doing so can result in problems of major magnitude.

DECEIT

The doctrine of deceit was established with purely commercial transactions in mind; it was not intended for application to design professionals. Nonetheless, attorneys have been able to use it for that purpose, especially in an effort to help secure a favorable verdict in the event negligence cannot be shown. For the most part, the cause for action arises from the same cause as express warranty: stating as fact something that may not be a fact.

Consider the case of Doug B. He was retained by a married couple to perform a subsurface site exploration of a building lot, in order to ascertain subsurface conditions and the type of foundation required for the house being planned. After performing his work, Doug submitted a report indicating that fill at the site varied between twelve and sixteen inches in depth. During excavation, however, as much as six feet of fill were encountered, necessitating use of a more costly foundation. Doug's clients sued him for both negligence and deceit, in hopes that either or both causes would be found valid, so they could recover their unanticipated costs.

Proving the Claim

To establish deceit, a plaintiff must show that there was:

- an assertion or statement, as a fact,
- of that which is not true,
- by one who has no reasonable ground for believing it to be true,
- made with the intent to induce recipients to alter their position.

In typical application, deceit applies to something like the purchase of a used car. You test-drive the vehicle, hear a noise you don't like, and decide not to purchase it. But then the sales representative asserts that the noise really is nothing at all (knowing the car may have a bad transmission), inducing you to change your mind. You buy the car, and, one week later, the transmission fails. Certainly Doug B. did not act in such a manner, but just as certainly he was found guilty of deceit. An examination of the criteria reveals why.

An Assertion or Statement, as a Fact Doug stated in his report that there were between twelve and sixteen inches of fill on the site. He did not equivocate at all. He later claimed the statement was an opinion, but, as the court noted, "He did not give his statement in the form of an opinion but as a representation of fact. His assertion was not a casual expression of belief, but a deliberate affirmation of the matters stated. . . . Moreover, even if defendant's statement was an opinion . . . his unequivocal statement necessarily implied that he knew the facts that justified his statement."

Of That Which Is Not True There can be no question about the fact that Doug's statement was untrue. There were six feet of fill on the site, not the twelve to sixteen inches Doug supposedly found.

By One Who Has No Reasonable Ground for Believing It To Be True Doug had no reasonable ground for believing his statement was true because he performed the tests negligently. At the time he probably did not know this, but that does not matter. In California, where this case occurred, it is not necessary to show actual knowledge of falsity; in fact, the statement can be made in complete good faith. The question is, did Doug have reasonable ground for believing his statement to be true? *No* is the answer, because the nature of the tests he performed—even if performed properly—could yield misleading results. *Any test or observation that is based on sampling leaves room for error.*

Made with the Intent to Induce Recipients to Alter Their Position As the court observed, "Defendant's intent to induce plaintiffs to alter their position can be inferred from the fact that he made the representation with the knowledge that the plaintiffs would act in reliance thereon." In other words, if someone is going to rely on the statement, it will be assumed by the courts that there is an "intent to induce the recipients to alter their position."

The Outcome As it so happened, the plaintiffs in this case were unable to recover for negligence. Even though Doug failed to abide by the standard of practice, there were no damages; the more expensive foundation would have been required in any event. Since Doug's actions imposed no additional cost, the court interpreted the suit against him as an attempt to gain unjust enrichment. The plaintiffs were able to prove their point insofar as deceit is concerned, because damages are not necessary to prevail. However, because there were no damages, the plaintiffs did not receive an award.

While Doug may have taken some pleasure from the judge's decision, he still had to pay a substantial amount of money, in the form of the time lost in researching the claim, conferring with his attorney and others, answering interrogatories, participating in depositions, and appearing in court, as well as that associated with aggravation and preoccupation. In addition, he had to pay for his attorney's time. It is also appropriate to note that Doug might have been sued for deceit even if he had performed the tests properly and they simply gave misleading results.

Defending the Claim

How do design professionals prevent claims for deceit when any statement they make may be construed as an "intent to induce the recipients to alter their position"; when it is extremely difficult to know whether or not there is "reasonable ground" for making the statement because it does not have to be shown that deceit was intentional? Two defenses should be considered, and both are used to prevent claims rather than respond to them. The first defense consists of performing the type and amount of work required in order to provide the findings that the client seeks. Doug was dealing with prospective homeowners who wanted to keep their prepurchase costs as low as possible. Doug obliged by doing just a minimal amount of work; in fact, too little work to provide the level of information sought. If the fee for doing the work properly will exceed the client's budget, the client must either modify its budget or modify its expectations of what a design professional's services will yield. If the client is unwilling to do either, then the design professional should probably refuse the commission, because it is bound to lead to dissatisfaction or, as in Doug's case, worse. *This guidance applies to all design professionals.* Observing the placement of half the rebars in a given area does not permit a design professional to certify anything about placement of *all* the rebars. Serious errors may have been made in placing those that were not observed. Accordingly, if the client wants a design professional to certify that all rebars were installed properly, the installation of all rebars *must* be observed. A client who wants it otherwise wants something for nothing, creating an untenable risk for the design professional who accedes to such a client's wishes. In essence, this risk transforms the design professional's statement into an insurance policy. If there is a problem with some of the rebars whose installation was not monitored, the design professional could be required to pay for correction of the problems, as well as any damage caused by the problem.

The second defense relates to the wording of any statements made: *State as fact only that which is fact.* If there is any room for doubt or any possibility for error, it is a design professional's responsibility to make it known, even if it is something the client does not want to hear. In Doug's case, action for deceit probably would have been prevented had his report concluded with something along the following lines.

- On November 8, 1990, we visited the proposed homesite and performed the following exploration activities: . . . .
- The following tests were performed: . . . .
- Based upon the exploration we conducted and the tests we performed, it is our professional opinion that. . . .
- Because borings indicate conditions only at the location of the borings, the subsurface structure was inferred by interpolation. Variation from the inferred subsurface structure may be expected. Should such variations be found, the findings and recommendations of this report no longer pertain unless we are given the opportunity to reevaluate them in light of new findings.

In a similar manner, one could say:

From the period November 8 through November 15, personnel of our firm monitored installation of rebars in the eight columns shown on the drawing attached. The manner in which approximately 50 percent of all rebars were installed was closely monitored. Based on these observations, it is our professional opinion that construction in these areas proceeded according to plans and specifications. Our observations do not preclude the possibility that some rebars may have been installed improperly, nor do our observations relieve the contractor from the contractor's responsibility to install all rebars in accordance with plans, specifications, and commonly accepted procedures used by the industry.

As a professional, it is your responsibility to inform clients of their risks. Clients who want an absolute statement must understand the work required to back that statement and the fees and expenses required for that work. If clients are unwilling to pay the fee required for the work, then they must be willing to accept a less absolute statement, as well as the additional risk associated with a less comprehensive service.

DEFAMATION

Caroline H. is a structural engineer. As part of her scope of services, it was her job to assist the project owner with contractor prequalification. One of the contractors expressing interest was A.B.M., Inc. Caroline had worked with the firm once before and found it to be less than satisfactory. She informed the owner about this, but the owner had also worked with A.B.M. and was satisfied with the company's performance, and thus decided to leave it on the list. When the bids came in, A.B.M. was lowest. Caroline, not wishing to work with A.B.M., gave the owner more specifics about her prior experience, relating facts both orally and in writing. She said the firm was incompetent and not at all hesitant about shifting the cost of its own errors to the owner or others. She also noted that A.B.M. had fraudulently attempted to charge for materials that were never delivered to the site. Upon learning all this, the owner decided to award the contract to the second-lowest bidder. A.B.M.'s president, Arthur M., contacted the owner to determine why his firm did not receive the award. The owner told him what Caroline had said. Soon thereafter, A.B.M. sued Caroline for defamation, libel, and slander.

While the foregoing scenario is hypothetical, it is an accurate portrayal of numerous similar suits, all of which make it continually more difficult for design professionals to serve as their clients' trusted advisors. Will A.B.M. win its case? Answering that question requires a review of the laws involved.

Proving the Claim

To win an action for defamation, there must be:

- a published statement
- exposing the plaintiff to public hatred and ridicule,
- made before witnesses who understand the statement to be derogatory,
- resulting in injury and damages being suffered by the plaintiff.

It is appropriate to examine these factors more closely.

A Published Statement *A published statement* means any transmission that comprises meaning between two or more parties. When defamation results from the spoken word, it is referred to as *slander*. When it results from something on paper, including drawings or caricatures, it is referred to as *libel*.

Exposing the Plaintiff to Public Hatred and Ridicule This is generally interpreted to mean a lessening of public opinion of the plaintiff to the point that the plaintiff suffers an injury of some type. In some states, the statement must contain a specific reference to the plaintiff. Most states require only that there can be no mistake about whom reference is made, whether or not the individual's name is used. In either case, the "public" can be a public of one person.

Made Before Witnesses Who Understand the Statement to Be Derogatory The point to note here is reliance on what the witnesses understand the statement to mean, *not* what the speaker or writer may have intended. In the case of Caroline H., there was only one witness, and Caroline's intent was clear. However, in statements quoted in newspapers, where the "witnesses" may comprise thousands of people, a far different situation arises. Even if 70 percent of the readers understand that the statement is not derogatory, the 30 percent who infer a derogatory intent will comprise a sufficient number to support this criterion for action.

Resulting in Injury and Damages Being Suffered by the Plaintiff A plaintiff may suffer two types of damages: *general* and/or *special*. Special damages are those related to the specific context of the defamation. In A.B.M.'s case, for example, the special damages would include loss of the profit that A.B.M. otherwise would have earned on the project. General damages are those assumed to result from the defamation, even if not proven with particularity. For example, statements made about an East Coast firm may make it almost impossible for that firm to get a project on the West Coast. In such instances, special damages cannot be shown; nonetheless, general damages can be presumed to exist. Courts may also impose *punitive damages* if it can be shown that the defamation was made with malice, that is, with the intent to harm the plaintiff. In establishing punitive damages, the court will consider the context in which the defamation occurred, the degree of malice, and similar factors.

In making a determination of defamation, the courts will look to tort law in the state, specifically as it relates to slander and libel. This law commonly will reflect its English Common Law heritage.

When the concept of tort action for defamation originated in sixteenth-century England, slander and libel were treated in the same way. Unless the case involved *defamation per se*, special damages had to be shown. Defamation per se included accusations of criminal activity, fraud or deceit, unchastity on the part of an unmarried woman, or "loathsome disease." In the case of defamation per se, general damages were assumed. Later, in seventeenth-century England, the tort of libel was separated from slander. This was due principally to the development of the printing press, which greatly expanded the "reach" of a printed statement. This distinction is still maintained in many states, in that a defamatory statement made orally on a radio or television broadcast would be actionable for libel, not slander.

The distinction between libel and slander is more than an historical curiosity. Many states treat the two differently. Except for slander per se (e.g., accusations of fraud or deceit), special damages must be proved before any general damages can be obtained. This reflects the understanding that slanderous remarks often are made in the heat of argument, whereas libel, involving the written word, usually requires premeditation. In the case of *A.B.M., Inc. v. Caroline H.*, Caroline apparently committed slander per se, by accusing A.B.M. of fraud. As a result, assuming the court sided with A.B.M., the latter would not have to prove special damages in order to collect general damages. In all likelihood, however, A.B.M. would seek both special and general damages.

In some states, libel is treated in much the same manner as slander. Most treat it differently, however, and often distinguish between *libel on its face* and *libel per quod*. Libel on its face is libel that is clear, without referring to the plaintiff by innuendo. This type of defamation usually entitles the defendant to general damages. *Libel per quod* generally requires the plaintiff to show special damages before being eligible for general damages. Because it is so difficult to prove special damages in such instances, "scandal sheet" tabloids in many states get away with extensive defamation through innuendo.

Defending the Claim

Several basic defenses are available to Caroline H. The most basic is the *absence of one of the criteria* needed for the tort to be upheld. In this instance, however, everything seems in place: There was a published statement (both written and oral) exposing the plaintiff to public hatred and ridicule, made before witnesses who understood the statement to be derogatory, resulting in injury and damage to A.B.M.

Another basic defense is *truthfulness*. In most states, a defamation claim will fail if the defendant can prove that everything said was true. Providing she has the necessary witnesses and documentation to corroborate her allegations, therefore, Caroline might have an absolute defense. Some states consider motives, however and, as such, telling the truth is not always a defense, as when it has been done in hopes of damaging someone else.

A third defense is *privilege*, in which it is shown that the statement, while it may have been defamatory, was privileged and therefore not actionable. One such privilege is associated with comments made on public issues, based on an analysis of facts. The theory is that the value of obtaining comment on public matters exceeds the value derived by protecting someone's good name. There is also the privilege of consent, in which the party defamed permits revelation of the defamatory materials. For example, a person who challenges dismissal from a position tacitly consents to any comments that may arise during discussion of the cause for dismissal.

Defamatory statements are also considered privileged when they are made to defend a person's own interests. For example, when a collection agency calls to obtain payment of a past due amount, a statement such as, "Their services were so terrible I had to hire someone else to do the work over," may fall within the context of this privilege.

The privilege of reply holds that a person who has been defamed by a statement may defame the other party in response. This is considered a "heat of passion" privilege that recognizes the basic human tendency to hit back. It likely would not apply if the defamatory "comeback" was made after there had been sufficient time for tempers to cool.

Finally, there are privileged communications, such as those between litigants and their attorneys, providing everything stated relates to the proceedings at issue.

In the case of Caroline, of course, none of the comments was privileged. How, then, do design professionals go about relating important information to their clients without fear of retribution? One technique, discussed in Chapter 5, involves creation of specific contract language that makes information about the performance of others confidential between design professionals and their clients. If clients release the information to someone else, then, according to the contract, clients would be required to indemnify design professionals from any action alleging defamation. In Caroline's case, had such a contract provision been in force, chances are the owner would not have told Arthur M. why A.B.M. was not selected or even that it was low bidder.

Another approach—and one that should be followed in any event—is, once again, sticking to easily provable facts, especially when the comments are made in writing. When comments are oral, there may be some "slack," in that it will be difficult for the parties to remember precisely what was said. Do not take license with this aspect of the law, however, because the resulting event—for example, not being selected for a project or being terminated—will suggest strongly that something defamatory must have been said, making proof of special damages a relatively simple matter.

UNFAIR COMPETITION

Barney B., a local civil engineer, was retained by Major Centers, Inc., to check and, where possible, upgrade the design of a new mall Major was about to build in Barney's area. The plans had been prepared by Retail Design Associates, P.C., a large firm located about 1,000 miles away. Major retained Barney to help assure the design complied with all local codes and ordinances, and otherwise would proceed with a minimum of difficulty, given local conditions.

Barney went at his job with vigor. He recommended storm sewers instead of street drainage, larger sanitary sewers, molded instead of square curbing, and so on. As a consequence, Retail Design Associates had to redo a number of drawings and specifications, causing late delivery on Retail Design Associates' part, as well as additional cost. Retail Design Associates sued Barney, claiming commercial disparagement and interference with contractual relations.

Will Retail Design Associates' case succeed? Answering this requires a review of the tort of unfair competition. It encompasses three specific types of action:

- commercial disparagement,
- interference with contractual relations, or
- interference with prospective economic advantage.

Commercial Disparagement

Four conditions must be present in order to prove commercial disparagement. They are:

1. a published statement that is false and injurious,

2. made with malice,
3. which interferes with plaintiff's commercial relationships, and
4. results in special damages.

Although commercial disparagement is closely akin to defamation, there are several major differences that make commercial disparagement far more difficult to prove. In defamation, an accusation of fraud or deceit is required. If it is proved, it would be considered defamation per se and the plaintiff would be entitled to general damages. In commercial disparagement actions, the charge must be ineptitude or inferior work, and it must be demonstrated that these comments resulted in special damages, that is, those stemming directly from the incident. In addition, defamation actions require defendants to show their statements were true, whereas actions for commercial disparagement require plaintiffs to show the statements were false. The plaintiff also must show in most (but not all) states that the statements were made with malice, something which is not required in defamation actions.

How did Barney fare with regard to the charge of commercial disparagement? The only defenses are missing elements (to prove the case) or truth. A review of the four conditions reveals the answer.

Condition 1: Published Statement That Is False and Injurious Publishing has the same meaning in disparagement as it has in defamation. Since one can assume Barney communicated with Major Centers, Inc., the existence of a published statement can be presumed. But did Barney say directly or by innuendo that Retail Design Associates was incompetent or habitually produced inferior work? There is no indication that he did. And, even if he had, would that statement be false?

Condition 2: Malice If there had been a series of communications between Barney and Major Centers, perhaps it would have revealed malice or an intent to cause harm. But no such communications existed, and, if they had, it would be up to the trier of fact to determine intent.

Condition 3: Interference with Plaintiff's Commercial Relationships There can be no question that Barney's work interfered with Retail Design Associates' commercial relationship with Major Centers: The project became more costly and was delayed. Nonetheless, while the relationship may have been strained, Retail Design Associates did not lose its client.

Condition 4: Special Damages Since Retail Design Associates did not lose its client, it is difficult to establish special damages. True, the firm did have to absorb the cost of redesign, but did this come about as a result of Barney's comments? Possibly they could have been considered special damages if Barney had said that ineptitude made the additional work necessary.

The Outcome In review, it becomes clear that Retail Design Associates cannot win an unfair competition case based on commercial disparagement because several elements are missing. No false and injurious statement was produced; no malice was proven; and there were no special damages.

Although Barney was victorious on this issue, he still lost, simply because he had to mount a defense. Chances are a claim would not have been filed at all had Barney only followed what used to be a commonly held ethical practice, namely, contacting the firm that did the original work before proceeding to review or modify it. This gets right back to the Golden Rule. Certainly you would like your peers to treat you in that manner; just as certainly you should treat your peers in the same manner. And, again, this professional "nicety" has significant practical impact. Barney could have informed a representative of Retail Design Associates that he would be reviewing and upgrading the plans and could have inquired about any special considerations of which he should have been aware. This would have eliminated the element of (unpleasant) surprise if the original designers were unaware of the owner's intentions. It would also have made Barney aware of certain considerations; for example, "We prefer the use of square molding to rounded," or "We have an updated set of plans you should see." After reviewing the work and deciding on the changes needed to satisfy code and to effect upgrading, Barney could have called Retail Design Associates again, this time to discuss the specific recommendations he would be making. Perhaps he would even have changed a few, based on the other firm's comments. But, in any event, he would have kept them informed and would have helped establish the kind of personal relationship that often makes suits undesirable. In addition, by keeping Retail Design Associates informed, they could have started working on the revised drawings sooner, thereby avoiding much unpleasantness. Today, of course, no association can require its members to act in this manner. It is obviously a professional approach, however, and just as obviously it can greatly reduce a professional's liability exposure.

Interference with Contractual Relations

The tort of contractual interference (malicious interference) comprises two basic elements. There must be:

- an intent to interfere with a known contractual right, and
- actual interference.

Actual interference is relatively easy to show. It typically involves termination of a contract. In Barney's case, of course, it resulted in a straining of the relationship.

It is far more difficult to demonstrate an *intent to interfere*. In some states, this must be the only intent, whereas, in most, malicious interference must be the major intent. In Barney's case, he may actually have wanted to interfere, with the hopes of having more work given to him at the expense of Retail Design Associates. In most states, it would be up to the jury to decide if this was his major intent. In certain others, of course, malicious interference would have to be the only cause. In a state abiding by the latter doctrine, then, Barney would be home free. Since he was retained to review and upgrade the plans and specifications, malicious interference could not possibly have been his sole intent. Part of his intent had to be fulfilling contractual requirements.

If the jury sides with Retail Design Associates and states that malicious interference was Barney's major intent, then special damages would be available; no general damages would be assumed. In some cases, special damages may also include *consequential damages*, that is, those damages incurred secondarily, as a result of the initial damages. For example, if as a result of Barney's interference Major decided to withhold a payment to Retail Design Associates and if that refusal to pay rendered Retail Design Associates unable to make a mortgage payment, resulting in the lien being foreclosed on its building, then the loss of business Retail Design Associates experienced, its additional administrative costs, and so on, all would be considered consequential damages. The court could also levy punitive damages, given the existence of malice.

In this claim, as in that for commercial disparagement, it is up to the plaintiff to prove malicious intent. Given the existence of Barney's contract, it would be difficult to do.

Interference with Prospective Economic Advantage

The tort of interference with prospective economic advantage is basically the same as the tort for interference with contractual relations. The basic difference is that no contract exists. The classic case in this area is *Tarelton v. McGawley*, a case that was resolved in England in the late eighteenth century. The plaintiff was seeking to develop trade off the African coast. As a canoe filled with natives approached his vessel, his competitor fired a cannon over the canoe's bow. This was particularly effective in discouraging the natives from trading with the plaintiff, but, on his return to England, the plaintiff filed suit. The court agreed that the defendant was not justified in his action, which resulted in loss of the plaintiff's prospective customers, thus comprising unfair competition. At one time, most design professional codes of ethics directly prohibited interference with prospective economic advantage, as well as other elements of unfair competition. In at least one instance, however, a member of one of the groups who had his membership revoked countersued and won. As a consequence, most codes of ethics today treat this subject in a somewhat oblique, largely unenforceable manner.

5
Contracts for Professional Services

A contract is a legally binding agreement that sets forth each party's responsibilities to the other. An effective contract also contemplates some of the typical contingencies that may occur and identifies the manner in which they will be handled. The purpose of this chapter is to provide some general information about contracts and the contract formation process, and to identify and discuss certain contract provisions that apply to some of the most significant contingencies.

The size of this chapter reflects the importance of contracts. In fact, no project should be pursued without a written agreement. The potential for disputes makes it essential to establish mutual responsibilities. Without a written contract, assignment of responsibilities may be left up to assumption; key provisions will be subject to quirks of memory. Committing understandings to writing helps prevent problems and is a key aspect of design professional–client communications. To help enhance this communication process, it is suggested that contracts be written in plain English rather than in the "legalese" that is commonly used in contracts prepared by attorneys. The assistance of attorneys still is essential, however, but they should be consulted principally for purposes of review.

Design professionals should have a full understanding of contracts, for these are the instruments that govern their responsibilities with respect to each commission they undertake.

CONTRACTS: AN OVERVIEW

A contract is a legally enforceable agreement that sets forth the obligations of each party to the other. Any violation of these obligations can expose the party committing the violation to sanctions of law.

Categorizing Contracts

Many terms are used to categorize contracts. Some of the most common are as follows.

Bilateral and Unilateral You will almost always be dealing with *bilateral contracts*, that is, contracts involving two parties who exchange promises. As an example, you promise to perform certain services, and the other party promises to pay for them. In a *unilateral contract*, one party requests the other party to act, but the other party is not obligated to do so. If the other does, however, then the party requesting action is required to do what was promised, that is, provide payment.

Enforceable, Void, Voidable, and Unenforceable An *enforceable contract* is one that comprises all elements necessary to bind the parties. If one of the parties reneges, the contract has been breached. The party injured by the breach may find it necessary to seek judicial relief. In granting relief, a trier of fact may assess damages against the breaching party or may require that party to complete the contract as originally agreed. The cost of obtaining judicial relief usually must be borne by the party seeking it, unless the contract specifies otherwise.

A contract is *void* when one of the elements required to bind it is missing, typically through oversight. This does not mean that you are without recourse if, after performing $10,000 worth of work, for example, you discover that the contract is void (*see discussion of implied contracts and quasicontracts, below*).

When a contract is *voidable*, one of the parties to it has the ability or legal right to call it void. This right could exist because the other party gained acceptance by using fraud, duress, or some other unacceptable means that would make enforcement of the contract contrary to public policy. If one party made a mistake, as by mislocating a decimal point, this, too, could make the contract voidable.

An *unenforceable* contract is similar to a voidable one, because something about it makes it statutorily impossible to enforce. For example, all states have a statute of limitations that applies to contracts. These statutes provide that, within a certain period of time after agreed-to services have been performed, the contract related to those services can no longer be enforced.

Express and Implied An *express contract* is one that results from a specific agreement between two parties, in which each has expressed obligations and expectations. An *implied contract* is one that is presumed or implied to exist by virtue of the parties' actions with respect to one another, especially if they are acting on the basis of custom (such as a custom in an industry) or precedent.

Oral, Written, and Verbal An *oral contract* is one that is entered into based on the spoken word; a *written contract* is in writing. A *verbal contract* is one based on words; that is, written or oral, and it also is used to indicate an oral agreement.

Quasicontracts A quasicontract exists when one party has been enriched by the other and the one who provided the services or products that caused enrichment seeks compensation. A quasicontract is similar to an implied contract in that an express understanding is not required, but it differs from an implied contract because it is not necessary to show that both parties involved conducted themselves as if a contract were in effect. For example, an architect may seek at least partial payment for inadvertently performing a study that had not been requested. Any payment made voluntarily by the client or as a result of judicial relief would be paid in the interest of fairness, assuming the service had some value to the client.

Binding a Contract

For a contract to be binding, it must comprise several elements. These include *agreement* (offer and acceptance), *consideration, legal form, authentication, competent parties,* and *legal purpose.*

Agreement Agreement exists when one party makes an offer and the other party accepts it. An *offer* is just that: one party's tender or proposal to perform an act or service in return for some form of consideration (e.g., payment). The offer may be unequivocal or, more commonly, it will be made with reservations. The offer will be considered outstanding and valid until it is accepted or rejected by the party to whom it is tendered, or until it is withdrawn by the person making it. Withdrawal is automatic when a reservation states that the offer will expire if not acted on within a certain period of time.

Acceptance means an acceptance of the offer by the party to whom the offer is made.

Consideration *Consideration* is defined as something of agreed-upon or perceived value given by either party to *bind the contract*, that is, to make it enforceable. The value of the consideration does not have to be substantial. Performing just a small amount of work may be enough to make the contract binding. If so, you would be permitted to enforce the contract's provisions.

Legal Form *Legal form* refers to the enforceability of the contract by virtue of its adherence to public policy. It is not uncommon for certain elements of a contract to violate law or interpretations of law. In some jurisdictions, only the offending clause is struck; in others, the entire contract may be considered void or unenforceable. To prevent the latter outcome, most contracts include a *severability clause.* This states that the voiding of any element of the contract affects only that element; all other provisions remain in force.

Authentication Authentication is an aspect of legal form. It in some cases requires a corporation to apply its corporate seal, a holdover from times when a wax seal was used. In most states, however, the signature of a company officer can serve as authentication, at least insofar as contracts are concerned. Some jurisdictions also require authentication through *attestation*, typically requiring the use of witnesses or notaries public.

Whenever you are operating in an unfamiliar jurisdiction, determine its authentication requirements.

Competent Parties Competent parties are individuals who are sane and, when representing business entities, have authority to do so. Most contracts include a brief provision requiring business representatives to attest that they are empowered to act on their companies' behalf.

Legal Purpose A contract written to facilitate illegal activity is not enforceable in courts of law or other recognized forums. "Illegal activity" may be more subtle than supposed. For example, unlicensed contractors have in some cases discovered that they are unable to enforce contracts, because offering to do the work without a license is illegal. Engineers and architects assumedly can be subject to the same judgments.

TYPICAL CONTRACT FORMATS

Several types of written contracts are commonly used. These include *conventional proposals, negotiated terms and conditions, special contracts for major projects, model contracts, multiple contracts, client-developed contracts, other types of written agreements,* and *oral agreements.*

Conventional Proposals

Contract formation frequently begins when a client requests a proposal. As discussed in Chapter 3, and assuming use of QBS, you would then meet with other design professionals and, ultimately, the client, to develop a workscope, general conditions, and a fee proposal.

General conditions describe nontechnical understandings that comprise the business context within which services will be rendered, such as mutual responsibilities relative to certain issues, when payments are due, and any limitations of liability. (The example provisions discussed below all relate to general conditions.) Most design professional firms have developed standard general conditions that they submit with their proposals.

A *fee proposal* sometimes lists the hours certain staff members are likely to expend in implementing services and their hourly rates, as well as the cost of specific operations, such as computer use and testing. Fee proposals often are accompanied by a general schedule of fees and costs, something that would be referred to should additional services be required during the course of the project.

In many cases, a proposal comprising a workscope, general conditions, and fee proposal will be submitted to the client with a cover letter that may also indicate the *performance schedule*. Otherwise, the schedule may be referenced in one of the three other elements, or comprise a fourth. Once the client accepts the proposal in writing, it becomes an agreement. In fact, this is a common occurrence when you and your client have worked together before and each of you has a good understanding of the other's methods of operation, expectations, and risk tolerance levels.

Negotiated Terms and Conditions

The proposals submitted by engineers and architects are sometimes subject to discussion, a process frequently called "contract negotiation." This term can be misleading, however, since "negotiation" often connotes dollar issues. While dollar issues are involved, discussion usually centers on the scope of services needed.

Once a final workscope has been agreed to, terms and conditions are discussed. The client may not agree to everything contained in standard general conditions and may seek to impose alternative and/or additional conditions. Risk assessment is a vital element of this process.

Design professionals should be intimately aware of the scope of services' impact on risk and thus the degree to which defensive measures included in general conditions can be relaxed or should be strengthened. Generally speaking, the extent and effect of defensive measures should be directly proportional to the client's willingness to assume risk. The more risk the client is willing to accept, the more the design professional should seek to use the contract and contract formation process to contemplate and assign responsibility for these risks should they materialize. The reason for this guidance is basic: The greater the risk, the more likely it is that problems will arise. A construction contract that contemplates such problems can help minimize their effect, by establishing who will do what in the event they occur.

In some cases, the proposal modifications required are so extensive that standard general conditions must be wholly rewritten. In other cases, they can be amended, as through an *addendum* that identifies additional conditions to which the parties agree.

Special Contracts for Major Projects

When a large and complex or otherwise risk-prone project is involved, standard general conditions seldom will suffice. While they may be submitted as a basis to initiate discussions, a more comprehensive agreement usually is needed. The larger firms that typically are engaged for projects of this type frequently use standard contracts of their own design, at least for certain types of recurring major commissions, for example, office buildings, bridges, shopping malls, or hazardous materials remediation. Some also have developed any number of standard contract clauses that can be selected to form a unique contract through word processing.

Model Contracts

A prime design professional will often employ a model contract when retaining interprofessionals. The models most commonly used are developed by associations. One of these is the Engineers' Joint Contract Documents Committee (EJCDC), an umbrella group comprising the American Consulting Engineers Council (ACEC), the American Society of Civil Engineers (ASCE), the National Society of Professional Engineers (NSPE), and the Construction Specifications Institute (CSI). The American Institute of Architects (AIA) also produces widely used model contracts.

The product of any association activity comprises compromises made to reflect the differing needs and viewpoints of the association's members. Many of these members modify association-developed materials to make them uniquely suited to their own needs. In any event, do not assume a model contract can be used "as is" simply because it has been developed by a recognized association. Obtain qualified legal assistance before assenting to its terms and conditions, at least to help assure conformance with the laws in the state with jurisdiction.

Multiple Contracts

Multiple contracting is referred to by several names, but all imply the same arrangement: The owner contracts separately and simultaneously with the prime and those who ordinarily would serve as interprofessionals. As with conventional approaches, the prime retains responsibility for design team coordination and still retains "leverage" by being in a position to approve or disapprove the owner's payments to the interprofessionals. Nonetheless, the approach provides a variety of benefits and is strongly recommended as a result of them.

One of the most significant benefits for interprofessionals is cash flow, since payments would be made directly to them. All too often, prime design professionals receive payment from an owner but then withhold payments to subcontracting design professionals, in order to gain use of their money. This practice can seriously erode team attitudes and thus benefits no one. Interprofessionals also gain direct access to the owner should the need arise, possibly because the prime insists on activities, procedures, or design approaches that are seriously flawed. Ethically, an interprofessional retained by the prime should not report to the owner, because the prime, not the owner, is the interprofessional's client. Realistically, interprofessionals are not generally constrained by this situation and will inform an owner when they believe problems are on the horizon. The problems usually must be very serious to suggest this procedure, however, because of the prime's leverage with respect to interprofessionals' progress payments and due to interprofessionals' usual desire to obtain more work from the same prime.

The prime design professional benefits from multiple contracting because it reduces exposure to "vicarious liability," that is claims that result from errors or omissions allegedly made by subcontracting design professionals. In addition, multiple contracting lessens bookkeeping requirements, and, in those states where gross receipts or sales taxes are applied to design professionals' services, it can lower the prime's tax burden as well, since the prime would no longer be a conduit for interprofessionals' fees.

The owner benefits from multiple contracting because it can promote better attitudes among design team members, it establishes a direct line of communication with interprofessionals, should it be needed, and it makes fees clear; that is, the owner learns how much each of the interprofessionals is being paid and how much the prime is receiving for design team coordination services. Regrettably, primes in some cases "beat down" the fees of their interprofessionals and impose a significant charge (not separately identified) for coordination. An astute owner usually wants to be assured that the fees paid to all parties for all their services are reasonable.

Some owners are reluctant to employ multiple contracting because they fear it will result in their having to deal with three or four design professionals, not just one. As noted above, this fear is not realistic; owners should be informed of the counteractive forces at work. It has also been noted that multiple contracting may create more bookkeeping requirements on the owner's part, but this is hardly a significant drawback in light of the benefits to be gained.

The multiple contracting concept can be implemented through one multiple contract that requires the signature of each of the design professionals (e.g., prime architect or civil engineer, structural engineer, mechanical and electrical engineer, and geotechnical engineer), or it may be implemented through separate contracts. As a consequence, it also is known as separate contracting or separate and multiple contracting.

Client-Developed Contracts

The need for legal review is particularly pronounced when clients use their own unique contracts, as often is the case with government entities and major corporations. Certain conditions of such contracts often tend to be one-sided, seeking to transfer client liability to the design professional. Engineers and architects too often will agree to such conditions because they fear that attempting to modify them will cause a major client to seek a more agreeable (and less astute) replacement. It is precisely this type of contract that often needs the most modification, since it could otherwise force you to accept a severe liability exposure that is not rightfully yours and that you are powerless to control. As difficult as it sometimes may be to change the attitudes of major clients, it can be even more difficult to win disputes with them after the fact.

Some of the contracts that owners develop take the form of *client-developed proposals, competitively priced proposals, purchase orders,* and *modified construction contracts.*

Client-Developed Proposals In some cases, the complete proposal format for a project is client developed and includes a scope of services, general conditions, and a "fill-in-the-blanks-type" fee proposal. Because the scope of services is developed unilaterally, it is based on assumptions, creating a greater likelihood of false expectations and misunderstandings. Client-developed proposals can also encourage submission of the lowest possible fee, a situation that can encourage an overly optimistic evaluation of the level of effort that actually will be required.

Review such requests for proposal (RFPs) or quotation (RFQs) closely before responding to them. Once appropriate blanks are filled in and the submission is signed, it becomes a contract upon the client's acceptance. While it may be possible to modify the scope of services and/or general conditions later, do not assume this can be done. If modifications are needed, speak with the client before making them. It may be possible to submit an altered proposal that will better serve the interests of both parties.

Competitively Priced Proposals In some instances, a client will ask for competitively priced proposals. The format used is essentially the same as that discussed above, except each respondent must develop a unilateral workscope. Ideally, this workscope would be the same as that developed through QBS. More often than not, however, there is pressure to keep fee (and thus workscope) to a minimum, in order to secure the engagement. If there is such pressure but the client has decided to pursue the engagement nonetheless, it is essential to delineate the workscope precisely, in order to help prevent misunderstandings about the services being offered.

Even though you may submit your own general conditions with such workscopes, clients may insist on using their own. Bear in mind that the scope of services and general conditions should complement one another for purposes of effective risk management.

Purchase Orders In some instances, a client-proffered contract will be nothing more than a standard purchase order, used to procure all types of supplies and services. Purchase orders are not appropriate for professional service agreements, and the client should be so informed. This can be done in person, over the phone, or via a well-worded letter. Choose the medium most suited to the people and circumstances involved. Emphasize that a purchase order does not serve the interests of either party.

Modified Construction Contracts Some clients use a slightly modified general construction contract to solicit professional services. These are typified by language that refers to a design professional as CONTRACTOR and by clauses that call for an architect or an engineer to obtain a performance bond and/or acquiesce to retainage provisions. Such contracts should be modified to reflect more fully the circumstances involved, for both parties' benefit.

Other Types of Written Agreements

No matter what type of written contract is involved, the scope of services, general conditions, performance schedule, and fee proposal all may be included in one continuous document, or the workscope, performance schedule, and fee proposal may be appended to, and referenced in, general conditions. Certain other materials also could be appended to the document and thus also become part of the agreement. Note that documents do not have to be appended physically to a contract to be part of it, as when other contracts or portions of them are *incorporated by reference*; for example, "DESIGN PROFESSIONAL shall be bound by the same terms and conditions as CLIENT, as identified in CLIENT's agreement with the owner, which is hereby made part of the AGREEMENT."

Note, too, that a contract can grow in size over time, as initial agreements are modified or expanded through addenda. Unless specifically stated otherwise, each addendum is subject to the same terms and conditions as other parts of the agreement.

Oral Agreements

You probably have at some point been told about "the good old days" when agreements were struck through oral understandings confirmed by a handshake. Although oral agreements still can be binding, it almost always is a mistake to rely on them for performance of professional services. The world we live in and the civil justice system that governs business dealings have become highly complex. You are responsible—and thus liable—not only to your client, but also to any other party who foreseeably could be damaged or injured as a consequence of your services. Extensive liability exposures are involved, and innumerable other concerns bear consideration. Should a dispute of some type arise, the parties involved will each be required to recollect what was said months—possibly years—before. Even the best-intentioned people have trouble remembering such detail, and, when it comes to pocketbook issues, many have difficulty remembering their best intentions.

Despite the problems associated with oral agreements, design professionals will sometimes begin work on smaller assignments after receiving an oral authorization. In such cases, they should present a written agreement for client acceptance within a day or so of project initiation. Those who begin work without a written agreement generally do so only as an accommodation to long-time clients for whom they have worked frequently in the past and who have signed their standard agreements many times before. As an alternative, it is possible to enter into a basic ordering agreement with a client who uses services on a somewhat regular basis. The agreement would set forth applicable rates and general conditions, and would stay in effect for a given period or until cancelled by either party.

Most design professionals agree that it would be foolhardy to take such an approach when a major, complex, or risk-prone project is involved, or when there is no significant prior relationship with the client.

Amendments Oral modifications of written agreements are themselves oral agreements. This fact is sometimes overlooked. Thus, design professionals who always insist on working with a written agreement will sometimes accept an oral amendment, exposing themselves to potential problems. Oral amendments should be immediately confirmed in writing, to help prevent any misunderstandings and to help prevent a claim that you did not do what you had orally agreed to do.

Fax machines can be particularly effective tools for oral amendments, especially when the oral amendments require a design professional to perform an additional service for an additional fee. A number of experienced engineers and architects report that they will not start the additional work until they reduce the amendment to writing, sign it themselves, and then fax it to the client for signature and return. The entire process can actually take less than thirty minutes.

THE BENEFITS OF FORMING AND HAVING A WRITTEN AGREEMENT

Although some of the benefits both parties derive from forming and having a written contract have already been alluded to, it is worthwhile to review them in more detail. Four of the most significant are *mutual understanding, establishing your own rules, sizing up,* and *identifying and allocating risk*, as follows.

Mutual Understanding

The process of developing a scope of services and general conditions requires in-depth communication that permits each party to derive more appreciation of the other's concerns. Better understanding promotes better contracts and more satisfying relationships.

Establishing Your Own Rules

When a contract is silent on certain issues, applicable tort provisions will be applied. For example, when a contract does not specify how a dispute will be resolved, litigation can be demanded by either party. Litigation is costly, convoluted, and slow, however, so parties to a contract often agree to use an alternative dispute resolution mechanism. Parties to a contract can also agree that neither will sue the other for consequential damages or that the time period during which either can initiate a claim against the other is shorter than otherwise allowed by the applicable statute of limitations. The ability to establish one's own rules is limited, of course. It does not permit measures that are against public policy, nor are extralegal requirements binding on third parties.

Sizing Up

It is important to understand your client's attitudes and motivations, because some may lead to avoidable problems. The process of contract formation allows you to assess an organization or an individual and to determine whether or not you want to provide services. Beware of any client who has little compunction about sacrificing quality or shifting liabilities to others.

Identifying and Allocating Risk

Your ability to evaluate a client's attitudes and motivations are enhanced when the subject of risk is discussed early in the relationship. Not too many years ago, such an approach was frowned upon; most design professionals' contracts ignored potential problems. In fact, some considered it unprofessional or undignified to discuss risks. Times have changed. Today, astute design professionals advocate a candid discussion of risks as a vital element of the contract formation process. In that way, risk management can be incorporated into the workscope, as well as general conditions.

DEALING WITH RISK

Of the three principal groups associated with a construction project—owners, design professionals, and contractors—design professionals benefit the least (in terms of profit), yet their risks can be the most substantial. They face claims from the other two parties, as well as from innumerable others seeking to hold them liable for damages allegedly arising from their errors, omissions, or other acts. In most cases, the risks involved are personal and may exist for an indefinite period. Despite the extent of these risks, many can be managed by contract, providing they are contemplated during the contract formation process.

Generally speaking, the process of contract formation permits you to deal with risk in at least eight specific ways.

First, you can offer to execute a commission in a more professional manner, to minimize the likelihood of problems resulting from inadequate communication, coordination, and cooperation. Quality-oriented clients will often authorize such an approach, because experience has taught them that it is the least expensive approach in the long term. Therefore, in order to reduce risk and to assess client attitudes, implement proposal-phase services professionally and offer to implement design-, construction-, and postconstruction-phase services just as professionally.

Second, thoroughly inform clients about their responsibilities. Use a contract's general conditions to require clients to cover any costs you otherwise would bear due to their negligence in fulfilling their responsibilities. Most clients will accept these provisions—indemnifications—because they are fair and reasonable.

Third, based on your own and others' experience, identify potential problems that you and your clients are powerless to prevent. Inform clients about these risk and the options available for managing them. Assuming they are not major, you may be willing to accept some. If you do, however, it may be appropriate to impose a fee premium in order to build the resources to pay for the problems that eventually will materialize, a concept discussed below. The client can also absorb some of these risks (typically through an indemnification), possibly transfer them via insurance, and pay you your normal fee. A combination of these two approaches is possible by relying on a risk allocation and limitation of liability agreement. All these concepts are discussed below.

Fourth, you can identify unpreventable risks that are so significant that you cannot possibly accept them, in part because they may be uninsurable. Typical of these are risks stemming from modification of existing structures or hazardous materials engineering. Let clients know they must accept such risks (via indemnifications) if you are to accept the assignment.

Fifth, you can use the contract to close certain loopholes that could otherwise become traps. For example, your agreement could include a list of services that you have explained and offered to the client and that the client has declined. This procedure would block an attempt to hold you liable for failing to perform a service the client told you was unwanted.

Sixth, you can use the contract to establish extralegal conditions that reduce certain risks which otherwise would exist by virtue of law. As examples, you and your client can specify a dispute resolution mechanism that is more effective than litigation, reduce the impact of a statute of limitations, or restrict either party's ability to sue the other for consequential damages.

Seventh, you can work to develop a contract whose language is so clear that the intent of the parties cannot be mistaken. Such a contract helps prevent misunderstandings. It also makes provisions easily understood by a trier of fact, such as a jury. This latter aspect in and of itself may discourage a plaintiff from pursuing an otherwise marginal claim.

Achieving these seven methods of reducing your exposure to risk hinges on your communication skills. Clients should be made cognizant of risks and techniques for dealing with them. If they are adamant about not taking prudent measures, not remunerating you to accept risk, or not accepting it themselves, then the ultimate risk-reduction method—*number eight*—must be considered: *Walk away*.

ISSUES TO CONSIDER IN CONTRACT FORMATION

You should be aware of several important issues before attempting to form a contract. These include *assumption of liability, professional liability insurance, disparate bargaining power, indemnifications, definitions, word selection, exculpatory wording*, and *the role of attorneys*.

Assumption of Liability

As a design professional, you are subject to the doctrine of professional negligence. Because professional negligence is a liability governed by *tort law*, it is a *tort liability*.

Tort liability is distinctly different from *contractual liability*, something that arises when a contract provision is breached. Tort liability and contract liability are not mutually exclusive, however. For example, if you contractually agree to act in a non-negligent manner, a negligent act would create *both* a contractual liability and a tort liability.

In forming a contract, avoid any provision that would obligate you to *assume liability*, that is, to do more than common law requires. For example, a contract provision that obligates you to perform "at the highest professional level" obligates you to abide by a standard far in excess of that established by the doctrine of professional negligence. Even if you do not commit an act of professional negligence, it could still be alleged that you did not perform "at the highest professional level," thus exposing you to a claim for breach of contract and a lawsuit to collect the damages caused by that breach.

Assumption of liability can also create severe insurability problems, because professional liability insurance only covers you for negligent acts. Failing to perform at the highest professional level is not necessarily a negligent act. Instead, it may be nothing more than a failure to deliver on a promise you made, something covered by few—if any—professional liability insurance policies.

Are you satisfied that you would never use incautious language? *Never?* What about a letter you or someone in your employ may prepare, which happens to say, "We traditionally have performed at the highest professional level, and now we offer that type of service to you." Later, during formation of the contract, that letter is appended to your agreement or included by reference.

How do you prevent such pitfalls? *Act professionally. Learn and educate.* In fact, most contract provisions that would cause you to assume liability are requested, not because your clients are devious, but rather because they are unaware of certain issues. Good clients always are willing to learn, especially since doing things a better way can increase their protection, by not invalidating your insurance. It can also help eliminate the false expectations that may arise when you agree to do that which you cannot reasonably accomplish. Failure to meet expectations creates frustration, and frustration often triggers a claim.

Perform professionally, and you create realistic expectations, make clients more aware of their actual risks, minimize your unnecessary risks, and help assure that those that do remain are covered by insurance or are otherwise manageable.

Professional Liability Insurance

Virtually all professional liability insurance policies contain a provision that specifically excludes coverage of liabilities assumed by contract. Most also contain certain other exclusions you should be familiar with, not only as a generally prudent business practice, but also because it gives you an important lever in contract negotiations. For example, when the client asks, "Why can't you operate at the highest professional level," your response might be the following:

> Mr. Smith, in my opinion, our firm is the best. Our people are concerned and dedicated, and we have a comprehensive program of quality assurance. But, if I agree by contract to perform at the highest level, I've committed to an obligation that exceeds what's required by law. That means I would have assumed a liability I'm ordinarily not exposed to, and that would be enough for my professional liability insurer to say, "We're sorry, but that's specifically excluded from coverage." Now, even though I think so highly of our firm, I'm not about to say we're perfect. Even the very best may commit errors or omissions. So, if I agree to perform at the highest level, and I don't, I'm not insured, and that's not good for me or you. I'd much rather just leave that provision out of the contract. The law says I'm liable if I commit a negligent act, and, if I do, I'm insured.

Since professional liability insurers essentially insist that you agree to do no more than is expected of you under tort law and since clients often insist that you carry professional liability insurance, you should be able to educate your clients to the point where they, too, insist that you promise no more than is expected. In the process, they will learn more about the true nature of their own risks. When they realize that actions, not words, provide protection, they may be far more interested in other services you can offer to reduce their risks.

Disparate Bargaining Power

Disparate bargaining power is a term used to connote one party's unfair advantage over the other, causing the other to accept unfair terms. It results in what some refer to as a *contract of adhesion*.

As an example, assume an architect is operating in an area where one major company or public agency is the source of most contracts and the client's standard agreement requires the architect to sign a document certifying that the project was build according to plans and specifications. The architect points out that such a document should not be signed, because all workers were not monitored all the time, and some may have "buried" an error or two.

The architect offers to say that the observed work was built according to plans and specifications or that, in the architect's opinion, based on observations made, the work was completed according to plans and specifications. The client's response is, "If you want this job, you'll accept the certification requirement in the contract. Otherwise we'll get someone else." The architect reluctantly accepts the provision but the next day writes a letter saying acceptance was forthcoming because the work is needed and clearly the client exercised its disparate bargaining power in coercing the architect to accept a provision that is wrong. Such a letter serves two purposes. First, it may encourage the client's representatives to reconsider the issue and change their minds. Second, if they don't change their minds, and should the architect be subject to a problem because of the clause in question, the client's exercise of disparate bargaining power can be used as a strong defense.

Naturally, there is always a risk that raising the disparate bargaining power issue will result in your not being retained for a project. However, if you are in contract negotiations, the client has already invested a significant amount of time. Starting anew with another consultant would cause delays, no matter what type of selection method is used, and the other consultant may share your outlook. As points to remember, then, bring up the disparate bargaining power issue when it is relevant, but emphasize it only after all other issues have been settled.

Indemnifications

Indemnifications comprise one of the most serious issues both you and your client will discuss during contract formation. In many instances, you both will have the same goal: agreeing to no indemnifications except those that make the other party liable. But the indemnifications you seek generally are far different from those sought by clients. For your position to be accepted, you have to explain it well to your client, and, to do that, you need to be familiar with the issue's background.

Client-Proposed Indemnifications The indemnifications typically proposed by clients originally were developed to protect property owners from claims filed against them as a result of actions of a party who had constructive use of the property. For example, if you rent a house, its owners would likely require you to indemnify them against claims arising from your use of the house. This is fair and reasonable, because you, not the owners, have use of the property.

Indemnifications are almost always required of construction contractors, and rightfully so. When they enter a site, they have constructive use of it: They are responsible for the means, methods, and sequencing of construction, as well as safety.

When engineers or architects are on a site, they do not have constructive use of it. They are there only to perform a service for the owner. They are fully responsible for the adequacy of that service and usually are required by law to compensate anyone damaged by their negligence, whether or not their contract says so (although a contract may limit their liability to a certain amount to certain parties).

If a client-proffered indemnification would require you to accept additional risk—more liability exposure than is required by law or custom—the question becomes: Whose risk would it otherwise be? Almost invariably, it is the client's, typically an owner client's portion of the overall risk that must be accepted in undertaking a project. By means of an indemnification, the client is attempting to transfer some of that risk to design professionals. In other words, the client is trying to use you as a source of "free" insurance. But is it really free?

If part of your service includes service as an insurer, you need to create the resources required to "make good," should the need arise, just as any insurer would do. The simplest way of doing this is by imposing a fee premium, that is, a flat amount or a percentage of fee that is charged in addition to the fee for "normal" services.

In establishing the fee premium, consider the same factors evaluated by insurance companies. For example, if it can be shown statistically (actuarially) that there is a 1 percent chance of a given type of risk materializing and that the settlement or judgment will be $100,000, the "cost" of each of every one-hundred such risks accepted is $1,000. More than $1,000 would have to be charged, however, to consider defense costs, including the value of staff time and fees charged by attorneys and experts.

Most insurers consider it prudent business practice to transfer some of their risks to reinsurers, mostly investors who function as insurers of insurance companies. While this permits insurers to reduce the cost of coverage that otherwise would prevail, they still must recover the cost of this risk transfer: the reinsurer's fee.

Engineers and architects can transfer their ordinary risks, but client risks that they accept via indemnification are not ordinary. They are extraordinary, and design professionals generally *cannot* transfer them, because insurance companies consider them untenable. Furthermore, by accepting extraordinary risk, design professionals face the possibility that they may lose their coverage even for ordinary risks, making the overall severity of risk that much more substantial. Clients need to understand that their risks are affected, too. Do they really want to transfer their risks to a design professional "insurer," when doing so may result in that insurer losing the means to insure anything at all?

An insurance company that does not charge enough for the risks it assumes cannot expect to stay in business very long. Design professionals who serve as their clients' insurers are subject to the same economic fundamentals. But the fees they charge to adequately fund extraordinary risk is likely to be far in excess of the premium charged by an insurer, for whom the risk would be ordinary. If design professionals are willing to accept such risks for a far smaller fee than an insurance company or for no fee at all, clients would be well advised to consider closely their engineers' or architects' ability to make good, should the need arise. If design professionals cannot adequately fund the risk, a client, who typically holds a far superior economic position, might be required to pay, irrespective of contract provisions.

Given the pitfalls, why do so many clients still try to transfer their risks to design professionals via indemnifications? The answer can be summed up in one word: attorneys. All too often the lawyers who prepare contracts on their clients' behalf look only to the words and not to the facts. Indemnifications look good on paper; they seem to get a client something for nothing, thus justifying the attorney's fee or salary. When these indemnifications must be exercised, however, they may be found worthless, simply because the design professional cannot make good. As such, the client may have to contend with a major liability exposure that otherwise could have been managed through insurance. This illustrates why clients are almost always best served by transferring their risks to insurers, through an owner's protective policy that protects them, their agents, and their consultants (such as your firm) from insurable losses arising from the work performed. While some liability exposures are not insurable in this way, the law generally creates an *implied indemnification*, which makes people or firms responsible for the consequences of their own negligent acts. In fact, some jurisdictions feel so strongly about this issue they have expressly prohibited certain types of indemnifications, such as those that would make a party liable for the damage caused by another party, even though that other party caused 99 percent of the problem.

Another major concern, discussed below, is determining what some client-developed indemnifications actually mean. Construction industry case law is littered with disputes whose central issue is the intent of an indemnification and, consequently, who is responsible for what. Each side to such a dispute is represented by attorneys who are fully confident that their interpretation is correct. Accordingly, any client-proffered indemnification must be scrutinized closely. To underscore this need, consider the dramatized case of *The City of San Diablo v. Delacourt Engineering*, discussed in *The ASFE Professional Liability Test Kit*. The unfortunate David J. Delacourt agreed to "hold harmless, indemnify, and defend the City from and against all claims or liability arising out of performance of the work herein." A problem arose, and Delacourt was sued, along with the city and the general contractor. The case against Delacourt and the contractor was dismissed, because negligence could not be proved. The city was not so fortunate, and, when its case was over, it sent Delacourt a bill for $30,000 as his portion of the city's defense costs. Although it had been demonstrated that Delacourt was not negligent, it was also shown that he had agreed to defend the city, and, unquestionably, nothing in the indemnification made Delacourt's agreement contingent on his negligence. Delacourt's professional liability insurer refused to participate in resolving the dispute about the bill or in paying the bill itself, since the problem had occurred due to Delacourt's acceptance of a contractual liability.

As you might be aware, public agencies, in particular, can be insistent about indemnifications. Their nature is such, however, that you may wish to try an "end-run approach" to resolving the problem. For example, if you are told you must accept the unfair provision as a condition of your being retained, you may wish to note for the record, as by a letter written to the agency in question, its procurement director, or other appropriate party, that the jurisdiction employed disparate bargaining power in forcing you to accept the clause. A government entity may be far less upset about such a notice than a private sector client, but the question remains: Will it work? This question feasibly might have to be answered in court, where you (probably absent participation by your professional liability insurer) would have to prove your point to a jury consisting of local taxpayers and/or a judge on the jurisdiction's payroll.

Another "end-run approach" could be tried when you are satisfied that the indemnification offered violates the governing jurisdiction's laws and thus is unenforceable. In that event, you could simply accept the clause as it stands, comfortable in the belief it is meaningless. Be extremely circumspect about doing this, because a number of jurisdictions have adopted the *doctrine of express negligence*. This holds that even a contract provision that is against public policy *will* be upheld when it is written in clear, unmistakable language. It would be particularly dangerous to offer clients a clause you know to be illegal, in hopes of leaving both of you satisfied, albeit for opposite reasons. *Recognize that rulings as to the meaning of a contract provision commonly go against the party who developed it.* Because you offer something unenforceable to the client, a court might rule that, in this case, it is enforceable. In situations such as these, obtain guidance from an experienced attorney.

Design Professional-Proposed Indemnifications Indemnifications proposed by design professionals generally require clients to accept risks that are rightfully theirs to begin with but that feasibly could be considered yours, unless language exists to the contrary. Certain other indemnifications require clients to accept risks that you are powerless to prevent or insure and that create liability all out of proportion to your hoped-for gain (i.e., profit). Such indemnifications thus become part of the consideration design professionals need in order to make their offers.

Any contract you prepare should have as few indemnifications as possible, *but* everything for which you seek indemnification should be included. This often can be accomplished by combining several related issues into one clause and writing one indemnification to cover all the issues.

In order not to alarm clients from the outset, by proposing a contract with many indemnifications, consider preparing a basic contract that omits issues calling for indemnification. Then, in your cover letter, you could say something along the following lines:

Dear Mr. Smith:
    I am pleased to enclose a basic contract for your review. Please note that the risk issues listed below have not been addressed.

(*Listing*)

Several approaches are available for dealing with each, and we should address these together. I will call you in a week to schedule an appointment, unless I hear from you before then.

In discussing the suggested clauses, the client may ask, "What's the indemnification for? Who's going to sue?" An effective response might be the following:

Let's hope no one does, and that's probably exactly what's going to happen. But let's be realistic. The courts have gotten out of hand. A person tries to commit suicide by jumping in front of a subway, but he lives and sues the subway system because the car didn't stop in time. Who'd be crazy enough to bring a suit like that? Evidently someone who's not so crazy, because the poor fellow who tried to kill himself won the case. But, even if he didn't win, someone had to spend time and money in defense, simply because the action was filed. Now, if I'm accused of negligence, I have to defend myself, and that's a risk I have to accept every time I accept an engagement. But why should I have to expose myself to a risk that has nothing to do with the quality of my design or that arises from events I cannot possibly control? It's your risk to begin with, and, if something goes wrong, you're probably going to be sued anyway. I'd be named simply because I'll pay X dollars just to get out of it. So, basically, your agreeing to indemnify me really doesn't increase your exposure. It just gives some legal extortionist one less person to sue.

Your attempt to deal intelligently with risk may be something new for some clients. They may say something such as, "I've worked with a lot of other designers, and they don't ask for any of these indemnifications in their contracts." An effective response might proceed along the following lines:

We've discussed the risks involved, and you see what I'm up against; I understand what you're up against. But I feel pretty comfortable in saying that you're an astute businessperson, and, if you were in my shoes, you'd probably be doing exactly what I'm doing. In my opinion, design professionals who aren't aware of these risks or who accept them anyway, even though they can't insure them or cover them with their own assets, are pretty poor businesspeople, and that kind of carelessness may affect their technical output, too. You know, you can't expect someone who's unconcerned about their own risk to be very concerned about yours.

You probably would be better off not dealing with clients who are adamantly opposed to the reasonable indemnifications you seek. The attitudes they convey suggest stubbornness, inflexibility, and lack of empathy, any one of which can aggravate a minor problem into a major liability that may be uninsured.

You also should be circumspect about clients who willingly accept all indemnifications without a murmur. Attitudes that are "too good to be true" may be exactly that. Similarly, it is vital for you to consider clients' ability to honor any indemnifications they accept. If they do not have the financial resources required or a means for transferring their responsibility through insurance, you could be left in an extremely uncomfortable position.

Definitions

The need to include definitions in a contract cannot be overemphasized. Clients may not question certain words or phrases, but you and your clients' understanding of them may be entirely different. As an example, consider the phrase "hazardous materials." Leaving the term undefined or using the definition included in federal guidelines may be wholly insufficient. State definitions also may exist, and these feasibly may differ from state agency to state agency. The impact of differing interpretations can be severe; a given material may not fall within a federal definition of hazardous, but it may be included within a state's definition.

Word Selection

The contract is a complete understand of the agreement, and all of it is conveyed by means of the written word. Just as it is essential to define certain words or phrases, so it is essential to use common words and phrases with care. This is particularly the case with absolute words, such as "all," "every," "none," and so on. We have gotten used to people using these words in a loose manner. If we hear "all," we assume it does not mean "absolutely without exception." But, in a contract, it does. Similarly, it is vital to avoid the confusion that can occur through use of pronouns, such as "whose," "who," "he," and "his." Undeniably, all of this is nitpicking. In fact, however, the failure to nitpick has cost consulting engineers and architects millions of dollars over the years, due to unfortunate word selection or inexact grammar.

Exculpatory Wording

Several of the example contract provisions given below were developed in response to actual claims made against design professionals. Many of these claims were based on inaccurate suppositions, given customs of the profession and construction industry. Nonetheless, design professionals have in many instances been found liable, in part because they did not explain certain limitations or customs to those who were allegedly not aware of them. By addressing a variety of these issues in writing, as is appropriate, clients become better informed. Nonetheless, some may be inclined to think or state that you are trying to limit your responsibility.

Clients should be given the real reason for your including these "please-don't-eat-the-daisies-type" provisions in your contract. In fact, you are not attempting to evade responsibility; you are merely trying to make crystal clear what your responsibilities do and do not comprise, as a matter of fact and tradition. In essence, nothing has changed, except that you are being more informative. Clients who state that another design professional never imposes such restrictions should be told that such restrictions always apply; some design professionals are simply more astute than others when it comes to informing the client of what they are.

You should make no attempt to get around the fact that your approach is based in part on the litigation that design professionals have been required to confront. Clients will understand this, because they are similarly affected by "litigious mania." To indicate how others have responded, simply refer to the owner's manual that comes with any late-model automobile, which will take great pains to indicate numerous "please-don't-eat-the-daisies-type" warnings. In essence, the tenor of the times demands that manufacturers and service providers advise people of eventualities and precautions that they should take or that they should consider by virtue of limitations of which they may otherwise be unaware. Clients also should be reminded that you are vulnerable to third-party suits as a consequence of your accepting the engagement, making contract caveats particularly necessary, should it be alleged that others have the right to file a claim because they are third-party beneficiaries of the contract.

The Role of Attorneys

Lawyers are advocates, obligated to do the best possible job for their clients. Many lawyers, perhaps most, evidently define "doing the best possible job" as getting the most they possibly can for their clients. It is partly for this reason that so many of the contracts you will be offered contain one-sided requirements that you cannot possibly meet, insure, or accept.

In fact, the principals involved in a contract know more about what is needed than lawyers, because these principals have to deal with reality, not words. You and your client should together develop your own language and not abdicate that responsibility to attorneys. However, attorneys should be called upon to review the language, to help ensure it conveys the necessary meaning in the event it is challenged, and they should advise you about alternative approaches and their implications. Attorneys should also review the document you develop for conformance with applicable laws and inclusion of all important issues. Likewise, they should consider the insurability of your contract, as should your insurance agent or insurer. But do not let lawyers write a contract for you.

You are an expert in your field. As such, you let your clients know what their options are, and, ultimately, you implement their decisions. Have your lawyer serve you in the same way.

NEGOTIATING A CONTRACT

Some may infer that contract negotiation is an adversarial procedure. This is not, and certainly should not be, the case. You are expected to serve as your client's trusted professional advisor; the contract negotiation phase of your relationship gives you an opportunity to demonstrate your professionalism. To do so, enter discussions with an intimate understanding of your *bottom line on risk*. Recognize the important interplay between your fee proposal, general conditions, scope of services, and performance schedule, so you are in a position to give and take. If the client's budget or schedule is so restrictive that you must reduce the extent of your services, your risk is increased. Assuming you want to be associated with the project, you may be able to counteract this increased risk through certain design techniques, for example, by providing redundancy and/or through provisions in your general conditions.

If the client is unwilling to accept certain defensive provisions included in your general conditions, will it be possible to charge a higher fee to fund the heightened risk? Will it be possible to take more time or expand the technical services offered, to help reduce the unknowns and the risks arising from them, and thus your need for defensive provisions?

Be candid and objective in discussing these issues with clients. Clients who are aware of your concerns and who want you to fulfill their needs will work with you to develop fair and equitable agreements. Candor and objectivity very often comprise the foundation on which lasting relationships are built.

No matter who the clients may be, always be prepared to decline the engagement if you cannot develop acceptable terms. Serious losses have befallen many design professionals who have relinquished too much. (As a general rule, people who want something in the worst way often get it just that way.) Recognize that your termination of negotiations does not necessarily mean you will not obtain the commission. Clients with whom you are unable to come to terms initially will sometimes be sufficiently impressed by your professionalism to rethink their position. They may call you the next day, or perhaps several weeks or even several months later, after interviewing other firms. Many design professionals have heard clients say, in effect, "We interviewed several other firms, and each was willing to accept our indemnification and some of the other provisions you wouldn't agree to. We looked into it and discovered that, while they were willing to accept the risks, they really had no way of dealing with them effectively. So we're coming back to you." On the other side of the coin, you may never hear from the clients again. Losing such clients may be a much better outcome than gaining them.

Several books have been written about negotiating contracts, and seminars are frequently given on the subject. These should be considered by anyone who needs experience in the techniques involved. Experience alone can be a harsh and costly instructor.

TYPICAL CLAUSES

The clauses that follow relate to several of the major issues you normally would address in your own contracts. They are presented for your reference and guidance *only*, not for use as is. Any model's wording should be reviewed carefully before being applied, at least to help assure compliance with laws in the state where the contract will be enforced. Note, too, that thoughts expressed in some of the provisions could be combined with other provisions. In other words, consider the issues presented, not for purposes of using the wording as is, but, rather, as examples of how issues can be dealt with. Also consider the explanations given, particularly in light of what already has been said about risk and tort law.

You will note that plain English is used and that the purpose of certain provisions is explained within the text of these provisions. These techniques are advocated to help promote understanding by your clients and by any third parties or triers of fact who ultimately may have to determine what was meant. This approach makes the contract longer than otherwise, and it fails to use some of the conventional language lawyers commonly employ. However, the tight, brief contracts lawyers typically develop all too often result in disputes, because the people who have to rely on them day-to-day are not certain of their meaning.

Despite use of plain English, the example provisions employ certain common conventions. For example, DESIGN PROFESSIONAL, CLIENT, and AGREEMENT are shown in capital letters. When a number is used, it first is spelled out, then shown as numerals in parentheses. This is done to help prevent any mistakes. *Recognize, however, that any numbers shown, such as 45 calendar days (as opposed to 30 or 60) or $50,000 (as opposed to $25,000 or $100,000), have been selected on an arbitrary basis, solely for purposes of illustration.*

As to other conventions, the word *days* is generally amplified by *calendar*, to help assure no one considers it a "workday." (If you are open on Saturday, is that a workday?) And, whenever the word *including* is used, it generally is followed by *but not limited to*, since courts generally interpret "including" to mean "only the following."

The indemnification used requires the client to pay for the value of the time you and your staff members spend in defending a suit. Unless this reimbursement is specifically called for, you probably will not obtain it. However, instead of repeating it time after time, you feasibly could include just one clause that states that, in computing the cost of legal defense as per any indemnification provisions, the value of your time and out-of-pocket expenses will be included.

As already noted, the wording you select for your indemnifications must be developed with deliberate care. Rely on an effective attorney who is thoroughly familiar with the law in the state where your contract will be interpreted, should a dispute arise. In some states, your contracts must explicitly exclude indemnification for your gross negligence and willful misconduct if the rest of the indemnification is to be upheld.

So there is no misunderstanding about one important issue, realize that any and all statements as to what you should or should not recognize in your fees, clients you should or should not deal with, and so on, are based on generally accepted concepts of prudent business behavior. No moral or professional obligations are implied or intended. In all cases, your fees, client acceptance measures, and related practices are your business and your business only.

The issues covered in the sample clauses that follow include:

- certification,
- consequential damages,
- construction cost estimates,
- construction monitoring,
- curing a breach,
- discovery of unanticipated hazardous materials,
- excluded services,
- freedom to report,
- indemnification,
- job site safety,
- limitation of liability,
- maintenance of service,
- ownership of instruments of service,
- record documents, and
- right to reject and/or stop work.

Each of these issues is important, but, in aggregate, they comprise but a fraction of the wide range of issues that must be considered. Design professionals should be conversant with all of them. Excellent guidance is available from a number of sources, including *The DPIC Companies Guide to Better Contracts* and the book on which it is based, *The ASFE Contract Reference Guide, Second Edition.*

Certification

A proposed contract may require you to certify that certain conditions existed during a certain phase of, and/or at the conclusion of, construction. For example:

Certification

Upon completion of the construction phase of the project, DESIGN PROFESSIONAL shall provide a written certification that construction was completed in accordance with approved drawings and specifications.

Whether or not the contract contains such wording, you may be required to certify certain conditions to satisfy requirements of a lender, insurer, surety, another design professional engaged by the owner, or a government agency. In some cases, your receipt of final payment will hinge on your willingness to sign a certification.

Problem *Certify* and derivative words (e.g., *certification* and *certificate*) are dangerously ambiguous. To some they mean "guarantee" or "warrant." To others they mean "declare" or "state." Insurers generally interpret *certify* as "state authoritatively," which is almost tantamount to "guarantee" or "warrant."

There is no problem with certifying facts. A design professional can certify that a representative visited a site on certain dates and that certain tests were performed or construction was observed. However, there is a problem when it comes to certifying any event or condition based on professional judgment or opinion, because either is subject to error, even when the work upon which it is based has been performed flawlessly.

Certifying as fact that which cannot possibly be known as fact exposes a design professional to strict liability for breach of warranty and/or deceit, and this liability can involve consequential and possibly even punitive damages. Certifying inferred conditions therefore creates a contractual liability that makes a design professional responsible for the fallability of human judgment, even though that judgment was prudent and based on work performed without error. Because the liability involved goes beyond that established through the doctrine of professional negligence, it is *tacitly or expressly excluded from professional liability insurance policies*. The existence of a certification can also cloud the issue when it comes to deciding whether a problem has arisen because of a negligent act or because of an assumption of liability.

Solution The certification issue is so important, and the use of certification documents is so widespread, that you should consider addressing the matter "head on," in language that makes the basis for your position absolutely clear, as follows:

Certification

During the course of construction, DESIGN PROFESSIONAL may be called upon to determine the degree to which certain design conditions have been achieved by contractors. In performance of this work, DESIGN PROFESSIONAL will use sampling procedures; that is, selected portions of the work will be subject to close review and/or testing, and the results observed will be inferred to exist in other areas not sampled. Although such sampling procedures shall be conducted by DESIGN PROFESSIONAL in accordance with commonly accepted procedures consistent with applicable standards of practice, CLIENT understands that such procedures indicate actual conditions only where and at the time sampling is performed and that, despite proper implementation of sampling and/or testing procedures, and despite proper interpretation of their results, DESIGN PROFESSIONAL cannot assure the existence of conditions that DESIGN PROFESSIONAL infers to exist. Since a certification that certain conditions exist comprises an assurance of such conditions' existence, CLIENT agrees that it would be improper for DESIGN PROFESSIONAL to certify that certain conditions exist when DESIGN PROFESSIONAL cannot assure they exist. Accordingly, CLIENT shall not require DESIGN PROFESSIONAL to sign any certification, no matter by whom requested, that would result in DESIGN PROFESSIONAL certifying the existence of conditions whose existence DESIGN PROFESSIONAL cannot assure. CLIENT also agrees that CLIENT shall not make resolution of any dispute with DESIGN PROFESSIONAL or payment of any amount due to DESIGN PROFESSIONAL in any way contingent upon DESIGN PROFESSIONAL's certification of the existence of conditions whose existence DESIGN PROFESSIONAL cannot assure.

Whether or not the clause is accepted, you still will have to deal with any certification requests that arise, such as those made by government agencies, lenders, insurers, or other parties. When this occurs, meet with the party involved and explain the problem. Your goal should be to develop wording that accurately represents the limitations, for example: "Design professional states [or declares] that, based upon certain tests design professional has performed and/or certain observations design professional has made, it is the design professional's professional opinion that. . . ."

In some cases, by virtue of the wording of a certain law or regulation, *certify* must be used. The goal in this event would be to obtain a definition of *certify*, preferably on the form being signed. Commonly used wording is:

Certify means to state or declare a professional opinion of conditions whose true properties cannot be known at the time such certification was made, despite appropriate professional evaluation. A design professional's certification of conditions in no way relieves any other party from meeting requirements imposed by contract or other means, including commonly accepted industry practices.

In some instances, law or regulation already contains a definition. For example, the Environmental Protection Agency defines *design professional certification* to mean the rendering of a professional opinion, and, reportedly, California statutes contain a similar provision.

Alternative Solutions If the client refuses to agree to an effective certification clause, several alternatives may be used. These include *client indemnifications, insurance through endorsement, self-insurance, unilateral modification of the certification document*, and *claiming disparate bargaining power*.

Client Indemnification A client indemnification is only as effective as the client's ability to honor its provisions when the need arises, a condition affecting all client indemnifications.

Insurance Through Endorsement Insuring the certification through an endorsement to your professional liability insurance policy is a possibility, but usually it is a remote one. Professional liability insurers are reluctant to accept the risks inherent in certifications. Even if an endorsement is available when needed, it may not be continued in the future. This would make it wise to consider insurance through endorsement only when the contract also calls for the client to indemnify you. In any event, the duration of the endorsement should be considered, as well as the method by which the client will pay for it over time.

Self-Insurance Self-insurance exists when you are unable to transfer the risk through an endorsement or indemnification and thus create a means of providing recovery, should it become necessary. Assuming you are willing to underwrite the risk, it could be funded through an appropriate fee premium.

Unilateral Modification of the Certification Document Some design professionals unilaterally modify certification documents they are asked to sign, as by footnoting *certify* (and others) whenever it appears and by defining the word properly, directly on the form. Even an agency that has accepted such modification in the past may not do so in the future, due to events as basic as personnel changes. In addition, even if the rewritten form is accepted, an attempt feasibly would be made to hold you to the original, unmodified wording, should a claim of some type arise. It might be argued that, by modifying the wording, you jeopardized public health and safety, something you are legally obligated to protect by virtue of your being granted a professional license.

Claiming Disparate Bargaining Power You could claim that a client exercised disparate bargaining power in obtaining your contractual agreement to issue a certification, providing you made known your reluctance to agree to the requirement during contract formation. Whether or not the contract includes a certification requirement, you should also resist signing any ill-advised certification and should ultimately agree to sign one only because disparate bargaining power was used, for example, that receiving payment was contingent upon agreeing to certify.

Not one of the foregoing alternatives is as effective as refusing to sign a certification, unless *certify* and its implications are properly defined. Any alternative that results in your willingly attesting to the existence of conditions you know may not exist could possibly result in allegations of breach of warranty or deceit, and even a challenge to your license.

A client who cannot understand why you should be so concerned about certifying the existence of conditions that may not exist or who does understand but nonetheless wants you to accept the risk for purposes of expediency is not the type of client with whom most reputable design professionals would want to be associated.

Consequential Damages

Consequential damages are those that result as a consequence of another event, as when severing a power line interrupts factory production or rupturing a water line causes flooding of basements.

Problem As a result of consequential damages, you could be held liable for damages wholly out of proportion to your fee.

Solution Because of the problems associated with consequential damages, it is advisable to include a general limitation of liability in your contract. Having this limitation should not discourage you from seeking additional protection, however. In this regard, the best solution may be a general limitation of liability clause that recognizes the potential for consequential damages or a separate clause that has the same effect. One that does this and also protects the client from *your* charges of consequential damages might read as follows:

Consequential Damages

CLIENT shall not be liable to DESIGN PROFESSIONAL and DESIGN PROFESSIONAL shall not be liable to CLIENT for any consequential damages incurred by either due to the fault of the other, regardless of the nature of this fault, or whether it was committed by the CLIENT or DESIGN PROFESSIONAL, their employees, agents, or subcontractors. Consequential damages include, but are not limited to, loss of use and loss of profit.

Some design professionals include one-sided consequential damages (or other) clauses in their agreements. These would preclude the client from seeking recovery from you but would allow you to seek damages from the client. In some states, such one-sided agreements are accepted. In others, they are voidable or mutuality is inferred; that is, if clients may not seek consequential damages from you, you may not seek consequential damages from them.

Construction Cost Estimates

Many contracts call for an engineer or an architect to provide a construction cost estimate, for example:

Estimate of Construction Cost

DESIGN PROFESSIONAL shall prepare an estimate of project costs and submit same to CLIENT for CLIENT's approval.

This obviously is information clients want and need, and obliging them seems harmless enough. But it isn't.

Problem An estimate often is treated as a *guaranteed maximum*. In fact, in some states, an automobile repair shop's estimate must be a guaranteed maximum by force of law. Design professionals are not necessarily in a good position to estimate costs, however, and their mistaken estimates can (understandably) upset clients and, thus, trigger claims.

Solution The best solution is to explain your limitations and to have clients rely on a cost estimator or contractor. If this is unacceptable, consider giving clients what they want, *but*—by virtue of contract language and discussion—

be sure they understand your limitations. Failure to discuss these limitations or put them into writing can lead to erroneous assumptions. A model clause that explains your limitations is given below. It is labeled OPINION OF PROBABLE CONSTRUCTION COSTS, *not* ESTIMATE, simply to avoid the interpretative problems *estimate* may create:

Opinion of Probable Construction Costs

DESIGN PROFESSIONAL shall submit to CLIENT an opinion of the probable cost required to construct work recommended by DESIGN PROFESSIONAL. DESIGN PROFESSIONAL is not a construction cost estimator or construction contractor, nor should DESIGN PROFESSIONAL's rendering an opinion of probable construction costs be considered equivalent to the nature and extent of service a construction cost estimator or construction contractor would provide. DESIGN PROFESSIONAL's opinion will be based solely upon DESIGN PROFESSIONAL's own experience with construction. This requires DESIGN PROFESSIONAL to make a number of assumptions as to actual conditions that will be encountered; the specific decisions of other design professionals; the means and methods of construction the contractor will employ; the cost and extent of labor, equipment, and materials the contractor will employ; contractor's techniques in determining price and market conditions at the time; and other factors over which DESIGN PROFESSIONAL has no control. Given the assumptions that must be made, DESIGN PROFESSIONAL cannot guarantee the accuracy of DESIGN PROFESSIONAL's opinion of probable cost, and, in recognition of that fact, CLIENT waives any claim against DESIGN PROFESSIONAL relative to the accuracy of DESIGN PROFESSIONAL's opinion of probable cost.

Some clients may object to the waiver, claiming it eliminates any liability on your part and therefore creates a disincentive to your doing the work well. In response, you should point out to the client that it does not necessarily eliminate liability, since others who rely on the cost opinion, such as a lender or an owner, could be damaged, claim negligence, and sue. You should also point out that it does not take the threat of a lawsuit to encourage professionalism on your part. You have a reputation to uphold, and, in the professional community, reputation is a paramount concern. Note, too, that all design professionals of whom you are aware, unless they also offer construction cost estimating services, are subject to the same limitations and just might not be as candid as you are.

Alternative Solutions If it will make the client happier, and assuming you can use it as a bargaining chip, offer to delete the waiver indicated above and end the clause with "... DESIGN PROFESSIONAL cannot guarantee the accuracy of DESIGN PROFESSIONAL's opinion of probable cost."

Another approach is to review the original client-proffered clause in an attempt to develop an acceptable alternative. The original clause says:

Estimate of Construction Cost

DESIGN PROFESSIONAL shall prepare an estimate of project costs and submit same to CLIENT for CLIENT's approval.

In rewriting it, note that you will be submitting an estimate only for construction of your portion of the work and that yours is an "engineering [or architectural] estimate," that is, an opinion of probable cost. Chances are, the client will then agree to the following wording:

Estimate of Construction Cost

DESIGN PROFESSIONAL shall prepare an engineering [or architectural] estimate of the probable cost of constructing work designed and specified by DESIGN PROFESSIONAL and shall submit said estimate to CLIENT.

When submitting your opinion of probable cost, note clearly in your cover letter that it is an engineering [or architectural] estimate, and define what that is per language given in the clause considered the best approach, and/or define it in the DEFINITIONS section of your agreement. In addition, if your contract lists services that will not be provided (see EXCLUDED SERVICES, below), you would include in that list "Construction Cost Estimates (Other than Engineering [or Architectural] Estimates of Probable Construction Cost)."

These safeguards should be enough to shield you from a significant source of needless liability problems. If clients absolutely insist on their original wording—and assuming such an attitude does not dissuade you from working with them—consider obtaining cost estimating services through subcontract. In that case, however, it may be necessary to integrate this consideration into any provisions that otherwise limit your freedom to subcontract and to change "DESIGN PROFESSIONAL shall prepare...." to "DESIGN PROFESSIONAL shall provide...."

Construction Monitoring

Virtually every major study of construction failures has agreed on one point: There is a pressing need for design professionals to monitor construction of their designs. Various case histories of professional liability losses demonstrate the same point.

Problem No set of plans or specifications can be constructed totally without interpretation of any type. A number of design-related decisions must be made by contractors, most under the designer's assumption that industry custom will be followed. But, in many cases, "custom" is subject to debate, sometimes—ultimately—before a trier of fact. Conflicts between the work of various designers also can occur, either because of inadequate coordination during the design phase or because of misinterpretation by a contractor in the field. In these and other instances, the impact of errors can be mitigated substantially if knowledgeable design professionals or their representatives are on the scene, to resolve errors, omissions, conflicts, and ambiguities or to create a field change to soften the dollar and time impact of someone else having done something wrong.

Some design professionals believe that "occasional site visits" are sufficient, because they are contacted quickly when their input is required. This position assumes that contractors and others on site will recognize a problem. If a problem goes undetected, however, the call will not come in and serious consequences could ensue. It may take years and thousands of dollars to prove your plans were not at fault.

Problems also can arise when field observation is performed by an independent firm that obtains the work because your firm does not offer the service or because the other firm proposes to perform it for a lower fee. In fact, persons who have no knowledge of the concepts and detail that underlie a set of plans and specifications, or the assumptions made by the designers, are far less likely to recognize problems, including those related to designer errors and omissions. Frequently, a construction monitor not affiliated with the designer's firm will "pass" an improper structural connection or other piece of completed work simply because it conforms to a drawing or detail. A more astute observer might note that the drawing was in error, that the wrong grade of steel was specified, the wrong size transformer was called for, and so on.

As discussed in Chapter 3, it is especially important for the prime design professional, structural engineer, and geotechnical engineer to maintain a full-time construction-phase monitoring presence; the prime, because of overall responsibility and liability; the structural engineer, because of the life-safety implications of the work involved; and the geotechnical engineer, because of the likelihood of unanticipated or changed conditions. The fee the owner is required to pay for this service usually is well worthwhile, due to the time-consuming problems that it can prevent. Insurance never is a substitute, especially so because the amount recovered—when and if anything is recovered at all—is significantly diminished by cost of recovery. Nonetheless, some clients seem to want things both ways. They are unwilling to pay the fee you need to perform construction monitoring, but they want you to take full responsibility for the problems that lack of construction monitoring, or lack of adequate construction monitoring, creates. At best, such clients are uninformed about the contingencies that can arise during construction, despite high-quality plans and specifications. Uninformed clients can be dangerous.

It is suggested strongly that construction monitoring should be a basic service, if not for all design professionals, then at least for primes, structural engineers, and geotechnical engineers. The service should be recommended to the client when the workscope is under development, both to facilitate personnel scheduling and to assess client attitudes. Explain why the service is needed and its value. If the client decides that construction monitoring is unnecessary or that another party can provide it, you either should obtain protection against the problems that could occur as a consequence of the client's decision or you should give serious consideration to refusing the commission. This applies whether the client is the owner or a prime design professional.

Solution Precepts embodied in the following clause possibly should be included in your contract:

Monitoring of Construction

CLIENT recognizes that construction monitoring is an essential element of a DESIGN PROFESSIONAL's service, provided to reduce problems during construction by facilitating detection of, and/or rapid response to, unanticipated or changed conditions, or errors or omissions committed by design professionals, contractors, materials providers, or others. CLIENT also recognizes that no party is as intimately familiar with DESIGN PROFESSIONAL's intents as DESIGN PROFESSIONAL's personnel, including those individuals whom DESIGN PROFESSIONAL prepares for and assigns to construction monitoring tasks. Accordingly, CLIENT agrees to retain DESIGN PROFESSIONAL to monitor construction of DESIGN PROFESSIONAL's portion of the project, and DESIGN PROFESSIONAL agrees to assign to the monitoring function persons qualified to observe and report on the quality of work performed by contractors, *et al.* (that is, the general contractor, subcontractors, subsubcontractors, materialmen, and others). CLIENT recognizes that construction monitoring is a technique employed to reduce the risk of problems arising during construction. CLIENT also recognizes that provision of construction monitoring by DESIGN PROFESSIONAL is not insurance and does not constitute a warranty or guarantee of any type. In no case shall performance of construction monitoring in any way alter the responsibility of contractors *et al.* for the quality of their work and for adhering to plans and specifications. Should CLIENT for any reason not retain DESIGN PROFESSIONAL to monitor construction, or should CLIENT unduly restrict DESIGN PROFESSIONAL's assignment of personnel to monitor construction, or should DESIGN PROFESSIONAL for any reason not perform construction monitoring during the full period of construction, DESIGN PROFESSIONAL shall not have the ability to perform a complete professional service. Should DESIGN PROFESSIONAL for any reason not have the ability to perform a complete professional service, CLIENT waives any claim against DESIGN PROFESSIONAL and agrees to indemnify, defend, and hold DESIGN PROFESSIONAL harmless from any claim or liability for injury or loss arising from problems during construction that allegedly result from findings, conclusions, recommendations, plans, or specifications developed by DESIGN PROFESSIONAL. CLIENT also agrees to compensate DESIGN PROFESSIONAL in defense of any such claim, with such compensation to be based upon DESIGN PROFESSIONAL's prevailing fee schedule and expense reimbursement policy.

Alternative Solutions Whenever a project involves innovative or complex design, or when there are significant risks associated with the project, such as those involving modernization and rehabilitation, a potential for encountering hazardous materials or contaminants, and so on, it is suggested that there is no alternative: Either you monitor the work, or you should not be responsible for it. Some firms even go farther than that, however, by saying that they either will monitor the work or not accept the commission.

If you are willing to perform the work but not to monitor its construction, and if the client will not accept the foregoing, you may wish to seek acceptance of a provision such as the following:

Responsibility for Design

CLIENT recognizes that it is impossible to include all construction details in plans and specifications, creating a need for field interpretation of plans and specifications. Since the intent of DESIGN PROFESSIONAL's design, specifications, and design recommendations is best understood by DESIGN PROFESSIONAL, DESIGN PROFESSIONAL cannot take any responsibility for the adequacy of DESIGN PROFESSIONAL's design, specifications, or design recommendations unless DESIGN PROFESSIONAL's work includes the construction monitoring necessary for field interpretation to determine whether or not the work performed is in substantial compliance with design, specifications, design recommendations, and their intent.

Although some design professionals are using such clauses, they have not yet been subject to any known challenge in court. Feasibly, they could be struck down, because they are almost tantamount to an indemnification that relieves design professionals from responsibility for their work, an indemnification that courts are loath to uphold. Accordingly, design professionals might be well advised to make it clear that construction monitoring is a vital element of a complete professional service and that they will retain responsibility for their professional acts, errors, or omissions but only to the extent that their impact would not have been mitigated through construction monitoring. Accordingly, a superior clause might read as follows:

Responsibility for Design

CLIENT recognizes that it is neither practical nor customary for DESIGN PROFESSIONAL to include all construction details in plans and specifications, creating a need for interpretation in the field by DESIGN PROFESSIONAL. CLIENT also recognizes that construction monitoring often helps DESIGN PROFESSIONAL to identify and correct quickly and at comparatively low cost any errors or omissions that are revealed through construction. For the foregoing reasons, construction monitoring is considered an essential element of a complete design professional service. Accordingly, if CLIENT decides that DESIGN PROFESSIONAL shall not provide construction monitoring, CLIENT agrees that DESIGN PROFESSIONAL shall be responsible for the consequences of DESIGN PROFESSIONAL's negligent acts, negligent errors, or negligent omissions only to the extent that DESIGN PROFESSIONAL's monitoring services would not have prevented or mitigated such consequences.

As can be seen, the second clause puts the client on notice that construction monitoring is part of the overall design professional service and explains why it is essential. The liability of the design professional is not eliminated altogether. Instead, the design professional remains liable for those problems that the client's decision to forgo construction monitoring would not have prevented. Realistically, of course, litigation might be required to determine exactly what the design professional's liability should be.

Although either of the foregoing two clauses might afford protection, having either upheld would probably depend on your calling the liability limitations to your client's attention. This usually is done by highlighting the provision, printing it in a larger or bolder typeface, or through some other appropriate means. Failing that, clients might allege that it was your intent to "sneak" something through, and the argument might be upheld, at least in part, because of your professional status.

Note, too, that neither clause affords as much protection to you as actually performing construction monitoring. As such, you should make it a point to offer this service, at least to gauge client attitudes. Clients who say, in effect, that they will decide on monitoring "later," perhaps because they have not dealt with you before, should be advised of the scheduling required to ensure that an adequate number of qualified personnel will be available when needed. If clients still will not agree to monitoring, chances are they have no intention of opting for it or do not understand the importance of scheduling and the problems that can arise when you are unable to assign those you want. In such cases, you may want to "stick to your guns" as to the waiver and indemnification provisions of the first clause indicated.

In the event that your client requests construction monitoring but later decides against it, alert the client to the impact of the decision with respect to contract provisions that will apply. If for any reason the provisions were not included in your contract or are not fully satisfactory, be sure a letter to the client makes it clear that you cannot be held liable for others' faulty interpretation of your intent, inability to recognize problems you would have caught, and so on. Some design professionals also suggest that construction documents furnished to contractors should make it clear that contractors are responsible for curing any and all construction defects at their own cost, including defects that the design professional or any other construction monitor failed to detect.

Curing a Breach

Despite the most carefully considered plans and best intentions, either party to a contract may breach the agreement, willfully or inadvertently.

Problem If the potential for breaching an agreement is not recognized in the agreement itself, attempts to remedy the breach could lead to a dispute that results in one of the parties not wanting to cure the problem. Various difficulties can arise from such attitudes.

Solution The best solution is to identify mutual responsibilities, should a breach occur, and what a breach does, and does not, imply. Such concepts are embraced in the following:

Curing a Breach

In the event either party believes the other has committed a material breach of this AGREEMENT, the party maintaining such a belief shall issue a termination notice to the other, identifying the facts as perceived, and both parties shall bargain in good faith to cure the causes for termination as stated in the termination notice. If such a cure can be effected prior to the date by which termination otherwise would be effective, both parties shall commit their understanding to writing, and termination shall not become effective. If in curing an actual or alleged breach either party shall waive any rights otherwise inuring by virtue of this AGREEMENT, such waiver shall not be construed to in any way affect future application of the provision involved or any other provision.

If your client will not accept this provision, consider not accepting the client.

Discovery of Unanticipated Hazardous Materials

Hazardous materials are commonly construed to mean wastes buried on site or ones that have migrated to a site from another source. These are being discovered ever more frequently at locations where they were not anticipated, particularly in urban areas. Civil, environmental, and other consulting engineers whose work commonly includes development and/or monitoring the construction of excavation plans, as well as architects who perform this service, should be aware of this potential. The best defense against problems is reliance on geotechnical or geoenvironmental consultants who are in a position to identify hazardous materials and know the proper response, should they be detected. When existing projects are concerned, architects and structural, mechanical, and electrical engineers may encounter unanticipated asbestos, PCBs, or other hazardous materials.

Problem Unless discovery of unanticipated hazardous materials is contemplated by the contract, you may incur a substantial—and probably uninsurable—claim from an injured employee, "sidewalk superintendent," or other party, and you may not be paid for whatever emergency response action you had to take. In addition, you should want to be assured that discovering unanticipated hazardous materials subjects the agreement to renegotiation or termination.

Solution A provision embracing terms and conditions similar to those of the following should be included in all contracts involving penetration of the earth's surface or work in, or with, existing structures.

Discovery of Unanticipated Hazardous Materials

Hazardous materials or certain types of hazardous materials may exist where there is no reason to believe they could, or should be, present. DESIGN PROFESSIONAL and CLIENT agree that the discovery of unanticipated hazardous materials constitutes a changed condition mandating a renegotiation of the scope of work or termination of services. DESIGN PROFESSIONAL and CLIENT also agree that the discovery of unanticipated hazardous materials may make it necessary for DESIGN PROFESSIONAL to take immediate measures to protect human health and safety, and/or the environment. DESIGN PROFESSIONAL agrees to notify CLIENT as soon as practically possible, should unanticipated hazardous materials or suspected hazardous materials be encountered. CLIENT encourages DESIGN PROFESSIONAL to take any and all measures that in DESIGN PROFESSIONAL's professional opinion are justified to preserve and to protect the health and safety of DESIGN PROFESSIONAL's personnel and the public, and/or the environment, and CLIENT agrees to compensate DESIGN PROFESSIONAL for the additional cost of such work. CLIENT agrees promptly to make to appropriate agencies whatever disclosures are required by statute or regulation. In addition, CLIENT waives any claim against DESIGN PROFESSIONAL and agrees to indemnify, defend, and hold DESIGN PROFESSIONAL harmless from any claim or liability for injury or loss arising from DESIGN PROFESSIONAL's encountering unanticipated hazardous materials or suspected hazardous materials, or arising from disclosures of information made by DESIGN PROFESSIONAL in a good-faith attempt to abide by applicable statutes or regulations, or otherwise preserve and protect the public and/or the environment. CLIENT also agrees to compensate DESIGN PROFESSIONAL for any time spent and expenses incurred by DESIGN PROFESSIONAL in defense of any such claim, with such compensation to be based upon DESIGN PROFESSIONAL's prevailing fee schedule and expense reimbursement policy.

Alternative Solutions Although different wording can be used, there is no alternative to the concept.

Excluded Services

In order to reduce their liability exposure and to enhance client awareness, some design professionals' agreements include a list of services that they have offered to the client but that the client has decided either are not necessary or will be obtained from another source.

Problem The problem that listing excluded services seeks to limit or eliminate is a claim by a client or third party that you did not make the client aware that certain services were available or may have been necessary. Since the client relied on your expertise, it may be alleged, you were negligent for not letting the client know what was needed. Note, however, that such a listing does far more than create a shield against a possible claim. It also helps alert the client to other services potentially applicable to the project and to the need to obtain from another source those necessary services that will not be obtained from you.

Solution A typical clause that identifies excluded services may read as follows:

> **Excluded Services**
>
> Services available from DESIGN PROFESSIONAL are not limited to those specified in the scope of services indicated herein. Other services that are available and applicable to CLIENT's project have been made known and explained to CLIENT. DESIGN PROFESSIONAL has indicated how each such service may affect CLIENT's risks, and, where DESIGN PROFESSIONAL has deemed a service necessary for successful project completion—from a technical standpoint, to comply with law, or for other reasons—DESIGN PROFESSIONAL has made such known to CLIENT and CLIENT has confirmed that, in CLIENT's opinion, such services are not necessary or, if CLIENT believes they are necessary, CLIENT has made, or shall make, arrangements to obtain them from a source other than DESIGN PROFESSIONAL. These excluded services include:
>
> *(Listing or annotated listing)*
>
> In that it would be unfair for DESIGN PROFESSIONAL to be exposed to liability for DESIGN PROFESSIONAL's failure to perform a service CLIENT has instructed DESIGN PROFESSIONAL not to perform, due to CLIENT's preference or desire to obtain such service from another source, CLIENT hereby waives any claim against DESIGN PROFESSIONAL and agrees to defend, indemnify, and hold DESIGN PROFESSIONAL harmless from any claim or liability for injury or loss allegedly arising from DESIGN PROFESSIONAL's failure to perform a service CLIENT has instructed DESIGN PROFESSIONAL not to perform. CLIENT further agrees to compensate DESIGN PROFESSIONAL for any time spent or expenses incurred by DESIGN PROFESSIONAL in defense of any such claim, in accordance with DESIGN PROFESSIONAL's prevailing fee schedule and expense reimbursement policy.

Alternative Solutions Two alternatives to the above are worthy of mention. The first is elimination of the waiver and indemnification if the client insists. The waiver in fact may not be necessary, and the need for the indemnification, given the list of excluded services, is questionable. Any related claim probably would be quickly dismissed.

The second alternative relates to the listing and your ability to asterisk or otherwise highlight any service you deem essential. This option may be particularly effective if the waiver and indemnification are dropped.

Freedom to Report

You are expected to serve as your client's professional advisor, candidly relating observations, opinions, and recommendations. Such candor can lead to problems.

Problem Contractors are frequently filing or threatening to file libel and/or slander suits in an attempt to influence an engineer's or an architect's report to the client, or to obtain recovery from the report's consequences. Typically, the reports involved advise that a given contractor should be excluded from a list of prequalified firms, a low bidder should not be relied upon by virtue of poor performance in the past, materials should be rejected due to unacceptable test results, or that a contractor should be replaced because of demonstrated inability or unwillingness to fulfill contract requirements.

Solution Be highly selective about the words you choose in reporting anything that could have a negative financial impact on another party. Report facts as facts, objectively, without embellishment or emotion; report opinions as opinions, carefully noted as such, and do your best to document everything you say. By taking this approach, you will make it difficult for the affected party to win a libel or slander suit against you.

Inform your client that the risks involved can be minimized through effective contractor selection, that is, reliance on negotiated contracting or, at least, prequalification. But your client must also recognize that you cannot act as an effective advisor if you face the threat of a lawsuit every time you state or imply something negative about someone.

The best solution might be a contract provision through which the client agrees to indemnify you for a slander or libel action brought against you by another party that the client has engaged directly or indirectly, and about whose performance or reputation you are required to render opinions and reports. Consider the following clause:

Freedom to Report

It is contemplated, that, during the course of this engagement, DESIGN PROFESSIONAL will be required to report on the past or current performance of others engaged or being considered for engagement directly or indirectly by CLIENT and to render opinions and advice in that regard. Those about whom reports and opinions are rendered may, as a consequence, initiate claims of libel or slander against DESIGN PROFESSIONAL. To help create an atmosphere in which DESIGN PROFESSIONAL feels free to communicate candidly, CLIENT agrees to waive any claim against DESIGN PROFESSIONAL and to defend, indemnify, and hold DESIGN PROFESSIONAL harmless from any claim or liability for injury or loss allegedly arising from professional opinions rendered by DESIGN PROFESSIONAL to CLIENT or CLIENT's agents, in DESIGN PROFESSIONAL's capacity as CLIENT's professional advisor. CLIENT further agrees to compensate DESIGN PROFESSIONAL for any time spent or expenses incurred by DESIGN PROFESSIONAL in defense of any such claim, in accordance with DESIGN PROFESSIONAL's prevailing fee schedule and expense reimbursement policy.

Alternative Solutions Another approach would be to treat certain information as confidential and for clients' eyes only. This information would comprise principally your opinions of certain others (usually contractors and subcontractors) and/or their work; your report and the envelope that contains it both would be clearly marked CONFIDENTIAL. Clients then would distribute the material at their own risk, and they would indemnify you for any claims that result, for example:

Freedom to Report

It is contemplated that, during the course of this engagement, DESIGN PROFESSIONAL will render opinions as to the past and current performance and capabilities of others engaged, as a service to CLIENT. All such opinions shall be rendered by DESIGN PROFESSIONAL in confidence and shall be marked CONFIDENTIAL. Such opinions are not intended for distribution by CLIENT. To help create an atmosphere in which DESIGN PROFESSIONAL feels free to communicate candidly to CLIENT, CLIENT waives any claim against DESIGN PROFESSIONAL and agrees to defend, indemnify, and hold DESIGN PROFESSIONAL harmless from any claim or liability for injury or loss allegedly arising from DESIGN PROFESSIONAL's written opinions of the past or current performance of another party retained directly or indirectly by CLIENT. CLIENT further agrees....

If the client is unwilling to accept the clause, or if you decide not to offer it, contractual protection will be unavailable. Accordingly, you should be sure to have insurance coverage for the exposure; be cautious in all your reporting, and instruct others in your firm likewise to use discretion.

Indemnification

Clients will often attempt to gain acceptance of an indemnification that transfers certain of their risks to design professionals.

Problem Several types of indemnification are used, and some are more onerous than others. The following is a *broad-form* clause:

Indemnification

DESIGN PROFESSIONAL agrees to hold harmless and indemnify CLIENT from any and all liability, including cost of defense, arising out of performance of the work described herein.

A broad-form indemnification is grossly unfair, because it would require you to cover your client's costs even when the problem involved has been caused *solely* by the client. In fact, this type of indemnification is in many states statutorily prohibited as being contrary to public policy and thus unenforceable. Nonetheless, as previously discussed, some jurisdictions have also adopted the doctrine of *express negligence*. This holds that even a provision that is against public policy *will* be enforced when it has been stated in clear, unequivocal language to which both parties have agreed.

An *intermediate form* indemnification requires you to hold harmless and indemnify your client when you contribute to the problem:

Indemnification

DESIGN PROFESSIONAL agrees to hold CLIENT harmless from and against liability arising out of DESIGN PROFESSIONAL's negligence, whether it be sole or in concert with others, in connection with the performance of the work described hereunder.

This may sound fair, but a close reading shows that you may be liable for 100 percent of your client's alleged damages, even though you caused only 1 percent of them. Note, however, that the state with jurisdiction may provide a statutory remedy that holds that indemnifications such as these automatically are interpreted under a *comparative negligence rule*, so the percentage of your financial contribution may not exceed your percentage of culpability; if you caused 1 percent of the problem, you contribute 1 percent of the cost. It may not be wise to rely solely on what the law states, however, because laws are subject to interpretation. In addition, some clients may be subject to a huge judgment, including exemplary or punitive damages, due to their role in creating the problem (willful neglect, for example), due to statutory provisions (treble damages), or due to application of a joint and several liability concept. In other words, even if the state with jurisdiction applies a comparative negligence rule, you may be called upon to pay 1 percent of $10 million, not including your costs of legal defense.

A *limited form* indemnification puts into writing what law requires in any event:

Indemnification

DESIGN PROFESSIONAL agrees to hold harmless and indemnify CLIENT from and against liability arising out of DESIGN PROFESSIONAL's negligent performance of the work.

Is this something to be concerned about? Probably, because putting a common law requirement into a contract switches liability from being solely tort to both tort *and* contractual. Whereas the requirement would be covered by your insurance were it solely tort, that is, not referenced in the agreement, including it in the contract creates a contractual liability that may be uninsurable.

Solution The best solution is learning the applicable laws in the state with jurisdiction and educating your clients. Assuming clients are reasonable, it should be possible to eliminate such indemnifications from your contracts. You are a professional and, as such, are governed by laws and standards established by society to protect your client.

Alternative Solutions If the client, for whatever reason, insists on an indemnification, have it examined by legal counsel with respect to applicable laws of the governing jurisdiction, to ascertain exactly what your exposure may be. You should also work with your professional liability insurer to assure provisions of the indemnification can be covered.

You may also want to consider asking clients about the type of protection they are seeking. Are they looking for you to pay more than your fair share? The answer will probably be no, and the same answer may be given to the question: Should I be required to pay any portion of exemplary, punitive, or treble damages assessed against my client because of my client's willful misconduct or similar activity? Assuming these are the questions asked and answers given, request that these understandings be committed to writing. For example:

> **Indemnification**
>
> DESIGN PROFESSIONAL agrees to hold harmless and indemnify CLIENT from and against liability arising from DESIGN PROFESSIONAL's negligent performance of the work. It is specifically understood and agreed that in no case shall DESIGN PROFESSIONAL be required to pay an amount disproportional to DESIGN PROFESSIONAL's culpability or any share of any amount levied to recognize more than actual economic damages.

If your contract includes indemnifications or limitations protecting you, as is suggested, it is important that these be referenced in the indemnifications, since otherwise an inherent conflict may exist. In that case, a modified clause may read as follows:

> **Indemnification**
>
> DESIGN PROFESSIONAL agrees to hold harmless and indemnify CLIENT from and against liability arising out of DESIGN PROFESSIONAL's negligent performance of the work, subject to any limitations, other indemnifications, or other provisions CLIENT and DESIGN PROFESSIONAL have agreed to.

The provision about disproportional payment, and so on ("It is specifically understood and agreed. . . .") could be added.

The various other alternatives you could apply are limitless, as are the various forms of indemnification.

Jobsite Safety

Jobsite safety is a vital issue; one you need to understand thoroughly, so you can explain it to your clients and, more importantly, so you can avoid any contract language that could be construed to make you liable for jobsite safety.

Problem Construction sites comprise one of the most dangerous of all workplaces. Jobsite safety responsibilities usually are assigned to general contractors, because they are the overall coordinators of the work and have constructive use of the site. Should one of the contractor's workers be injured, the worker's remedy generally is limited to workers' compensation insurance, which protects the contractor. In a few states, design professionals are also protected by the contractor's workers' compensation insurance. This is not the case in most states, and, in many, the maximum workers' compensation benefits for certain types of injuries—even fatal ones—are significantly below those won through highly publicized jury awards. This can set into motion a search for "deep pockets" and an attempt to somehow impose responsibility on a source other than someone protected through workers' compensation insurance, for example, *you*. As such, you must do your utmost to ensure that nothing in your contract can be construed to imply that you are responsible for jobsite safety, as through acceptance of stop-work authority. *You should under no circumstances accept a clause such as the following*:

Jobsite Safety

DESIGN PROFESSIONAL shall be responsible for any losses or injuries that occur at the jobsite due to unsafe conditions that resulted from the plans and specifications provided by DESIGN PROFESSIONAL.

Solution Clients should be informed of the jobsite safety issue in any event. Specifically relative to the clause above, however, they should be told that they are asking for something that already is established by law, because design professionals owe a duty of care to anyone who foreseeably could be damaged by their professional acts. But your work does not exist in a vacuum, nor do you customarily tell the contractor how to achieve the results you specify. *The clause could be interpreted to impose responsibilities that exceed those normally required by law* and thus could void any professional liability insurance coverage you otherwise might have. The best solution is elimination of the clause or use of an alternative clause that provides specific limitations, so that there is no confusion as to who is responsible for what:

Jobsite Safety

Insofar as jobsite safety is concerned, DESIGN PROFESSIONAL is responsible solely for DESIGN PROFESSIONAL's own and DESIGN PROFESSIONAL's employees' activities on the jobsite, but this provision shall not be construed to relieve Owner or any construction contractors from their responsibilities for maintaining a safe jobsite. Neither the professional activities of DESIGN PROFESSIONAL nor the presence of DESIGN PROFESSIONAL or DESIGN PROFESSIONAL's employees and subcontractors shall be construed to imply DESIGN PROFESSIONAL has any responsibility for methods of work performance, superintendence, sequencing of construction, or safety in, on, or about the jobsite. CLIENT agrees that the General Contractor is solely responsible for jobsite safety and warrants that this intent shall be made evident in the Owner's agreement with the General Contractor. CLIENT also warrants that DESIGN PROFESSIONAL shall be made an additional insured under the General Contractor's general liability insurance policy.

To help facilitate the various stipulations of this clause, you may wish to give your clients sample language for use in the owner-general contractor agreement, for example:

> CONTRACTOR agrees to waive any claim against OWNER and OWNER's agents, architects, engineers, and their employees acting within the scope of their duties and to defend, indemnify, and save them harmless from any claim or liability for injury or loss that allegedly may arise from CONTRACTOR's performance of the work described herein, but not including the sole negligence of OWNER, OWNER's agents, architects, and engineers, or employees. CONTRACTOR will require any and all SUBCONTRACTORS to conform with provisions of this clause prior to commencing any work and agrees to insure this clause in conformity with the insurance provisions of this contract.

The insurance protection clause may read (at least in part):

> CONTRACTOR shall require CONTRACTOR's insurance carrier to add OWNER, OWNER's professional consultants, and their agents as additional insureds under CONTRACTOR's general liability insurance policy with respect to services performed by CONTRACTOR for OWNER.

Your clients should not be averse to these stipulations, since they help define and clearly limit resolution of potential problems. Failing that, the liability actually may boomerang back to the owner, as in a situation in which it is found that you are liable, but the owner is made to pay because you were acting as the owner's agent (regardless of the contract's wording) and the owner's pockets are much deeper than yours, because your insurance may not provide coverage.

Note: When hazardous-materials engineering projects are involved, you may wish to add the following to your provisions:

> In the event DESIGN PROFESSIONAL expressly assumes health and safety responsibilities for toxic or other concerns specified, the acceptance of such responsibility is not and shall not be deemed an acceptance of responsibility for any other health and safety requirements, such as those relating to excavating, trenching, drilling, or backfilling.

Alternative Solutions Alternative solutions often involve compromises that weaken the concepts that comprise the best approach. The weaker your protection, the greater your risk will be. Note, too, that your risk also will be affected by the nature of construction to be used and the type of project. As an example, multiple prime construction contracting tends to create significant safety exposures, because it is sometimes difficult to establish who is in charge of what. This also makes it more difficult to fix blame when a safety risk materializes, encouraging "shotgun" claims. Fast-track construction also creates a greater likelihood of problems, as does construction performed by contractors whose bids are low because they employ poor quality control and safety procedures, or they are unaware of conditions that make better-than-customary procedures essential.

Because construction site safety issues are among the most important of all, you may be well advised to refuse an engagement when you believe safety matters will not be managed effectively.

Limitation of Liability

Limitation of liability has for centuries been an operative contracting principle that recognizes that providers of a service obtain a small benefit (their profit) in the process of helping their clients (or customers) achieve a much larger benefit. For purposes of fairness, the risks visited upon service providers and those obtaining the service should be proportional to their benefits. Without this approach, clients would have little motivation to minimize their own risks or to authorize design professionals to perform quality-oriented services. As a consequence, clients would seek to have work performed as cheaply as possible, comfortable in the belief that, should anything go wrong, design professionals (and their insurers) would serve as a source of recovery, that is, free insurance.

The foregoing scenario should sound familiar because, to a very real extent, it represents exactly what has happened. Design professionals' desire to obtain commissions has resulted in a continuing reduction of service and increased willingness to accept more risk. Do clients therefore get something for nothing? No, because macroeconomic forces are at work. As a consequence of cheap engineering, continually more projects are subject to delays, overruns, and disputes, all of which increase owners' costs. These same problems have also resulted in sky-high rates for professional liability insurance, something else owners pay for through the overhead portion of design professionals' fees. However, if the past is a prologue, insurance might ultimately become unaffordable or unavailable, resulting either in few design professional firms being available to offer service or in owners having to self-insure the risks associated with professional negligence. Precisely this outcome was recognized more than a millenium ago by the Babylonian merchants who are credited with originating the concept of limiting service providers' (in their case, shippers') liability to a maximum fixed amount. The merchants reduced their risk by selecting providers with care and by insisting that providers employ measures designed to reduce the potential for losses.

The concept of limiting a service provider's liability has been in effect for many years in the United States. Anyone who travels on business sees it frequently. Airlines limit their liability for losing passengers' luggage. Hotels limit their liability if items are stolen from a room. These limitations are, or should be, made conspicuous to users of a service, so they will be aware of them and thus will be encouraged to take appropriate protective action, for example, storing their valuables in a hotel's safety deposit box instead of leaving them in a room.

The limitation-of-liability principle was first applied to design professionals in the early 1970s. ASFE developed a contractual provision calling for geotechnical engineers' liability to the owner and construction contractors to be limited to $50,000 (or some other fixed amount) or the fee, whichever was higher. Geotechnical engineers at the time were seriously threatened by the extraordinary number of claims being filed against them. Because they had not effectively communicated information about their risks, their clients were expecting perfect results in circumstances in which such results were virtually impossible to obtain.

ASFE's experience has shown that limitation of liability for design professionals reduces architects' and engineers' risks, not only by capping their exposure, but also because of the client-consultant communication that discussion of the concept engenders. All too often, projects are begun in a spirit of blind optimism, because project principals are unaware of their risks. By discussing risks with your clients—something a limitation of liability clause can virtually force you to do—the client becomes more aware of the potential pitfalls and the additional services or service amplifications you can apply to reduce risks, including the risk of errors or omissions. Such a discussion can lead to a better understanding between design professionals and their clients, while encouraging more professional execution of the commission, for the benefit of all.

Problem As effective as limitation of liability has been, some clients still object to it, viewing it solely as a device to reduce design professionals' liability exposure at their expense. There is reason to believe that the term "limitation of liability" induces this narrow focus, obscuring the full range of benefits the concept can establish.

Solution In order to encourage clients' acceptance of limitation of liability and thereby to gain its use as an effective risk management tool, the concept is being broadened, codified, and retitled. The most popular new name for it seems to be *risk allocation*, referred to below as *RA*. The concept is based on four others. Design professionals should be familiar with each of them, in order to explain RA and its benefit to their clients.

First, whether or not an "RC cap" (i.e., limitation of liability) is applied, a design professional's liability is *always limited*, the limit being defined by the amount of money a design professional has available to satisfy claims, awards, or settlements. As such, an RA cap does not impose a limit where none formerly existed; it merely establishes what the limit will be before, rather than after, a problem arises.

Second, design professionals are exposed to risk whenever they accept a commission. Some of these risks can be reduced by transferring them to others, conservative design procedures, more professional proposal-phase activities, more comprehensive workscopes, and general conditions that establish specific liability limitations (typically through waivers of liability or risk-transferring indemnifications). Despite these procedures, risk will still exist. Accordingly, as discussed before, the overall fee proposed for performing services on clients' behalf should include an allowance for risk funding, permitting design professionals to establish the reserves needed to cover their costs when risks materialize.

Many design professionals who have used the limitation-of-liability concept make no secret about the need to fund risk. Typically, their general conditions include a limitation of liability, and, if a client decides to strike the provision, an additional fee—typically, a fixed percentage of the overall fee—is imposed. The RA approach is more effective. As indicated by the suggested provision, given below, the amount of the limit—the RA cap—would initially be left blank, an approach that forces discussion of the concept. During discussion, the design professional would explain the concept and related issues, noting that the fee would be *reduced* when the RA provision is applied. As such, *RA provides a benefit to the client*, not just to the design professional. (While this is true for limitation of liability, the RA approach makes it far more evident.)

Third, the extent of risk associated with a project depends on the nature of the project and the client, as well as the degree to which other risk management options can be applied. For example, condominium projects are particularly risk-prone because the engineers and architects engaged owe a duty of care to all unit owners and their successors, and this duty may exist for an indefinite period, due to design professionals' personal liability exposure and lack of a statute of repose. Single-family housing developments are highly risk-prone for the same reasons, as are projects controlled by committees of volunteers, among many others. The greater the risk imposed by the nature of the project, the higher is the level of risk funding required.

Risks also are affected by the attitudes of clients. Some are far more willing to take risks than others. Generally speaking, clients who are not particularly risk-sensitive impose additional risks on their design professionals; risk-funding levels should recognize this factor, too.

The manner in which the commission will be executed affects risk. The quality associated with proposal-phase services is often eroded when fee-bidding is used, but, due to attitudes of the prime design professional, these services may not be particularly effective even when QBS is used. Likewise, the number and extent of services provided, and the care associated with each, will influence risk and risk-funding requirements.

In essence, then, the risk associated with a project depends on a number of variables and the interrelationships between them. These variables and their interrelationships must be carefully assessed in order to establish an appropriate risk-funding premium. The premium should also recognize the fact that nothing can be done to eliminate risk altogether. Even a blanket indemnification is limited by a client's ability and willingness to honor it.

Fourth, as noted above, the preferred RA contract provision would include blanks, virtually forcing discussion of the issue and, as a consequence, the entire subject of risk. In addition, to help assure that the clause is conspicuous (something that may be required to assure its being upheld by a court of law), it should be highlighted with a marker, set or typed in larger or boldface print, in all capital letters, or otherwise made evident.

In discussing risks and their impact on fees, design professionals should note that the RA approach gives their clients new and valuable options. If clients want no RA cap or a higher one than you prefer, the impact on risk-funding requirements could be reduced through specific limitations included in the general conditions, provision of additional services, or amplification of those services already agreed to, or a combination of these. While increasing the level or amount of service would result in a higher fee, the increase involved may be comparable to that associated with the higher risk funding required if the services are not provided.

In essence, many clients need to be educated about these issues and how risk is affected by the interplay between them. As a consequence, even if you are unable to gain client acceptance of an RA cap, you will more than likely be able to reduce risks or at least learn enough about client attitudes to suggest that the necessary risk preventive might be refusing the commission. Although all of these outcomes are available through use of a limitation-of-liability approach, they often are not attained because the name of the concept focuses (misleadingly) only on the design professional's benefit and the provision itself does not contain any blanks, obviating its discussion.

The preferred RA provision given below assumes a contract between the prime design professional and the owner. Note that astute primes will not accept a limitation on an interprofessional's liability without first obtaining a similar limitation from their own clients, that is, owners. Accordingly, when you are acting as an interprofessional, you should be sure to educate the prime about limitation of liability and RA when such education obviously is needed.

The preferred RA provision is as follows:

Risk Allocation

OWNER recognizes that DESIGN PROFESSIONAL's fee includes an allowance for funding a variety of risks that are imposed on DESIGN PROFESSIONAL by virtue of DESIGN PROFESSIONAL's involvement in, and association with, OWNER's project. One of these risks stems from DESIGN PROFESSIONAL's potential for human error. In order for OWNER to obtain a reduction in fee as a result of DESIGN PROFESSIONAL's ability to set aside a smaller allowance for risk funding, OWNER agrees to limit DESIGN PROFESSIONAL's liability to OWNER and all construction contractors arising from DESIGN PROFESSIONAL's negligent professional acts, negligent errors, or negligent omissions, such that the total aggregate liability of DESIGN PROFESSIONAL to all those named shall not exceed $_____. OWNER further agrees to require of the contractor and contractor's subcontractors an identical limitation of DESIGN PROFESSIONAL's liability for damages suffered by the contractor or contractor's subcontractors arising from DESIGN PROFES-

SIONAL's professional acts, errors, or omissions. Neither the contractor nor any of contractor's subcontractors assumes any liability for damages to others that may arise on account of DESIGN PROFESSIONAL's professional acts, errors, or omissions, except as otherwise stipulated herein.

Alternative Solutions   A number of alternative solutions are available. Through one, a dollar figure would be inserted, instead of a blank. Using the conventional approach, and assuming a $50,000 RA cap, you would insert the following language:

> ...to all those named shall not exceed $50,000 or the DESIGN PROFESSIONAL's total fee for the services rendered on this project, whichever is greater. (If OWNER wishes to discuss higher limits and the charges involved, OWNER should speak with DESIGN PROFESSIONAL.) OWNER further agrees....

Inclusion of language indicating that higher limits are available is essential, because it makes clear that the RA cap identified was set *by the client*. This aspect of the provision is deemed essential to help ensure its being upheld. Otherwise, a court might strike the provision as being unenforceable, since, as a general rule, judges are somewhat opposed to limitations of liability in professional service agreements. For this reason, too, it is essential to make the entire provision physically obvious, through highlighting or other means.

Another approach, although it is not as highly recommended as either of the two given above, would involve use of the "standard" wording formerly advocated by ASFE. It is similar to that given above, but the differences should make it obvious why ASFE now prefers the RA approach:

> **Limitation of Liability**
>
> OWNER agrees to limit DESIGN PROFESSIONAL's liability to OWNER and all construction contractors arising from DESIGN PROFESSIONAL's professional acts, errors, or omissions, such that the total aggregate liability of DESIGN PROFESSIONAL to all those named shall not exceed $50,000 or DESIGN PROFESSIONAL's total fee for the services rendered on this project, whichever is greater. (If OWNER wishes to discuss higher limits and the charges involved, OWNER should speak with DESIGN PROFESSIONAL.) CLIENT further agrees to require of the contractor and contractor's subcontractors an identical limitation of DESIGN PROFESSIONAL's liability for damages suffered by the contractor or contractor's subcontractors arising from DESIGN PROFESSIONAL's professional acts, errors, or omissions. Neither the contractor nor any of contractor's subcontractors assumes any liability for damages to others that may arise on account of DESIGN PROFESSIONAL's professional acts, errors, or omissions, except as otherwise stipulated herein.

All of the forgoing provisions make the design professional liable for third-party claims. Some architects and engineers attempt to shift this liability exposure to the owner and contractors through such wording as the following:

Limitation of Liability

OWNER agrees to limit DESIGN PROFESSIONAL's liability to OWNER and all third parties arising from DESIGN PROFESSIONAL's professional acts, errors, or omissions, such that the total aggregate liability of DESIGN PROFESSIONAL to all those named shall not exceed $50,000 or DESIGN PROFESSIONAL's total fee for the services rendered on this project, whichever is greater. (If OWNER wishes to discuss higher limits and the charges involved, OWNER should speak with DESIGN PROFESSIONAL.) OWNER further agrees to require of the contractor and contractor's subcontractors an identical limitation of DESIGN PROFESSIONAL's liability for damages suffered by the contractor, contractor's subcontractors, and all other third parties arising from DESIGN PROFESSIONAL's professional acts, errors, or omissions.

In one court case, a defendant design professional claimed that failure to deliver plans and specifications by a contractually mandated date was covered by the limitation-of-liability clause also included in the contract. The owner was a public agency that lost a substantial amount of federal funding due to the design professional's lateness. The owner claimed that the late delivery was a breach of contract, not professional negligence, and thus was not affected by the limitation of liability. The court agreed. In response, some attorneys have suggested that limitation-of-liability provisions should be broadened to include breach of contract. Such an approach may not withstand a court challenge, potentially voiding the entire clause. Furthermore, depending on laws in the state where the contract would be enforced, a court may be at liberty to say the provision also applies to the client. A limitation-of-liability or risk allocation provision in such cases thus could begin as follows:

Limitation of Liability

To the greatest extent permitted by law, OWNER agrees to limit DESIGN PROFESSIONAL's liability to OWNER and all third parties arising from DESIGN PROFESSIONAL's professional acts, errors, or omissions, or breach of this AGREEMENT, such that total aggregate liability of DESIGN PROFESSIONAL shall not exceed $_____. . . .

Innumerable other RA and limitation-of-liability clauses can be developed, depending on the wording preferred by the client and/or the design professional, as well as the specific issues covered by the limitation. In all cases, however, some type of clause should be included and should be made conspicuous, at least to spur discussion of risk. In fact, if this is all the provision accomplishes, its purpose will have been in large part fulfilled.

Maintenance of Service

When architects or (more typically) engineers are serving as subconsultants to a prime consultant, including project or construction managers, there is a potential for the termination of the prime's contract with the owner.

Problem In some instances, the termination is the result of, or creates, a dispute between the prime and the owner. The owner—your client's client—may wish to retain your services or have them retained by a replacement prime, to help assure continuity. Or, as has sometimes occurred, the project will be under construction and your services are being employed for construction monitoring. If your services must be terminated by virtue of the prime's decision (often to gain leverage or in retaliation), someone else may have to take on your work, creating problems for both of you with respect to interpretations of your plans and specifications. When you are the design professional of record for a specific aspect of construction that a jurisdiction requires the design professional of record to monitor, the situation can become even more difficult.

Solution The best solution is a contract provision between you and your client that permits you to carry on your work in the event the owner-prime contract is dissolved, so your services do not become a "bargaining chip" between the prime and the owner. Such a provision may read as follows:

Maintenance of Service to Client's Client

In the event that CLIENT's contract with CLIENT's client is terminated for any reason, DESIGN PROFESSIONAL shall be at liberty to continue services on CLIENT's client's behalf, through agreement developed between DESIGN PROFESSIONAL and CLIENT's client. DESIGN PROFESSIONAL's services to CLIENT shall be terminated as specified elsewhere in this AGREEMENT, but DESIGN PROFESSIONAL shall not be constrained from entering into an agreement with, or performing services for, CLIENT's client immediately upon termination of CLIENT's agreement with CLIENT's client, even in the event that termination of this AGREEMENT shall not have been formalized.

Alternative Solutions Although the concept embraced by the forgoing provision can be worded differently, there really is no alternative to the concept itself. If a client will not agree to the provision, you would be well advised to reconsider accepting the commission.

Ownership of Instruments of Service

Plans, specifications, reports, boring logs, and so on are instruments of service, not products, and thus should remain the property of the professional who prepared them. Nonetheless, some clients believe they should own instruments of service, because they paid for them. In truth, however, they are not paying for them; they are paying for service.

Problem Many clients, especially government owner clients, will require you to provide all copies of everything and allow you to keep one set for yourself. A typical client-developed provision may read as follows:

Ownership of Documents

All documents, including, but not limited to, plans and specifications, shall become the property of CLIENT. Upon completion of the work, DESIGN PROFESSIONAL shall deliver to CLIENT all copies of each and every document pertaining to the work. DESIGN PROFESSIONAL may retain one set of these documents for DESIGN PROFESSIONAL's files.

Unauthorized reuse is the most serious problem that usually can arise from compliance with such a provision, creating the potential for a claim against you, because your plans and specifications were applied under different circumstances and, because of these differences, the plans and/or specs were in error. Even if your work was perfect for the intended application and even though you did not authorize any other application, you would still have to mount a defense, and, in fact, you might lose.

Unauthorized reuse is not the only problem with which you should be concerned. Another can arise in those jurisdictions where instruments of service are treated as products. Any defects in them—errors or omissions—might be treated as product defects, thus invoking the doctrine of strict liability rather than the doctrine of professional negligence, obviating professional liability insurance coverage. In addition, with specific reference to the clause given above, be wary of any absolute word. In this case, the word *all* ("*All* documents . . . pertaining to the work. . . .") creates a condition that probably is not meant, since it would encompass agreements with subconsultants and subcontractors, notes and other proprietary information, and so on.

Solution Work with clients; educate them; attempt to obtain an effective outcome. The correct understanding should be as follows:

Ownership of Instruments of Service

All reports, plans, specifications, field data, field notes, laboratory test data, calculations, estimates, and other documents prepared by DESIGN PROFESSIONAL are instruments of service, not products, and shall remain the property of DESIGN PROFESSIONAL. DESIGN PROFESSIONAL shall retain these records for a period of _____ (_____) years following submission of DESIGN PROFESSIONAL's final invoice, during which period they will be made available to CLIENT at all reasonable times.

As an added element, you could copyright all, or some, of these materials. Obtaining a copyright is simple. It may involve no more than merely printing copyright wording on your instruments of service.

Alternative Solutions The retention period limit included may be set statutorily or may be established by the design professional. In cases in which engineers and architects are at liberty to set a retention period, there has never been agreement as to what the period should be. Some set it based upon the statute of repose, but such statutes are subject to change. In some cases, a relatively short retention period is involved, such as one or two years, and any extension beyond that time would be at the client's expense; for example,

> ... to CLIENT at all reasonable times. If CLIENT wishes DESIGN PROFESSIONAL to retain documents for a longer period of time, CLIENT shall so specify in advance, in writing, and shall pay in a timely manner all charges agreed to for DESIGN PROFESSIONAL's maintenance of such documents beyond the time period otherwise prevailing.

If it is your practice to maintain documents on microfilm or other media that permits you to discard the original materials, this, too, should be referenced. (Courts generally require the best available evidence, and that usually means the original documents when they are available.) Such reference can be made in two places: first, in wording that indicates that you will retain documents ("... DESIGN PROFESSIONAL shall retain these records *on microfilm* for a period of...."), and again relative to other than conventional maintenance:

> ... to CLIENT at all reasonable times. If CLIENT wishes DESIGN PROFESSIONAL to retain documents for a longer period of time or in a format other than microfilm, CLIENT shall so specify in advance, in writing, and shall pay in a timely manner all charges agreed to for DESIGN PROFESSIONAL's maintenance of such documents beyond the time period otherwise prevailing.

In the case of clients who insist on retaining ownership of instruments of service, it should be possible to strike a compromise whereby you furnish the materials to them but they indemnify you against unauthorized reuse. A clause embodying this concept is as follows:

> **Ownership of Instruments of Service**
>
> CLIENT acknowledges that DESIGN PROFESSIONAL's reports, plans, specifications, field data, field notes, laboratory test data, calculations, estimates, and other similar documents are instruments of professional service, not products. Although ownership of such documents normally is retained by DESIGN PROFESSIONAL, they nonetheless shall, in this instance, become the property of CLIENT. CLIENT recognizes that instruments of service should, under no circumstances, be reused without written authorization of DESIGN PROFESSIONAL to do so. Such authorization is essential, because it requires DESIGN PROFESSIONAL to evaluate the documents' applicability, given new circumstances, not the least of which is passage of time. Accordingly, in return for DESIGN PROFESSIONAL's relinquishment of ownership, CLIENT agrees to waive any claim against DESIGN PROFESSIONAL and to defend, indemnify, and hold DESIGN PROFESSIONAL harmless from any claim or liability for injury or loss allegedly arising from unauthorized reuse of DESIGN PROFESSIONAL's instruments of service. CLIENT further agrees to compensate DESIGN PROFESSIONAL for any time spent or expenses incurred by DESIGN

PROFESSIONAL in defense of any such claim, in accordance with DESIGN PROFESSIONAL's prevailing fee schedule and expense reimbursement policy.

In the event that instruments of service may be considered products in the state with jurisdiction, it may be advisable to expand the indemnification to read as follows:

> ... allegedly arising from reuse of DESIGN PROFESSIONAL's instruments of service or allegedly arising from considering DESIGN PROFESSIONAL's instruments of service products. CLIENT further....

As an added safeguard, you might consider adding conspicuous limitations language to each page of any reports, plans, specifications, and so on that you submit, using wording such as "This material is subject to limitations identified in Section _____, page _____." You may also wish to copyright materials, despite a clause giving ownership to the client, principally as a device to prevent unauthorized reuse.

Record Documents

Design professionals are frequently required to provide record documents, indicating what was actually built. A typical client-developed owner-prime design professional provision may read as follows:

Record Documents

Upon completion of the work, DESIGN PROFESSIONAL shall compile for, and deliver to, OWNER a complete set of record documents. This set of documents shall consist of corrected specifications and as-built drawings showing the location of all work.

Problem Record drawings are based, in large part, on information furnished by others. In most instances, it is impractical to verify the accuracy of this information. In some cases, it is impossible to verify it, and, more often than not, full accuracy is a rarity. However, unless this situation is clarified in writing, you could be held liable for losses arising from errors in record documents, regardless of the source of the error. The losses involved could be major. For example, consider the possible consequences of an error involving the location of a concealed power main in the event a cut-through becomes necessary. Relying on your "as-built" drawings, a contractor is instructed to begin cutting in what is indicated to be an appropriate location, but, as it so happens, this is precisely the location where the power main actually lies.

Solution The best solution comprises three elements. First, any contract provision referring to record documents should make their potential inaccuracies clear.

Second, refrain from using the terms "corrected specifications" and "as-built drawings." *Corrected* indicates "correct" or "without error." *As built* connotes that the plans represent the work as it was actually built. Although those in the construction industry may recognize the true meanings of these terms, someone new to the industry or otherwise unsophisticated may be relying on the plans and specifications, and/or a lay trier of fact may be called upon to render a decision. Accordingly, refer to "record specifications" and "record drawings."

Third, *each page* of your record plans and specifications should contain a notice or warning that they may comprise inaccuracies for which you are not responsible.

It is also important to clarify the ownership of these documents, as already noted. An appropriate contract provision may read as follows:

> **Record Documents**
>
> Upon completion of the work, DESIGN PROFESSIONAL shall compile for, and deliver to, CLIENT a complete set of record documents conforming to information furnished to DESIGN PROFESSIONAL by construction contractors. This set of documents shall consist of record specifications and record drawings showing the reported location of work. In that record documents are based on information provided by others, DESIGN PROFESSIONAL cannot, and does not, warrant their accuracy.

Language appropriate for *prominent notice* on each page of record specifications and drawings might be the following:

> This record document has been prepared, in part, based upon information furnished by others. While this information is believed to be reliable, the design professional cannot assure its accuracy and thus is not responsible for the accuracy of this document or for any errors or omissions that may have been incorporated into it as a result. Those relying on this record document are advised to obtain independent verification of its accuracy before applying it for any purpose.

Alternative Solution Several alternative solutions exist. The first would be to delete the provision from any client-developed agreement but to still include the warning in record plans and specifications.

If a client objects for some reason to the last sentence of the preferred provision ("In that record documents are based on information furnished by others, DESIGN PROFESSIONAL cannot and does not warrant their accuracy."), the sentence feasibly could be deleted. However, this may give one cause to wonder why a client would object to a statement of the obvious, inserted to help ensure that there will be no misunderstandings about an important issue. If the preferred language must be weakened for any reason, it would be appropriate to include definitions of record documents, record plans, and record specifications in the DEFINITIONS section of your agreement—an approach that might be valuable in any event.

You feasibly may want to strengthen the preferred provision by adding a waiver and indemnification protecting you in the event that your warnings in record documents are ignored.

Right to Reject and/or Stop Work

The client may want you to reject a contractor's work or to stop the work if corrections are not made. A clause such as the following may be offered to you:

Right to Reject Work

DESIGN PROFESSIONAL shall have authority to reject work that does not conform to contract documents. If rejected work is not promptly corrected, DESIGN PROFESSIONAL shall stop the work.

Problem Your role on a construction site is intended to be passive: You observe or monitor. Rejecting or stopping work is active and creates an immediate exposure in terms of, "You should have done..." whatever it is someone else failed to do.

Playing a passive role does not mean lack of involvement or concern. From the contractor's point of view, the phrase, "I'll have to recommend that the client reject this work," has almost the same force and effect as, "I'm rejecting the work." As such, in almost all cases, work rejection can be accomplished in a passive role just as in an active one, at least insofar as prompt correction is concerned. Few contractors will want to have work rejection occur later by the owner, forcing considerably more rework.

Stop-work authority is something else again. *Having the right to stop work can be, and often is, construed as the duty to stop work.* In fact, in the clause above, this duty is specific. Does it apply just to the way in which the contractor is doing something, or does it apply to jobsite safety as well? As you and your client must be aware, courts often look for deep pockets—no matter how remote the argument—to right a perceived wrong.

Another problem with stop-work authority is the consequential damage that can result from delays. It may be worth more to a client to accept poor work and open a new facility on time than to have the work done well and open late. A client who questions the wisdom of your stop-work authority could then sue you for the consequential economic losses suffered as a result of the delay. As such, given everything at stake, the client—*and only the client*—should make the decision about stopping work.

Solution The best solution is a clause that clearly states what it is you will be doing and what your responsibilities are. Consider the following:

Rejection of Work

DESIGN PROFESSIONAL may recommend to CLIENT rejection of work that, in DESIGN PROFESSIONAL's professional opinion, does not conform with DESIGN PROFESSIONAL's recommendations, specifications, or design.

Through this approach, it would be up to the client to reject or to stop work. You can provide all the input necessary for formulation of a decision.

Alternative Solutions If a client insists that you have work rejection authority, it may be acceptable; check with legal counsel. If the client insists that you take on stop-work authority, however, insist on a full waiver and indemnification. This would amount to adding the following wording to the client-provided wording given on page 146:

> ... corrected, DESIGN PROFESSIONAL shall stop work. CLIENT agrees to waive any claim against DESIGN PROFESSIONAL and to defend, indemnify, and hold DESIGN PROFESSIONAL harmless from any claim or liability for injury or loss allegedly arising from DESIGN PROFESSIONAL's stopping the work or failure to stop the work. CLIENT further agrees to compensate DESIGN PROFESSIONAL for any time spent or expenses incurred by DESIGN PROFESSIONAL in defense of any such claim, in accordance with DESIGN PROFESSIONAL's prevailing fee schedule and expense reimbursement policy.

6

Professional Liability Insurance

Professional liability insurance has become particularly important in recent years, due to the size and frequency of claims filed against engineers and architects. These developments have made it difficult for uninsured design professionals to obtain work and, at the same time, have substantially elevated the cost of coverage. This chapter examines the role that professional liability insurance has assumed, the industry that provides it, key aspects of coverage, and factors that affect its availability and cost. Design professionals need this understanding, not only to further their own risk management activities, but also to educate their clients, many of whom fail to recognize just what it is that professional liability insurance does, and does not, do. Because these clients, and not just a few engineers and architects, do not understand these matters well, many impose or accept significant, and often needless, risks, from the mistaken belief they are "covered."

THE ROLE OF PROFESSIONAL LIABILITY INSURANCE

When professional liability insurance was first introduced in the mid-1950s, it was offered principally to protect the assets of insureds. In the event damages had to be paid, as the result of a court's or an arbitrator's award, or as the consequence of a settlement, most of the money paid came from the insurer, leaving design professionals' and/or their firms' wealth basically intact. Engineers' and architects' expenses were confined largely to the deductible, that is, the initial, self-insured "layer" of coverage.

A Source of Full Recovery

Over time, due in part to client demands, it was recognized that professional liability insurance could also benefit clients and other third parties as a source of recovery whose size exceeded whatever amounts design professionals could otherwise bring to bear. As a consequence, liability insurance became a source of full recovery, abetting the notion that ours should be a risk-free society supported by a huge, vastly wealthy, and wholly impersonal insurance industry that can easily afford a loss and that can actually regard the headlines created by a major award as advertising. If reaching into deep pockets required unfair treatment of those only marginally at fault, so be it, was the belief. After all, the insurance was there, not so much to protect practitioners as to protect society, or so it seemed.

A Substitute for Professionalism

Design professionals also are "there" to protect society by conducting themselves and their work in a manner that will preserve and protect human health and safety. In that professional liability insurance could do the same, or at least make a damaged party whole, it began to be used as a substitute for the high degree of professionalism that engineers and architects had historically applied in the execution of their commissions. Clients regarded the insurance as something they were paying for in any event—something that could justify the less extensive workscopes they relied on to achieve design fee reductions. While undoubtedly many design professionals balked at this concept at first, enough ultimately concurred to initiate a trend that today affects all.

A Marketing Prerequisite

As relaxed standards of practice contributed to continually more claims and disputes, professional liability insurance—arguably the cause of the problem—became recognized as the cure. Uninsured firms and those without adequate coverage thus find it difficult to obtain work, making professional liability insurance a marketing prerequisite. In situations in which few firms can obtain the coverage, as for certain geoenvironmental projects, those who do have it can enjoy a tremendous advantage. And some firms go to great lengths to obtain it; for example, by securing bank letters of credit based upon their net assets and then applying the letters of credit to reinsure policies that otherwise would not be offered. If a claim is made, of course, the firm would have to cover any losses experienced by the insurer. Although this is nothing more than a roundabout approach to self-insurance, there is a major difference: The firm using it is able to furnish a certificate of coverage—something few of its competitors can do. For the same reason, groups of firms have banded together to form their own insurance companies, which, for all intents and purposes, are little more than shells, susceptible to collapse after two or three major claims. Many clients do not seem to care, however, demonstrating a worrisome naivete about the insurance industry and what actually may transpire once claims are made. Design professionals who use this approach are not naive; just weary. They have grown tired of trying to defend their lack of coverage. It has become far easier and more effective to mouth the password—"I'm insured"—which gains them access to new commissions. To their credit, most of the firms that obtain this marketing-oriented "see-through" protection are fully aware of its fragility. Accordingly, many take advantage of the supply-demand imbalance and insist upon executing their commissions in a professional manner, to reduce the likelihood that their insurance coverage—basically, their own assets—will have to be called upon.

A Cause of Barratry

The huge and impersonal resources created by professional liability insurance have made engineers and architects the targets of those attorneys who are drawn to deep pockets, as moths to a flame. Often relying on hired-gun experts, they don advocates' robes to foment claims where none should exist, a practice called *barratry*. Somewhat ironically, most such claims are satisfied not by insurers but, rather, by design professionals, in part because they have increased their policies' deductibles to moderate premiums. Most also must pay legal and expert witness fees from their own pockets, and, in all cases, the time spent to fight a claim—justified or not—is nonreimbursable, except to the extent that a contractual provision may intervene.

A Project Fundamental

The concept of professional liability insurance is becoming ingrained in American society, due in large measure to the courts' penchant for dispensing social justice, that is, making the damaged whole. As such, concepts of "vicarious liability" and "joint and several liability" are being applied in such a manner that those who retain uninsured providers run the risk of being considered indemnitors, especially when they have deep pockets of their own. In one recent case, a sewer authority serving several jurisdictions was required to pay damages to a couple who had $25,000 worth of trees wrongfully cut down by a contractor the authority had retained. The couple had originally sued the contractor. The contractor was uninsured, however, and filed for bankruptcy. The claim was reinstituted against the sewer authority. In its contracts with the jurisdictions it served, the sewer authority said it would evaluate the liability insurance coverage of its contractors; it did not say it would use only insured contractors. Nonetheless, the court determined that review implied coverage would be assured, and, since the couple was considered a third-party beneficiary of that contract, the authority was ordered to make restitution.

As courts continue to seek, and to find, any plausible excuse to compensate victims, professional liability insurance will become increasingly more necessary—a societal requirement imposed on those who perform, or cause performance of, activities that could foreseeably damage others.

A Lever for Reasonableness

Clients frequently try to achieve a secure outcome inexpensively, by having their design professionals assume liability for certain risks that could otherwise be mitigated through more professional service. For the most part, this approach requires design professionals to assume contractual liabilities, such as those associated with guarantees, certifications, or indemnifications. Since these contractual liabilities are uninsurable, however, a doubly dangerous situation emerges. The work necessary to reduce certain risks is not performed, and there is no insurance coverage for the exposure. Design professionals certainly do not want to be in such a position, and, given legal trends, clients should be fearful of it as well. After all, liability could revert to clients, in whole or in part. Since client positions are sometimes based on a misunderstanding of professional liability insurance and the false sense of security arising from that misunderstanding, explaining insurance facts to such clients can result in your insurance becoming a lever for obtaining reasonableness. Clients should not want to be the cause of your being uninsured, especially when the situation is highly risk-prone. Accordingly, for your own sake, and as part of your service as a trusted professional adviser, you should inform your clients about the realities involved. By doing so, you should be able to eliminate contractual liabilities in order to maintain insurance coverage and to enhance your workscope in order to reduce unnecessary risks.

Large private-sector clients and those in the public sector may be particularly difficult to convince. They often are dealing with contracts that have been prepared by their legal staffs, and, more often than not, these contracts have been agreed to many times before by your competitors, if not your own firm. You must therefore demonstrate that you are more astute than your competitors or even than you used to be. Assuming they have experienced no problems in the past, explain what they could run into and that, up to this point, they have been lucky. Since some of these clients will greet your observations with skepticism, ask your insurance agent or insurer for assistance. Conscientious insurers generally will be willing to write a letter, and, in some cases, they will also send a regional claims representative to meet with you, your client, and your client's attorneys.

In cases in which little or no professional liability insurance is available, as for work associated with hazardous materials, the need for insurance, and the inability to attain it, could very likely invoke the *doctrine of peculiar risk*. This doctrine holds that, when extraordinary risks prevail (i.e., the inability to obtain the insurance society requires), the client must authorize extraordinary safeguards. If these safeguards are not authorized, then the client must accept all or a significant share of the liability that results when these risks materialize. As such, being unable to attain professional liability insurance can also serve as a lever for reasonableness, providing the client is informed.

THE PROFESSIONAL LIABILITY INSURANCE INDUSTRY

To understand professional liability insurance, it is important to understand the industry that provides it.

The Principals

The insurance industry consists of four principal parties: insureds, insurance agents, insurers, and reinsurers.

Insureds Insureds are those to whom a policy's coverage is extended. In the case of design professionals, it is common to consider a firm as the insured, but this is not really the case. Although a firm may pay for, and derive protection from, a policy, those who are covered by it are the "named insureds," that is, the registered professional engineers and architects who assume responsibility for the firm's work products by placing their seals on their instruments of service. Coverage is provided in this manner because, by law, design professionals are individually responsible for their work. This includes the work performed by others, such as drafters, whom design professionals supervise.

The individual coverage aspects of professional liability insurance can create knotty problems for engineers and architects who are alleged to have committed a negligent act while with a firm that no longer exists or that is now uninsured. Unless these individuals have coverage from some other source, they may have to face the claim alone. Accordingly, it is vital for you to understand what type of protection is availble to you before you retire from, or otherwise leave, a firm or active practice, or take a position with another organization.

Insurance Agents Insurance agents sell insurance and obtain a commission from each sale. Some specialize in just one type of policy, but, more commonly, they offer a variety of coverages from one or several insurers. Careful selection of an insurance agent is always advisable, because effective agents do far more than sell. With regard to professional liability insurance, they should at least provide guidance in completing the application form, because many of the questions asked are subject to interpretation. Answers given affect the premium and, in some cases, determine whether or not coverage will be offered. Agents can also give advice on techniques that firms can use to lower the cost of their coverage and should be able to pass along information about new policies or policy provisions, as well as loss prevention measures you can use. Insurance agents also are among the first people called when a claim is filed or when a situation suggests that a claim probably will be filed. As such, they should be available to provide guidance when needed.

In selecting an agent, it is suggested that you ask peers about those whom they use and what their experience has been. An agent who serves a number of design professionals and who is expert in professional liability insurance matters can be an extremely valuable asset.

Insurers   Insurers are the companies that issue policies. They establish what their policies do and do not cover, limits and deductibles, and the premium that must be paid. They also handle any claims that are made. Those responsible for these functions are actuaries, underwriters, and claims managers.

Actuaries   Actuaries compute the odds that a given risk will materialize and the ultimate probable cost of that risk. In doing so, they rely on the law of averages (probability theory), determined through analysis of historical data.

When an insurer refuses to provide coverage for a certain type of exposure—as has occurred in recent years—the decision usually results from guidance provided by actuaries, typically because they know the risks are huge but are unable to calculate them reliably. Refusing to cover a risk is a major step for insurers; it is a decision not to sell their product. It is similar to a design professional's decision not to accept a commission—something that is done with great reluctance.

As you probably are aware, most insurers (as of January 1991) exclude pollution coverage from their professional liability and commercial general liability (CGL) policies. At one time, insurers offered this coverage without significant restraint, until they were inundated with major claims resulting from long-term damage. At that time, policies were modified to cover only those losses caused by a sudden release of pollutants. In one case, however, a court determined that sudden-release coverage applied to a long-term problem because the long-term problem clearly had to be caused by innumerable sudden discharges. This being the case, the court also ruled that each such sudden discharge had to be covered to the full limits of the policy. Given the court's attitude, and fearing (with good reason) that it would be emulated in other cases, the industry decided to terminate pollution coverage altogether.

Underwriters   Underwriters evaluate each applicant to determine the extent to which various risks (identified by actuaries) are likely to affect it and, accordingly, what the firm's premium should be, assuming its application is accepted. In some cases, evaluations are based exclusively on a review of an applicant's written responses to questions on the application. In others, written responses are used mostly to determine whether the application should be rejected or studied further. If further study is indicated, representatives of the insurer will actually visit the firm, to conduct an in-depth underwriting review.

While many design professionals have been hard hit by rapidly spiralling insurance costs, most are in a position to reduce these costs, as we will discuss.

Claims Managers When everything is running smoothly for insureds, the only contact they may have with their insurers is an occasional bill or newsletter. The true test of an insurer occurs when a claim or preclaim situation emerges. In fact, does the insurer offer preclaim counseling? If so, by whom? What are the person's qualifications, capabilities, and attitudes? How quickly and how effectively does the company respond to a preclaim situation? To an actual claim? Who are the lawyers that the insurer recommends to their insureds? What are their reputations? Are they known as professional and knowledgeable, or as something less than that? And what is their attitude? Is their principal concern "doing right" by the insured, or is it "doing right" by the insurer? By all means, speak with insureds who have actually had to go through a claim or preclaim experience with an insurer. To purchase insurance based solely on the premium could be a serious mistake. If an insurance agent is unwilling to identify appropriate people with whom you can check, casual discussion at association meetings should be of value.

Reinsurers As their name implies, *reinsurers* insure insurers. The manner in which they operate differs from insurers', in that they often purchase certain "layers" of risk. For purposes of illustration, assume an insurer has issued a policy with a $1 million limit and a $50,000 deductible. The insurer retains the $50,000–$250,000 layer and then transfers the $250,000–$500,000 layer to Reinsurer A, and the $500,000–$1 million layer to Reinsurer B. The amount that the insurer pays to the reinsurer is determined by underwriting factors similar to those used by the insurer.

Reinsurance helps bring stability to the insurance industry, because it diffuses major losses. A comparatively small insurer that faced a $100 million loss on its own would feasibly have to declare bankruptcy. In actuality, however, its losses would be limited only to the risks it decided to retain or was unable to have others accept.

Insurers determine how much risk they will retain and how much they will transfer to reinsurers, based on a number of variables. The more that is transferred to reinsurers, however, the more the insurer's rates are determined by factors over which the insurer has little or no control. And the reinsurance market can be volatile, especially so since as many as half the firms worldwide are not long-term participants. They enter the market when investment opportunities look good; they leave when other opportunities beckon. Those who are new to the industry tend to accept the "high-end" risks because, historically, these involve those layers that are the least likely to be penetrated and thus require the least resources to accept. In recent years, however, penetration of the uppermost layers has become far more common, forcing many of the newcomers to leave the industry or, in some cases, to go out of business.

The long-term reinsurers, many of which are insurance company subsidiaries, understand the market well and are prepared to accept some of the "bumps and bruises" that occur from time to time. However, when they are alone in the market, the amount of capital available is greatly reduced, thus reducing the amount of premium that can be written, that is, the capacity of the market. Capacity fluctuations have a direct bearing on the price you pay for coverage, as we will discuss.

The Cyclical Nature of Capacity

Capacity fluctuations affect liability insurance costs in accord with the basic laws of supply and demand. When capacity is so high that supply exceeds demand, premiums moderate and a "buyer's market" prevails. When demand exceeds supply, however, the situation is reversed: The cost of insurance rises, and a "seller's market" is created.

Historically, capacity rises and falls in cycles. In the late 1970s and early 1980s, insurers operated at the peak of a capacity cycle, and professional liability insurance was relatively inexpensive. By the mid-1980s, insurers were operating in the trough of a capacity cycle, and insurance rates were high. A review of the factors associated with this cycling provides excellent insight into the inner workings of the industry.

Generally speaking, insurers and reinsurers have two sources of income: underwriting income, derived from its insureds' premium payments, and income derived from investments in the stock and bond markets, buildings, mortgages, loans, and so on.

When the investment market is strong, conventional investors generally try to obtain as much money as they can to increase their involvement, typically by borrowing money from banks or other lenders. Although investors incur an interest expense when they borrow funds, their calculations show that their investments should generate enough income to pay the interest and still generate a profit.

Insurance companies do not have to rely on lenders to obtain investment funding. They already have a ready source of income: cash from premium payments. Thus, when the investment market is strong, they take steps to increase their underwriting income. They can do this by lowering their rates to attract insureds from other insurers and/or by relaxing their underwriting standards to accept risks they might otherwise refuse. Insurers recognize that either practice will increase underwriting losses, but that is the price that has to be paid—tantamount to interest—in order to increase participation in the investment market. Their calculations show that, despite increased underwriting losses, they will still enjoy healthy profits, due to the income they can earn by investing the increased premium dollar cash flow.

It is important to note that insurers' underwriting losses are subject to certain accounting requirements imposed by regulatory agencies. When a claim is filed, the circumstances surrounding it must be closely analyzed to determine how much it will ultimately cost the insurer. Even though actual payout may not occur until three to five years (or more) after the claim has been filed, the insurer must immediately set aside an appropriate amount in "loss reserves" to assure its ability to "make good" when the need arises. The amount set aside must approximate the full amount likely to be paid out. These loss reserves then are invested.

In the late 1970s and early 1980s, investment opportunities were particularly strong because there was such a high rate of inflation in the United States. Even relatively safe investments were providing double-digit returns. These high rates attracted numerous investors to the reinsurance market. By accepting risks that would not require payouts for three, four, or five years later, they derived immediate income to fund their investment ambitions. As a consequence, the insurance industry's capacity was increased; insurance rates and underwriting standards fell.

When the United States altered its monetary policies in the early 1980s, inflation was dramatically reduced and high-yielding, safe investment opportunities all but disappeared. Reinsurers left the market in search of greener pastures, and capacity was drastically reduced. But this was not the only problem facing the insurance industry. Due to the low rates offered, insureds had been encouraged to increase their coverage. Low underwriting standards had made insurance available to those who previously could not obtain it. As a consequence, claims rose to record levels, fueled in large measure by society's growing litigiousness. This situation caused insurers and state regulators to reexamine loss reserves, and many discovered an alarming situation. Due to inflation and other factors, they had seriously underestimated the cost of many losss. Since high-yielding investments were not available to help make up the difference, insurers immediately had to transfer vast amounts of money from capital to loss reserves. Since this significantly diminished the money available for investment, insurers had to look to underwriting income, rather than investment income, as their principal source of profit. This shift resulted in a precipitous rise in the cost of professional liability insurance, as well as more selective underwriting standards.

Some observers claim that the rapid rise in liability insurance rates experienced in the mid- to late 1980s indicates that liability insurers are mismanaged. In truth, however, they are simply reacting to competitive pressures with which all business entities must contend. An insurer that seeks to attain stability by not lowering its premiums and underwriting standards when others do will quickly find itself losing business. In fact, the situation insurers face is not too dissimilar from that faced by design professionals. When the demand for engineering and architectural services is strong, design professionals are in a much better position to insist upon professional execution of their commissions. When a buyer's market prevails, even some of the best-known firms will agree to relatively flimsy workscopes.

Those who claim that insurance companies are mismanaged also contend that they should be subject to more regulation, such as that that affects electric, gas, and other utilities. As it now stands, regulation is effected on a state-by-state basis, but the level of control is primarily product-based. State insurance regulators strive to assure that a company is legitimate and can make good on any policies it sells. By contrast, public utility commissions take a close look at everything a regulated utility does and thus become far more involved in review of management philosophy and decisions. Insurance companies could be controlled to that extent only through a federal apparatus, but, given the size, strength, and lobbying capabilities of the insurance industry, federal control is not likely to become a reality in the foreseeable future. In addition, federal control could not be easily exercised over reinsurers, because many are headquartered outside the United States. As a consequence, the professional liability insurance market will probably continue to operate in a largely unregulated manner, and, in responding to economic trends, it is likely to remain somewhat volatile. That being the case, design professionals and their clients cannot be assured that the professional liability coverage in force when they begain a project will be available when they finish it, let alone two or three years later, when claims are most likely to be filed. As such, you should never contractually agree to provide certain coverages in the future, nor should you look upon insurance as a substitute for high quality. High quality lasts for a lifetime. Your insurance may disappear next week, especially if you place your faith in an insurer that is not in the market for the long term or that low-balls premiums in hopes of building market share.

ASPECTS OF COVERAGE

Design professionals should be intimately familiar with what their professional liability insurance policies do, and do not, cover, as well as any time-related or other limitations that may exist. This understanding is essential for implementation of overall risk management strategies and to help ensure that clients do not overvalue the protection policies provide.

Establishing Overall Protection

Professional liability insurance is intended to create a source of recovery, in the event you commit an act of professional negligence. As already noted, most professional liability policies exclude coverage for any contractual liabilities you may assume, and, by understanding the implications of contractual liability, you may be able to dissuade clients from imposing them. It should be noted, of course, that professional liability insurance is not the only type of coverage firms carry. Commercial general liability (CGL) coverage is also obtained, with the nature of coverage varying from policy to policy. While some of these may cover slander or libel claims, few, if any, provide protection from contractual liability or breach of contract claims.

Your professional liability insurance agent usually is in a good position to review the full extent of your needs and to identify alternative means for obtaining the protection you require, given the nature of your practice. Experienced risk management consultants may also be able to help.

As a general rule, it is advisable to obtain professional and commercial general liability insurance from the same company, when practical, to afford protection in the event of claims that might be construed as being covered by either policy. In such cases, which are rare, both insurers might say that it is the other's responsibility to cover the potential loss, leaving the design professional caught in the middle.

In cases of exposures that cannot be covered by most policies, such as those relating to pollutants, toxics, and asbestos, it would be foolhardy for a design professional to absorb the risks. The exposures in question generally should be transferred to the client through a full or partial indemnification. (A partial indemnification is a limitation of liability, such as that established for professional negligence. Limitations of liability do not have to be restricted to that application, however.)

Whenever risk is being transferred to a client, design professionals should first determine whether the client is in a position to honor the indemnification, should the need arise. When the client is a transitory corporation that will likely be dismantled once a project has been completed, the head of such a group should accept personal responsibility for the exposure or otherwise create a long-term source of recovery, possibly through a bond, letter of credit, or a prepaid insurance policy.

Claims-Made and Related Provisions

Until recently, most CGL policies and even a few professional liability insurance policies provided *occurrence-based* coverage. Occurrence-based policies cover claims arising from occurrences that transpire while the policy is in force, even if the claim is filed years later, when the insured is covered by another insurer or none at all. Today, many commercial general liability insurance policies and virtually all professional liability policies are written on a *claims-made* basis. This means that the policy covers only those claims that are filed while the policy is in force.

Prior Acts (Retroactive) Coverage Most professional liability insurance policies provide coverage for "prior acts," that is, services that were performed before the policy went into force. Some prior acts or retroactive coverage is unlimited, meaning the policy covers any professional liability claim made while the policy is in effect, even if the claim is based on an allegedly negligent act committed ten or twenty years before. Other policies cover no prior acts at all, or, more commonly, they exclude from coverage claims arising from acts committed before some prior date. A claims-made policy with a retroactive date of December 31, 1984, would thus provide coverage for claims made while the policy is in force, *providing* the act giving rise to the claim occurred after December 31, 1984.

Design professionals should be aware of any prior acts limitations in their policies and institute appropriate safeguards, including development of self-insurance, if necessary.

Aggregate Coverage It is important for engineers, architects, and their clients to understand the limitations created by aggregate coverage. Very simply, it means that the limits of the policy apply for the term of the policy, which is almost always one year. In other words, a policy can provide no protection from claims after its limits have been exhausted during its term. Virtually all professional liability insurance policies provide aggregate coverage, creating another reason why professional liability insurance can never be a substitute for high-quality performance. Even though a firm has a million-dollar policy, it may only have a hundred dollars worth of coverage left.

Although the limit of coverage offered by professional liability insurance policies is aggregate, deductibles usually are applied on a per claim basis. In other words, if three claims had to be paid, each for $75,000, and your policy had a $50,000 deductible, you or your firm would pay $150,000 ($50,000 × 3) and your insurer would pay $75,000 [($75,000 − $50,000) × 3].

Cost of Defense The extent to which a professional liability insurance policy covers defense costs is yet another important issue. Not too many years ago, it was possible to obtain a cost of defense option that typically paid from 80 to 100 percent of attorneys' and experts' bills. This type of coverage was particularly beneficial for the design professions, because it encouraged many engineers and architects to fight meritless claims or those seeking monetary awards grossly out of proportion to actual damages. This coverage proved to be costly, and now it is provided, for the most part, only with policies that have low deductibles, for example, from $10,000 to $25,000. As a consequence, design professionals are once again opting to buy their way out of meritless or exaggerated claims, simply because it is less costly than proving the claims' invalidity. Note, however, that most professional policies still recognize the cost of attorneys and experts, by permitting insureds to apply it to their deductibles. However, defense costs paid by the insurer may also be applied to the aggregate limit. Defense costs can add up quickly and thus can affect a policyholder's behavior during the course of a claim.

Extended Discovery Extended discovery is coverage for claims filed after a claims-made policy has been cancelled or terminated, providing the event giving rise to the claim occurred while the policy was in force. Extended discovery, or "tail," is not something that insurers offer voluntarily. It is required by some state regulators and otherwise may be available (for a specific amount of time) on a negotiated, policy-by-policy basis.

Educating Clients Many clients are woefully unfamiliar with claims-made and other aspects of professional liability insurance—something they make clear in contracts they prepare. For example, some clients require an engineer or an architect to name them as additional insureds on a policy—something that almost never can be done.

Some clients may require their design professionals to maintain a policy in force during the course of the contract, possibly from the mistaken belief that the insurance is occurrence-based rather than claims-made. In fact, some of the most devastating problems—third-party claims—arise after the design professionals' work has been completed. For this reason, more astute clients will require design professionals to maintain coverage for one, two, or more years after their work has been completed. Design professionals should never agree to such provisions, however, due to the volatility of the market. In fact, the insurance may be too costly or otherwise unacceptable.

Client education often is required to obtain reasonable terms and conditions, and to acquaint clients with their risks, thus encouraging alternative measures to reduce it. Insurance agents can often provide assistance.

Endorsements and Riders

Endorsements and riders comprise policy amendments or additions that provide protection that would not otherwise be available. Whenever it appears that a contractual provision or assigned responsibility may not be covered, it is prudent to ascertain whether or not coverage would be in force by virtue of the standard policy and, if not, whether an endorsement or a rider might be available. If protection can be had through a rider or an endorsement, you should learn how much it will cost and the length of time for which it may be available. Most design professionals require their clients to pay for the cost of endorsements or riders that client demands make necessary. For the most part, professional liability insurers are extremely reluctant to provide endorsements or riders. All too often, they create exposures that their insureds are powerless to control or to prevent, making them virtually impossible to underwrite.

Project Insurance

Project-based professional liability insurance has been available for many years but has recently gained new popularity. As typically offered, all the design professionals selected for the project are covered by one professional liability policy that applies solely to the project in question. In essence, these policies function as vehicles to prepay claims likely to be made over time, except that the cost comes out of the current project budget. This coverage provides much more protection than conventional practice policies, because it remains in place for a given period of time, subject to aggregate reductions. (As claims are satisfied, the amount of coverage remaining is reduced.)

The nature of these policies is such that owners should be willing to pay for them, given the better protection afforded. Nonetheless, covered design professionals could contribute, since the gross fees derived from the project for which a project-based policy is obtained are not usually counted in the overall gross fees that underwriters consider in setting rates for practice policies. In other words, when project-based policies are in place, a firm's conventional liability insurance premium should be lower than it otherwise would be.

Project-based policies give owners more flexibility with respect to risk management, because they can be assured of the coverage they prefer, for the amount of time preferred. These policies also permit insurers to become more involved in projects, to obtain insight relative to risks, and possibly to impose certain requirements as a condition of coverage. For the most part, these requirements would help establish higher levels of communication, coordination, and cooperation than might otherwise have been available.

SELF-INSURANCE

Self-insurance and "going bare" are not the same. "Going bare" implies that a firm has no insurance and has taken no particular steps to create a source of recovery or to reduce risk exposures. Self-insurance implies that a firm is somehow creating a means for recovery through its own financial resources and may be taking measures to minimize potential losses, for its own sake and others'.

All design professionals are partially self-insured, by virtue of the deductible associated with their policies, and because they usually must absorb the value of the time required to defend a claim. Other types of self-insurance also are used. For example, as mentioned previously, some firms obtain a letter of credit backed by their net assets and then apply this letter of credit to reinsure a policy issued by a conventional insurer. Although this maneuver is performed principally to support marketing activities, the firm, in fact, is putting its assets at risk. A similar approach is also used by some particularly large firms that need a substantial amount of protection. Typically, they will obtain a conventional policy that affords protection up to, say, $5 million, and then will apply a letter of credit to cover an additional $5–$10 million layer.

Self-insurance of a sort can also be obtained through contract provisions that transfer risks to the client or others, in full or in part, through limitations of liability or full indemnifications. Self-insurance also can be provided through development of a cash reserve for use in satisfying claims or covering the cost of legal defense. Reserves established in this manner are not subject to any particular tax privilege, however, and thus are seldom developed. In essence, they create a tax exposure that often is more troublesome than the reserve is worth. Nonetheless, it might be possible to establish reserves through various investments (e.g., common stocks) or other techniques.

In all cases, of course, the best insurance of all is professional execution of a commission, as well as professional conduct in virtually all other activities that a firm pursues, including contract preparation, human resources management, marketing, and promotion, among others. Generally speaking, firms that are self-insured are far more attuned to the preventive value of quality-oriented performance, because they have so much at stake.

INSURANCE COMPANY FORMATION

Self-insurance should not be confused with establishment of an insurance company or pool by a group of practitioners. This approach was pioneered in the design professions by members of ASFE who founded Terra Insurance, Ltd. For more than fifteen years, Terra functioned by reinsuring policies issued to its insureds by insurers already organized to do business throughout the United States. As a consequence of new laws in the United States, it now is far easier to establish a "captive insurers" such as Terra. For this reason, the original Terra group has become Terra Insurance Company (A Risk Retention Group), headquartered in Vermont. The original Terra Insurance, Ltd., headquartered in Bermuda, was sold to an insured who operates the company as a reinsurer. Risk retention groups are small mutual-type insurance companies owned by their insureds and generally are immune from state regulations. Many organizations are looking into this potential, often citing Terra—and its comparatively low rates—as an example of what can be done. It should be recognized, however, that Terra's rates have not resulted from its captive status but, rather, from its insistence on quality-oriented performance from its insureds. All firms that apply for coverage are subject to extremely close underwriting review.

As of January 1991, the growth spurt in captives that began in 1987 has begun to ebb. Nonetheless, what occurred is certainly worth recalling, because it likely will recur sometime in the future. In particular, prospective insureds need to be highly circumspect about rates that may be too good to be true. All companies have to compete for available reinsurance capacity. A new insurer with relatively little experience or capital generally must pay more for reinsurance than others, assuming reinsurance is available at all.

All insurers also require appropriate staffing, in order to be able to respond to routine inquiries, handle functions such as claims management and underwriting, and so forth. For the most part, the overhead created will be the same on a per insured basis, no matter who provides these services, unless there is a relaxation of quality. Given liability insurance's strategic position, however, relaxed quality cannot be easily tolerated. If anything, the quality offered should exceed that available elsewhere, something that requires more expense, not less.

The notion that established insurance companies all tend to be large and successful as a result of huge profits is not accurate in all cases. While this may be true for certain types of insurers—life insurers, in particular—it does not necessarily apply to professional liability "lines." In fact, most professional liability insurers have, for many years, offered their coverage at or even below their underwriting cost; profits have been derived from investments. As such, if a new risk retention group or other insurer offers exceedingly low rates, it could be because of certain coverage restrictions, for example, no coverage for prior acts or contributory coverage whereby insureds pay a percentage of the claims cost over and above the deductible. "There's no such thing as free lunch" applies in this situation, as in most others. Look very closely. Have your insurance agent take a close look, too. Be aware of the pitfalls before rushing ahead. If your firm believes that insurance is needed more for marketing than any other purpose, beware. When this attitude permeates, it can lead to an all too casual approach that fosters problems that cannot be resolved by a certificate of coverage only. Remember: High-quality professional work and professionalism in all activities are essential, even when a firm has the best insurance money can buy.

In some cases, involvement in a captive insurer may be inspired by the desire to make money, since any net profits ultimately revert to the insureds and owners, but the smaller size of these captives limits the investment potential and also creates much greater risk. Just a few major losses could force a company out of business, especially if it loses the backing of reinsurers. One also has to contemplate the cyclical nature of capacity. Cheap coverage is no great "claim to fame" when a "name brand" costs just a little more or—when apples are compared to apples—a little less.

ESTABLISHING PREMIUMS

The factors that underwriters typically consider in establishing a firm's premium include the policy's limit and deductible, the firm's gross fees, the disciplines practiced by the named insureds, the geographic area served, the nature of the projects involved, the extent of subcontracting that a firm performs, prior claims against the firm, and the effectiveness of a firm's management. The two major A-E professional liability insurers also consider a firm's participation in certain programs and provide premium credits for appropriate loss prevention and quality enhancement activity.

It is worthwhile to review how these factors affect premiums and discuss techniques for lowering premiums without sacrificing protection.

Policy Limits and Deductible

The more protection a policy provides, the more expensive it is. To minimize the cost of coverage, therefore, design professionals should keep their protection to the lowest practical amount. This amount must be determined through a deliberate review of the circumstances each confronts, including personal preferences. Peace of mind can be an extremely important factor.

In evaluating insurance requirements, the first consideration should be risk. How likely is it that a claim will be filed? How much might that claim actually cost? What are the chances of more than one claim being filed during a policy period? Insurance agents can provide helpful guidance in this and other respects, but some agents' assistance may be suspect because their commissions are directly proportional to the amount of premium paid. While most professional liability insurance agents offer guidance in a professional manner, by giving honest advice, the potential for prejudice does exist. Several agents should be evaluated before one is selected.

In considering your risks, be certain to separate the wheat from the chaff. For example, mechanical engineers face potentially huge claims based upon allegations that their design resulted in a "sick building" or that legionella-contaminated water was spread from a cooling tower they designed. It is doubtful that either such claim would be coverable under a conventional professional liability insurance policy, however, because the risks probably would be excepted by the policy's absolute pollution exclusion. Assuming that to be the case, there would be no point to increasing a policy's limits to cover the risk. (Whenever there are doubts about coverage, obtain a definitive written response from the insurer.) The policy's prior acts provisions comprise another important consideration. An electrical engineer who specified PCB-filled transformers ten years ago would be unable to obtain coverage for the potential liability if the prior acts protection is not available or because the nature of the risk is considered "pollution-related" and, thus, not coverable.

Consider, too that courts generally are reluctant to force individuals or organizations to pay damages in excess of their insurance limits. As such, having a lower limit could result in less exposure rather than more. But this is not necessarily true in all cases, and—if you are unwilling to make a "leap of faith"—peace of mind could be obtained only through higher limits.

It also is appropriate to consider the original concept of insurance protection and determine how much could be lost. What are your and your firm's assets worth? For the most part, the more you stand to lose, the more protection you need. Some design professionals attempt to reduce the amount of risk by draining a firm of its assets and/or by using trust funds, foreign bank accounts, and similar stratagems to make themselves "litigation-proof." These often work out much better in theory than in reality, and, to a very real extent, the cost of hiding one's assets often exceeds the amount required to insure them.

Client requirements also comprise an issue that merits consideration. If many of them require you to carry a $1 million limit, then that may be the appropriate limit to obtain. If a $1 million limit is required only on occasion, however, it may be more appropriate to obtain a policy with a $500,000 limit and to obtain additional protection for given projects through project insurance or an endorsement.

For the most part, raising a policy's limit is not nearly as costly as lowering its deductible, because the lower end of protection is far more likely to be penetrated than the higher end. Particularly for those firms that must pay relatively high premiums by virtue of their size, discipline mix, or other factors, increasing the deductible can save a considerable amount of money. The premium dollars saved could then be put into an investment instrument to build a reserve that, over time, would equal the amount of the deductible's increase. Given the problems associated with cash reserves, however, some firms are opting to use variable life insurance policies as a reserve medium. These permit relatively rapid build-up of cash values on a tax-deferred basis. This type of approach may be most suitable for firms that seldom, if ever, experience a claim. In any event, it should be used *only with appropriate guidance.* In some cases, for example, a higher deductible may eliminate the insurer's participation in the cost of legal defense. Note, too, that insurers usually are concerned about an insured's ability to meet its deductible. Frequently, a deductible increase can be obtained only with an insurer's permission.

Gross Fees

The larger a firm's gross fees, the higher its premium will be for equivalent coverage, since higher gross fees suggest more risk due to more and/or larger projects. Assuming all other factors are equal, however, a firm whose gross fees are $2 million does not necessarily pay twice the amount as a firm whose gross fees are $1 million. The impact of fees proceeds on a sliding scale. As gross fees increase, each additional fee dollar has a lesser effect on the premium.

In establishing gross fees, an underwriter generally considers those projected for the current year and those actually earned for each of several prior years. A number of techniques are available for adjusting these figures, depending on the insurer's requirements. For example, certain types of work, such as development of manuals, may impose relatively little liability exposure. As such, fees received for this work may be excludable, thus lowering past years' totals. Your insurance agent should be able to provide assistance in this respect.

Insofar as future years are concerned, it may be possible to establish separate firms for certain activities, for example: a firm of design professionals; a separate firm that performs nonprofessional tasks, such as field work (e.g., surveying, drilling, construction monitoring), laboratory work, or drafting; and a third for involvement in high-risk activities, such as asbestos and other hazardous materials. The wisdom of taking this type of action depends on a number of factors, including procedures established by professional liability insurers. If significant savings will result, however, it may be worthwhile.

Primes can achieve lower gross fees by relying on the multiple contracting approach just discussed. Most insurers reduce premiums when a firm subcontracts a portion of its work, however, so the savings may not be substantial. Nonetheless, it is always worthwhile to consider prospective savings, something an insurance agent should be able to help with.

Relying on project-based professional liability insurance creates yet another option for lowering gross, since fees for projects covered in this manner usually are excluded from gross fee calculations. Although relying principally on project insurance would result in higher premiums than would be paid for conventional practice coverage, increasing reliance on project coverage could result in significant overall savings in cases in which clients treat the premiums as reimbursables or pay for them directly.

Disciplines Practiced

Certain disciplines are far more risk-prone than others; the experience of a large group performing the same discipline thus has an important bearing on premiums. In the late 1980s, for example, structural engineering was considered particularly dangerous from an insurer's point of view, resulting in major premium increases even for structurals who had practiced for many years on a claims-free basis. Firms that specialize in structural engineering have relatively few options. This is not necessarily the case for multidisciplinary firms, however, and, as already noted, the potential for creating a separate firm that pursues high-risk disciplines only should be considered. In some cases the cost of insuring the two firms separately may be less than the cost of insuring them together, particularly when a third, nonprofessional services firm is established to serve both.

Geographic Area

The area where a firm practices can have an impact on premiums because claims are more common in some areas (particularly urban areas) than others. There also are variances with respect to the cost of legal services and of statutes of repose. A firm with just one office or several offices all operating in the same geographic market can do little to alter its area of operations. Larger firms may have different choices, however, as when it comes to opening branches or merging with, or acquiring, other firms.

Project Types

The types of projects that a firm works on is an important underwriting concern. Generally speaking, any firms' risks are increased when it accepts fast-track projects, those that rely on multiple-prime construction contractors, or those that employ construction managers. Condominium projects also are considered high-risk.

Architects' and structural engineers' most risk-prone projects tend to be high-rise buildings, parking garages, and sports arenas, with structurals also facing higher risks when they design bridges. Bridges are also risk-prone projects for civil engineers, as are trestles, piers, docks, marinas, dams, tunnels, and waste treatment plants. Hospitals are particularly troublesome for mechanical and electrical engineers.

Firms that specialize in risk-prone projects can do little to alter their project mix. This is not the case for others, however. Generally speaking, they can strive to obtain projects that impose less risk through careful development and implementation of a business plan. In some cases, the insurance savings derived from altering the project mix may be more than sufficient to cover the cost of the effort required to make the change.

Subcontracting

Subcontracting usually can lower a firm's insurance costs, much as reduction of gross fees. Insurers' attitudes may vary in this respect (as well as others, of course): Some may treat subcontracted work as if the work were paid for directly by the owner, through a multiple contract arrangement; others may not. In either case, however, a firm will be exposed to "vicarious liability," that is, the liability associated with the subcontractor's work.

Claims Experience

A firm's claims experience usually is based on the insurer's claims experience with that firm. If the firm is applying for coverage for the first time, therefore, its management practices may be weighted more heavily than otherwise. If a firm has been with the same insurer for several years, however, and has never had a claim against it, its premium will be lower than an identical firm's which has experienced several claims in the same period of time.

Management

The quality of a firm's management relative to liability loss prevention is established by an underwriter's "judgment call," often with input from the insurance agent. Firms that are particularly well managed, that prepare effective contracts, that strive to execute their commissions in a professional manner, and so on, are well advised to bring such efforts to an insurer's attention.

Premium Credits

The DPIC Companies, a major professional liability insurer, pioneered the concept of encouraging loss-prevention-oriented practices by providing premium credits. For example, firms that regularly use limitations of liability in their contracts may earn annual premium savings of as much as 20 percent. A 10-percent credit is offered for participation in general loss prevention education programs sponsored by DPIC. CNA, another major professional liability insurer, offers similar programs, but not for use of limitation of liability.

Overall Impact

Table 6-1, furnished by the DPIC Companies, indicates the extent to which some of the different factors mentioned can affect insurance premiums,

TABLE 6-1: The Impact of Discretionary Variables on Professional Liability Insurance Premiums (Shown Cumulatively)*

	Firm A (Best Case)	Firm B (Worst Case)
Basic Premium	$13,960	$13,960
Subcontracting		
Subcontracts 25%	$12,564	-----------
Subcontracts 0%	-----------	$13,960
Project Mix		
Low Hazard	$ 9,423	-----------
High Hazard	-----------	$17,450
Experience		
No Claims	$ 8,481	-----------
Above-Average Claims	-----------	$19,195
Management		
Excellent	$ 6,361	-----------
Needs Work	-----------	$23,994
Liability Loss Prevention Education Program		
Participates	$ 5,725	-----------
Does Not Participate	-----------	$23,994
ACEC Peer Review Program		
Reviewed	$ 5,438	-----------
Unreviewed	-----------	$23,994
Limitation of Liability		
Use	$ 4,351	-----------
Does Not Use	-----------	$23,994
Total Premium*	$ 4,351	$23,994

*Assumes two firms practicing the same disciplines in the same area and earning the same gross fees.

assuming two firms that practice the same discipline in the same area and earn the same amount of gross fees. As can be seen, for the hypothetical case involved, the premiums paid could vary by more than 450 percent, on a best-case versus worst-case basis.

WHEN A CLAIM IS FILED

It is regrettable that words alone are insufficient to convey the heartache that design professionals experience when claims are filed against them, because it forces all too many to learn the hard way. Those who have done so can readily explain the fallacy of regarding professional liability insurance as a safety net. By the time that net is reached, extraordinary damage can be done.

In some cases, an engineer or an architect will expect a claim. In those instances, it is prudent to react to the first symptoms, often by contacting the insurance agent. In many instances, however, the claim will be completely unanticipated. As a consequence, research must be performed quickly. Since everyone on staff will usually have other priorities to attend to, performing the research creates scheduling nightmares. People have to be pulled off jobs to review files, ascertain the whereabouts of former employees, attend staff meetings, and otherwise attempt to establish facts. By the time many get back to their work, they are preoccupied with the claim, tired, and unable to concentrate. In short, claims often force people to carry on under precisely the type of conditions that make errors and omissions most likely. This necessitates—or should necessitate—more double-checking than usual, increasing costs and often creating delays, irritating clients and other design team members, and creating new pressures for all involved.

Once the facts are established, meetings must be conducted with an attorney, usually one selected by the insurer. If the firm's principals do not like the attorney, time must be spent finding another one, assuming the insurer concurs. (In some instances, a question of coverage may arise, and it could be necessary to hire a second attorney to help ensure that coverage is obtained. It is not even outside the realm of possibility that you might have to sue the insurer in order to force coverage.) Experts also have to be retained, and both the attorneys and the experts must be thoroughly briefed.

As the case moves forward and more becomes known, a variety of emotions often come to the fore: anger (sometimes hatred), self-doubt, and guilt. Some professionals can be depressed for weeks or months, with their malaise being aggravated still more by the incessant stream of bills from attorneys, experts, testing laboratories, and others.

By the time initial reactions finally cool down and the pace returns to a semblance of normalcy, interrogatories will arrive, along with subpoenas for records and depositions. Extensive amounts of time must be spent, and all the emotions that had finally been walled up break through once again.

The value of the time and productivity lost in these proceedings can be extraordinary, often amounting to $50,000 or more, and none of it is recapturable. Out-of-pocket expenses can also rise to that level or higher, depending on the services required and the professional liability policy's stipulations. Of course, in some cases, the professional liability policy may not apply at all. The negligence alleged may actually arise from a contractual liability, or the event in question may have occurred outside the policy's retroactive period. In such instances, emotionalism can reach a fever pitch, disrupting one's home life as well as one's business life, a potential outcome even when insurance is in force.

When it comes right down to it, as those with experience will aver, professional liability insurance does little if anything to prevent or even to ease the emotional trauma that claims can evoke, and it is wholly unable to restore lost time, lost client goodwill, and lost employee loyalty and confidence. In fact, most engineers and architects would be far better off practicing their professions as though they were uninsured, because—when it comes to what often are the most damaging losses of all—they really have no coverage.

7
Professional Risk Management

Risk is something we all face every day. Most risks do not materialize because we know what they are and take appropriate measures to avoid them. For example, prudent drivers know it is not enough to obey traffic signals; they also have to drive defensively, to avoid accidents caused by those who are less alert.

This chapter examines general approaches to risk management in professional practice, as well as a variety of specific procedures learned from others' hindsight, as reported in case histories of typical loss experiences. In reviewing specific methods, do not lose sight of the forest for the trees.

The most effective risk preventive is a professional approach to practice. This is not limited to professionalism in execution of technical efforts. In fact, many claims arise because of misunderstandings or unrealistic expectations, problems that often can be avoided through better professional-professional and client-professional communication from the outset of a project. As such, if *all* of a firm's technical and managerial responsibilities are approached and executed in a professional manner, the likelihood of problems—risks materializing—will be greatly reduced. In fact, as ASFE experience has documented, the most effective professional liability loss prevention activities are those that enhance the quality of professional practice. Be that as it may, some design professionals seemingly have adopted the attitude that laws and lawyers have made them victims, and little if anything can be done to change this situation. Undeniably, engineers, architects, and other professionals have become targets in recent years. Nonetheless, their victimization in large measure is something of their own doing. Their management practices are, in many cases, antiquated or barely existent. Many accept clients who are known to sue at "the drop of a hat" or who have gained a reputation for slow payments or no payments at all. Instead of formal contracts, subconsultants and even primes rely on letter agreements that state little more than, "We will do the work." Too often, projects proceed with something far less than a complete service and with so little coordination that problems during construction are virtually inevitable. And this list does not even begin to scratch the surface.

It does not have to be this way, as numerous successful, professionally operated firms have demonstrated. And a key to turning things around is becoming aware of the most prevalent types of risks, evaluating prospective engagements with these risks in mind, and creating an effective contract. Even the best professional-services agreement is not enough to manage risk altogether, however, and design professionals must recognize this situation as a fact of professional life; one that encourages them to pursue professionalism in all their activities and to respond quickly to symptoms that are known to signal "problems ahead."

DEALING WITH RISK IN GENERAL

There are two general approaches to dealing with risk, as already discussed: transferring it or retaining it.

Risk Transfer

Risk can be transferred to some other party through insurance or indemnification. Historically, design professionals have put too much faith in risk transfer, leaving them vulnerable to exposures they assumed were others' responsibilities.

Insurance Professional liability insurance is the most common risk transfer mechanism. As noted in the previous chapter, however, its coverage is far from absolute. Many exposures are not covered at all; those that are covered are covered only in part. Design professionals must contend with a policy's deductible on their own; some of the most serious losses—those associated with attitudes and time expenditures—likewise must be absorbed. As such, even under the best of circumstances, insurance affords only a partial transfer of risk. To think otherwise is to disregard reality.

Indemnifications Risk may also be transferred through complete or partial indemnifications, through which the client accepts certain risks, particularly those that are created by the nature of the project and that engineers or architects are powerless to control.

Indemnifications take one of two forms: full or partial. A partial indemnification typically comprises a limitation of liability, through which certain risks are shared. In the event that a risk materializes, the expense involved is allocated as per a contract's general conditions. As fair and equitable as risk allocation procedures may be, some clients simply will not accept them or, at least, those aspects of them that involve capping a design professional's liability. This is particularly the case when public-sector clients are involved. Public agencies and private-sector clients alike are even more reluctant about granting full indemnifications. As such, irrespective of the quality of protection afforded by full or partial indemnifications, they can be difficult to obtain, especially so because clients often seek to transfer their risks to design professionals, also by relying on indemnifications.

When indemnifications are obtained, it is appropriate to consider what might occur, should they have to be exercised. Especially when there is a potential for a major loss, some clients will contest the legality of the indemnification. While limitations of liability have generally withstood such tests, judicial decisions indicate they will be "strictly construed," that is, they will be applied in as narrow a manner as possible.

Courts are even more averse to enforcing full indemnifications, necessitating the use of language that makes absolutely clear what is, and is not, being indemnified. Even when such clarity is established, however, a question may arise as to the indemnifying party's ability to "make good." A client, such as a corporate entity, could dissolve to avoid a loss or could declare bankruptcy as a consequence of a potential loss, or any one of many other circumstances could materialize to prevent or impair an indemnification from being honored.

In short, indemnifications, like insurance, are imperfect risk transfer mechanisms. As such, commissions should be pursued as if indemnifications and insurance did not exist at all, because, in reality, they may be of little or no help when it comes time to seek their protective relief.

Risk Retention

It is fundamentally impossible to transfer all risks, even those that are known to exist, by virtue of the legal system that governs us. Design professionals are subject to a variety of claims they can do little to prevent. However, steps can be taken to soften the impact of these and other claims, should they arise.

One of the general approaches to lessening the impact of claims is establishing loss reserves by including a given amount of risk funding in fees. Loss reserves can prevent a firm from becoming crippled by a claim and can lessen the anxiety some engineers and architects experience when they consider the potential for claims.

Establishing loss reserves may be easier said than done. The need to submit a competitive fee can encourage design professionals to eliminate risk-funding allowances from their proposals or bids. And, if reserves can be established, they may be used for purposes other than those intended. Even when they are available, however, loss reserves generally tend to be limited, and, as with any other source of deep pockets, they can do little about reimbursing a firm for lost time and are next to useless in preventing the emotional upset claims can create. In short, while having loss reserves can be of great value, do not overrely on them. They are not a substitute for other measures or the proper attitude.

Note, too, that the risks to which design professionals are exposed go far beyond those that emanate from the commissions they accept. These risks also include those common to all other business entities, such as claims filed by disgruntled personnel or those stemming from alleged violations of federal, state, or local regulations. These potential problems, as almost all others, can be prevented from materializing through effective risk management, especially that form of risk management that is derived from more professional execution of business and professional responsibilities. In essence, claims filed against design professionals allege that they somehow failed to fulfill responsibilities that are theirs by virtue of agreement, custom, or law. Professionalism requires engineers and architects to be fully acquainted with their responsibilities and to execute them properly—procedures tantamount to understanding and abiding by the "rules of the road." Professionalism also requires "defensive driving," that is, the additional steps, such as effective communication, that can be implemented to help minimize the problems that can occur when other parties are not as astute.

The following discussion identifies some of the many additional steps that can be taken with specific regard to risks associated with execution of commissions.

SELECT CLIENTS AND PROJECTS WITH CARE

Well-managed organizations operate with a business plan that identifies the types of clients and projects that should be pursued in order to achieve business and professional goals. Risk factors should be considered closely when developing a business plan, given that certain types of clients and projects are far more risk-prone than others. This is not to say that firms must concentrate their business activities among clients and projects that present minimum risks. It does mean that engineers and architects should be fully familiar with the types of risks involved, to help assure that their plans call for prudent procedures to help manage those risks. This is particularly the case when a plan calls for penetration of markets new to the firm, because they present substantial opportunities. For example, the hazardous-materials remediation market is one that affords tremendous potential for many firms, but relatively few are equipped to become involved without internal expansion, that is, the addition of qualified staff who are familiar with the technical "ins and outs" of the work, as well as compliance with various regulations affecting personnel safety, the specialized contract provisions needed, and so on. Bear in mind, too, that risk is relative. Dealing effectively with clients and/or projects that are inherently risk-prone may present far less risk than a less concerned approach to clients and/or projects that traditionally pose low levels of risk.

Whether or not a firm is operating through a business plan, it is a basic rule of professional practice that a given type of client or project should not be pursued unless the firm is fully familiar with the inherent risks involved, permitting application of those project-specific risk-reduction techniques associated with contract formation. Workscopes, general conditions, schedules, and fees are not the only concerns requiring consideration, however. If the technical or other requirements imposed by a project exceed a firm's competence, the firm should not pursue the client or project type involved until it somehow obtains the competence needed to afford professional execution of the work.

No project is immune to risk. Even relatively routine commissions can impose severe losses, often because they are approached in too casual a manner. Accordingly, every project should be approached with a risk-oriented outlook. Assuming you are aware of the inherent risks associated with the nature of the work, your specific risk assessment should begin with an evaluation of your client and, when interprofessional work is involved, the project's owner as well.

The first question to ask is: Who is the client? If the individual or individuals are already well known to you, people you have worked with before, the question is answered. Note, however, that a person's behavior is influenced by far more than inherent traits. Developers who have been quality-oriented in the past may simply not have the funds or time required to implement their preferences on their next project. Similarly, the prime professionals you work with must at times bend their preferences to meet the specific demands of a given project or owner. Organizations are likewise subject to such variability, especially due to the different outlooks of those who represent them. The methods that have been used by company A in the past may be ignored altogether when a different project representative is involved or even when the same project representative reports to a different company CEO or supervisor. Do not assume that you are intimately familiar with how a specific client or individual will act on a specific project; circumstances that prevailed in the past may no longer be in effect. To obtain a full answer to your questions, ask them in a direct manner. Given prior relationships, accurate answers should be forthcoming. *Do not rely on assumptions.*

When the client is someone you have not worked with before, determine how the person got your name. This gives you an opportunity to thank the referral source and to evaluate the marketing communications techniques you are using, or could be using, to help assure more business through the same channels. Speaking with the referral source also gives you the opportunity to learn more about the client. Has the referral source worked with the client before? If so, will the source be working with the client on this project? If not, why not? What are some of the other projects with which the client has been involved? Who are some of the other design professionals that the client has retained? A surprisingly large number of design professionals seem reluctant to ask these questions, almost as though they are afraid of learning that a project or a client really is too good to be true. While it may be good advice not to look a gift horse in the mouth, realize that a client is not a gift. You will be required to extend a great deal of credit to the client and to accept significant risks as a condition of service. And, as anyone involved with horses will tell you, you should never let any horse into the barn without first having it thoroughly tested.

If you are working on an interprofessional basis, the same precepts hold true. If your client is working for an owner by whom the firm has not previously been engaged, has your client performed any research into the owner's background? If not, the client should be encouraged to do so, or both of you should do so together.

Assuming your research reveals nothing to suggest you should avoid a given client and/or owner, you should obtain credit information from the client directly. In fact, private-sector clients have expressed a great deal of surprise that the prime design professionals with whom they deal seldom ask for financial information and references. Design professionals report that, when they do request such information, those they ask it of often are pleased, saying in essence, "It's somewhat reassuring that you apparently do know something about business." Many engineers and architects must provide much more information when taking out a $1,500 loan than they seek when extending $15,000 in credit.

Once you believe the client's commission should be accepted, you should propose an agreement adequate to support professional execution of the work. The process of contract formation will then yield important information about specific client attitudes. If one of these is the willingness to lower quality in order to reduce fees, proceed with caution; lowered quality always results in heightened risk. Be particularly sensitive to statements that imply, "If you treat us right on the fee this time, there will be a lot more work in the future." Clients who offer this "come-on" will more often than not find another firm "next time," to whom they will make the identical "pitch."

INSIST ON AN ADEQUATE FEE

Human nature is such that, with few exceptions, all of the work proposed, even if foreshortened in order to offer a lower fee, will not be pursued as it should be if it will mean a financial shortfall. When it is recognized that money will be lost, corners are cut and, as a consequence, major risks are created. All too often the corner-cutting occurs after the initial design work has been completed or nearly completed, resulting in less time being expended for activities such as internal peer plan review, shop drawing review, or construction monitoring. As such, you should strive to make your fee inviolate. If it must be lowered, a quid pro quo should be forthcoming. This could take the form of a reduced workscope and/or the client's assumption of greater risk, through a full or partial indemnification. In some cases, however, due to the nature of the client, the fee may have been inflated to begin with, to create "room" for negotiation. Fee inflation for purposes of negotiation is a commercial tactic. As a professional, you are expected to serve as your client's trusted advisor. If you are willing to lower your fee without asking for anything in return, you obviously were not asking for your "best" fee, suggesting that you were attempting to take advantage of the client. An astute client will be sensitive to this, and, as a consequence, fee inflation may result in your not being able to obtain the commission or might destroy any hope of establishing a relationship founded on mutual trust. This does not mean you should refrain from charging particularly good clients a lower rate if you want to; it does mean that the lower rate should be requested at the outset.

Some architects and engineers consider it reasonable to offer a lower fee to new clients who are in a position to offer legitimate opportunities for future assignments. Lack of familiarity with a client's methods almost always necessitates the investment of more time than customary, however, so a customary fee often represents a smaller profit margin. When a long-time client requests a lower fee, be cautious. If you must oblige, be prepared to lose money on the project before you cut quality. The money lost by charging too little will almost invariably be far less than what you could lose by doing too little.

In all cases, *avoid undue optimism*. Always be prepared to say no to a client when an appropriate balance cannot be obtained between workscope, general conditions, fee, and schedule. Turning down a project has saved many firms from disaster.

PROVIDE QUALITY CONTROL

As already discussed, the quality control inherent in professional execution of a commission comprises a vital aspect of any project. Unquestionably, techniques applied to prevent errors or omissions, or to quickly counteract those that emerge, are among the most important of all risk management procedures. Flawless work does not prevent claims of negligence, however, nor does it prevent those claims from going undefended or even unrewarded. As such, while it is essential to implement an appropriate degree of quality control, reliance on quality control as the sole risk management method would be an error. Claims arise from more than technical shortfalls; nontechnical risks must therefore be managed as well.

APPLY REALISTIC ASSIGNMENT AND SCHEDULING PROCEDURES

Every effort should be made to assign specific tasks to those most capable of performing them well. When circumstances prevent this outcome, the person who receives the assignment should work closely with a more experienced individual, to obtain satisfactory task fulfillment and to enhance the capabilities of the less experienced employee. Case histories show that, in order to offer a lower fee or to "cut losses," some firms require less experienced personnel to perform without oversight or assign experienced individuals who are instructed to "cut corners." As these case histories demonstrate, either approach can comprise an act of utter foolishness.

Insofar as scheduling is concerned, design professionals are often under pressure to "do it yesterday." Although unrealistic scheduling requests can be resisted somewhat through contract wording that dilutes promises, a client still may expect an impossibly fast turnaround. Displeased clients will sometimes search for, and find, almost any excuse to justify a claim. To avoid this possibility, be candid with your clients. Let them know how much time it will take to perform work well. *Do not agree to deadlines you know you cannot meet.* The delays you create will at best frustrate clients and dash their expectations; they may also lead to significant financial losses that you may have to cover and whose size may be many times greater than the profit to be derived from accepting the work. Inability to meet deadlines can also give your firm an unfortunate reputation, and, when obtaining a commission is in part dependent on how well you have satisfied other clients, being late on one project could mean the loss of considerable future business.

If your firm is busy, but you nonetheless decide to accept additional work, be prepared to provide whatever it takes to have the work performed well, on time. This could necessitate the hiring of additional personnel and reassignment of certain existing staff members to supervisory positions, or even reliance on other firms to perform certain elements of the work, with additional quality control being performed by your own staff. Either of these approaches creates certain risks. So, too, does reliance on overtime, since longer hours result in lower productivity and the greater likelihood of errors and omissions. As a general rule, then, strive to set realistic deadlines and manage the work force in such a way that deadlines can be met. Do not sacrifice quality in order to accelerate performance. Remember: Someone who wants something in the worst way gets it just that way.

REACT QUICKLY TO SYMPTOMS OF PROBLEMS

Problems seldom just happen; they evolve. In the process, subtle signals will be given that everything is not working as smoothly as desired. Effective risk management requires comprehensive staff education to help assure these symptoms are recognized and reported promptly.

Everyone Has a Role

To establish an effective risk management program, *all* firm employees must be educated about certain basics. Key among these is a "what-the-heck-we're-insured" attitude, which has no place in a professional organization. It leads to claims; claims affect a firm's financial performance, and financial performance affects every member of the staff. The firm may be unable to obtain the equipment it planned to acquire; it may have to cut back on bonuses, if not eliminate them altogether; or it may have to forgo contributions to retirement plans. In other words, if only for selfish reasons, *everyone* has a vested interest in loss prevention. As such, *everyone* should receive and follow loss prevention instruction.

Although some of the general motivational guidance given will be the same for all personnel, most of the specific loss prevention instruction will be tailored to the nature of the work that people perform. Since many staff members perform a variety of tasks, it is necessary to identify each of the tasks. For example, most people answer a firm's telephone at one time or another, so all should be familiar with proper telephone etiquette. Why? Consider what can happen when improper telephone techniques are used, as when a client calls to inform you of a minor problem but is put on hold and forgotten. By the time you are able to get back to the client, the client is angry. This affects the client's view of the problem's severity and the demands imposed relative to action you should take and when. Thus, in answering the phone, as in countless other activities, employees should be told how to do things properly and *why* they should be done properly.

Documentation

All staff members need to appreciate the importance of documentation, because the existence of effective documentation can often result in claims being dropped or in not being made at all. This often occurs because the basis of a claim frequently is, "We never told you to do that," or, "You said something else at the time."

As a matter of routine procedure, notes should be taken whenever any member of your staff speaks with a client or anyone else connected with a project. Each note should indicate who was involved, when the conversation took place, and the substance of the conversation. As appropriate, notes should be turned into confirming letters or memos, and these should be sent to all affected parties: those involved in the conversation, others inside the firm, and persons outside the firm. Such techniques not only help create a strong defense posture but also help prevent exactly the kind of misunderstandings or communication breakdowns that can lead to problems and claims. The file copy of each memo or letter should also indicate to whom copies and blind copies were sent and when.

Feedback Network

The ability to react quickly to early symptoms of problems also relies on establishment of a feedback network, a need closely tied to documentation. For example, when someone involved with a project calls to speak with you when you are not in and seems angry, you should be fully informed. The telephone call memo (documentation) you receive should indicate that the caller seemed upset, so you can be sure to get back to the caller. If you are away for several days, you should make sure that someone else will receive the note, or a copy of it, so an immediate inquiry can be made. Likewise, if something seems "fishy" on the jobsite, whoever has that feeling should know to report it to superiors, so the situation can be evaluated. If a problem is found, the firm must be protected. But protection does not necessarily imply action based solely on self-interest, assuming one can label self-defense as self-interest. Rather, it often means action taken principally in the interest of the project. For example, the problem may include a situation over which your firm has no control. If the client does not do anything about the problem after you document its existence, little more can be done. In other cases, however, more must be done, including notification of authorities on a confidential or anonymous basis.

Typical Symptoms

Here follows an annotated listing of some of the typical indications that latent problems may exist. Appropriate staff members should be trained to recognize and deal with them effectively.

Report Followed by Silence Geotechnical and geoenvironmental engineers customarily issue reports of findings and recommendations on most projects they accept. Other design professionals issue similar reports, as when the project involves an existing structure. It usually is assumed that, within a period of time after the report has been issued, the client will get back to you for the next stage of work. For purposes of marketing alone, the client should be contacted within a certain period of time to determine the project's status. Loss prevention also dictates a call. All too often a client will not like what a report states and will select another firm to take the project forward, often by cutting corners to minimize costs. This creates more risk and thus a greater likelihood that something will go wrong. If it does, the firm issuing the original report will probably be sued.

Effective loss prevention also dictates that reports should include wording that affords protection in the event the submitting firm is "disengaged" after the report has been submitted, despite a contract that calls for "follow-on" services. Accordingly, if a project will be delayed or if there is a possibility that another firm has been, or will be, called in, the report should be reviewed immediately. If protective wording was not included or if something said could be misinterpreted or misapplied, the client should be advised at once. The client should be told that the report was prepared based on certain assumptions and that, since these assumptions are now invalid, certain precautions should be observed.

The key, of course, is follow-up. If you do not hear from the client within a reasonable period of time, call. If the client does not return the call, assume the project has been given to someone else. This may be a faulty assumption, of course, but it is always best to err on the side of safety. Likewise, it is always best to consider various eventualities in preparing and transmitting a report, including whatever assumptions are behind it.

Recommendations Are Not Implemented During the course of a project, or even at its outset, you may be in a position to make certain recommendations about effective procedures. Whenever your recommendations are not followed, determine why; do not guess, and do not be afraid to make a few waves. For example, assume you are the electrical engineer and have recommended or even specified installation of high-pressure sodium lamps of a given wattage, in order to achieve a certain maintained footcandle level in a parking lot. During a visit to the site, you notice that incandescent fixtures and lamps are being installed. You know they will be costly to maintain well and that the necessary maintenance probably will not be provided. As a result, the footcandle level necessary to help assure safety will probably not be achieved, increasing the risk of accidents or assaults and a third-party claim against your firm. Your client is the architect. You call and ask, "Why was the substitution made?" The architect has no clue, because the substitution was approved by the construction manager. With the architect's permission, you then write a letter to the owner, pointing out potential problems. This should be done in a wholly noninflammatory, matter-of-fact manner, noting that the recommendation was not followed and identifying potential problems that could result. Concerned owners should appreciate being told about such changes, because they affect their liability, too. As such, your letter could lead to enhanced safety. If that does not occur, however, and if problems later do result, your letter could serve as a powerful defense.

Risks Accepted to Cut Costs As soon as it becomes apparent that an owner or a client is willing to accept more risk as a means to cut costs, design professionals should become highly circumspect about everything they have done, are doing, and will be doing on the project. Heightened risks make problems more likely, and any problem is likely to result in claims against design professionals. Alert others to potential problems; use letters, memoranda, and other written means to provide solid advice, for the benefit of the project, the client, and yourself.

Problems Passed Along Some clients, including some prime design professional clients, adopt an attitude that a problem or a potential problem is nothing to worry about so long as it can be passed along to someone else. This may be another design professional ("Let him worry about interpreting your report"), a contractor, or the people to whom a building will be sold. Remember: As a design professional, you are responsible to more than just your client.

Advise your client and other appropriate parties of problems that could occur and preventive measures that should be applied. If need be, inform those to whom problems may be passed, so they can be on alert. Do not allow someone else's foolishness or greed to create a risk exposure for you.

Accelerated Scheduling   When it becomes necessary to work fast, recall the axiom, "haste makes waste." If the need for an accelerated pace results from poor scheduling on your part and if a delay is not acceptable, every effort should be made to comply with requirements without taking shortcuts. If this will mean less profit, or even a loss, so be it. Whoever set the schedule, or whoever was charged with meeting it, will learn a valuable lesson: The client will remain satisfied, and risks will be reduced.

When clients demand a foreshortened schedule after another has been agreed to, they should be willing to pay the additional costs associated with meeting it. It usually is wise to determine why the new deadline is needed. If clients are under some type of new constraint requiring them to move faster, chances are someone will somehow create a problem that would have otherwise been avoided. Stay on your toes.

Highly Unusual Unanticipated Conditions   Geotechnical and geoenvironmental engineers are those most likely to encounter highly unusual unanticipated conditions, but others are not immune. Clients do not like learning about these, but they will be even more upset to learn about them after a delay. The client should be immediately informed about unanticipated conditions when they are discovered, so remedial measures can be taken quickly. But be cautious: When highly unusual unanticipated conditions will result in a significant unanticipated expense, an effort may be made to cut corners elsewhere. Cutting corners often is a shortcut to problems.

Unqualified Contractor Is Selected   Many of the third-party claims lodged against design professionals result from errors made by contractors. Design professionals often are sued because there is no way to collect from contractors or to collect from them sufficiently. Design professionals also are subject to suits from contractors who allege that their own errors or omissions were somehow caused by a design professional.

Unqualified or unscrupulous contractors are able to obtain work by being the lowest bidders. Very often their bids are lowest because they have failed to create an adequate allowance for certain contingencies. This frequently occurs when contractors are not from the area where construction will occur and thus fail to account for a long rainy season, the need for imported fill, certain local code requirements, and similar factors.

As already discussed, techniques are available to assess owners' attitudes in these matters before work commences. If the retention of an unqualified firm is anticipated, the chances of problems are high. Appropriate defensive measures must be taken, and these must be pursued in such a way as to minimize any charges of libel or defamation. If your contract does not include field observation, it is essential that it does include recognition of the problems that feasibly could occur when observation is not provided. If your contract does include monitoring, but that element of the work is terminated, it is essential that you inform the client in writing of the impact on your responsibilities. While you naturally do not wish to offend clients, you should not accept the risk imposed by their own unwise actions.

Problems with Client's Field Representative It is not uncommon for disagreements to arise between a design professional's field personnel and those of the client. Your field personnel should be aware of this, and they should also be instructed to use diplomacy in order to establish rapport. Failing that, a client's field representative may attempt to stir up problems, as by exaggerating minor ones. If such negative relations do arise, the best approach may be to replace your field representative, assuming the situation has been caused by something along the lines of a personality clash. Sometimes much more is at work. For whatever reason—to reduce expenses or to shed less light on cover-ups—your field representatives may be given less work, may be told to be someplace else when critically important work is being performed, and so on. Your field representatives should be instructed to inform you of any such incidents at once, so that they can be reported promptly to your client. The same applies when the client's field representative is a foolish person with many years' experience, who permits work to be performed without adequate monitoring or who approves alternates without adequate consultation or review.

Problems Disappear In some cases, projects begun by particularly unscrupulous contractors will be plagued by problems early on, and then, almost overnight, the problems will vanish. Does this mean such contractors have decided that the aggravation is not worth it and that they should simply "toe the line"? Possibly. But in some cases it could also mean that someone is being paid to look the other way, and that "someone" could be one of your employees. When a situation is too good to be true, it often is. In such cases, therefore, it would be effective to perform a quality control "spot check." If it seems as though a cover-up is ongoing, do whatever you possibly can to determine the truth and to double-check everything previously approved on site.

Client Accepts Marginal Work Some owners are willing to accept marginal or even subpar work as long as it results in their projects opening on time. Although such owners may have excellent reasons for doing this, it does not mean they (or subsequent owners) will have a positive attitude when it comes time to effect repairs later. When clients do not require contractors to correct problems, it may be necessary to write a memo or a letter acknowledging the discussion in which the decision was made known to you. If the marginal work results in code violations, stronger action may be necessary.

Client Loses Interest In some cases, the head of a client or owner organization will be actively involved in a project but will then be reassigned or, for whatever reason, will lose interest, necessitating that someone else within the firm take over. When this occurs, it usually is prudent to meet with the new person in charge, to review the project's background, current concerns, and related issues. Even if such a meeting is held, however, people should be prepared for changes, as well as less attention from the top—something that typically occurs when those assigned to the project already have their hands full.

Delays When a project is subject to delays, during any stage, attempt to determine the cause. Frequently, it is the result of something going wrong in the area of financial resources. When this occurs, it may be necessary to cut corners or otherwise take action that deteriorates quality. Even when the cause of a delay seems somewhat benign, be aware that delays often result in time having to be made up later or in corners having to be cut, in order to compensate for lost time or increased costs.

Silence Instead of Dissatisfaction In some cases, you may be prepared to hear from an angry client, but, instead, you hear nothing at all. This may occur after construction bids have come in significantly higher than anticipated. The silence could mean the client was prepared for the significant variance and is not bothered by it. The odds of this happening are remote, however, as are the odds of the client not voicing displeasure over anything else you assume would cause upset.

When you are the one who will be in the position of delivering the bad news, be sure you understand why the undesirable situation occurred. If you discover that it occurred because of something you did, speak with your insurer or agent, possibly also with peers or others, to obtain suggestions on how to soften the blow, even if contract wording may protect you.

If the news comes directly to the client from another source and you learn about it, do not wait for the client to contact you. Client silence could mean that a suit against you already is being prepared. Express your concern. Personal diplomacy on your part could clear up a costly misunderstanding or, at least, mitigate the loss. Do not under any circumstances remain on the sidelines hoping the problem will go away. It won't, and a head-in-the-sand attitude will only make it worse. Note, too, that any client refusal to allow you to look into the problem, to permit rapid response to it, could result in the problem becoming worse quickly. By putting your desire to investigate and assist into writing, you will document your professional attitude and your ability to mitigate damage—something that could prove valuable, should a client seek to recover damages that result from the client's own delayed response to the problem.

Silence After Dissatisfaction Personal diplomacy also is necessary when a client voices dissatisfaction and then is silent. This situation almost always precedes the filing of a suit, so it is essential to maintain open lines of communication. Do not simply assume you have committed an error or omission; do not assume the client does not want to hear from you. Try to meet with the client as soon as possible; do your utmost to convince the client that you will do your best to make good if it is shown you are at fault. Again, putting your sentiments into writing may be of great value later.

Gut Reactions In some circumstances, you or some other member of your firm may have an uneasy feeling about a certain individual or situation. These are instinctual or intuitive reactions that all too often are ignored, because a project should be highly profitable, will look good on your experience record, and so on. *Always obey your gut reactions.* If something seems wrong, chances are something *is* wrong. As a case in point, every one of the above-mentioned symptoms was drawn from actual loss experiences reported by members of ASFE, and, in almost every instance, the professionals reporting the cases related that they felt there was something going wrong but that they failed to react quickly enough, effectively enough, or at all. They simply hoped they were wrong, for any number of reasons that seemed reasonable at the time.

REACTING TO PROBLEMS AND SYMPTOMS

As should be quickly recognized from the preceding discussion, documentation is vitally important, as is recognition of symptoms and reporting them to appropriate parties. By creating an effective contract that anticipates some of these problems, many can be handled in something other than a state of agitation. This is important, because actions taken or statements made while in a state of alarm can come back to haunt you. Always do your best to assume a somewhat dispassionate professional demeanor, even when the problem may have been caused by something you or someone else on your staff did, or did not, do. Deal with the client or other parties in a calm, confident manner; ignore words uttered in the heat of passion. Always do your best to apply personal diplomacy, in hopes of establishing a sense of calm and purpose, and do not give up your efforts until they clearly are of no further use. To the best of your ability, attempt to conduct yourself as if a jury were looking over your shoulder. It very well might have to, and, for this reason, it is essential to document your attitude, so the results of childish, unprofessional behavior do not land on your doorstep. It is not easy to do, and in some cases comprises a fundamental test of your professionalism. Do not get overly enamored of your style, however. Remaining cool in a crisis does not necessarily eliminate the crisis. Understand your own limits. Be quick to call on appropriate assistance, including that available from peers who can come in to investigate and establish valuable findings.

ESTABLISHING A RISK MANAGEMENT PROGRAM

Every firm should have one person who is assigned the task of risk management. In smaller firms, the time required may be such to make risk management a part-time job, something performed along with other responsibilities. In larger firms, however, the risk manager may head up a staff. The work involved may not be called risk management, however. Instead, it might be called quality control. Whatever its name, risk management should include more than technical quality control, because more than technical issues are involved. Risk management implies quality control in *everything* a firm does, including attempts to prevent claims and losses that are not really the fault of the firm.

A firm that proceeds without an effective risk manager has not infused quality into the risk management function. This lapse in professionalism could be damaging, if not fatal.

8

Dispute Resolution

Litigation—the filing of a lawsuit—is the most common form of formal dispute resolution. It is "formal" because litigation proceeds in accordance with a variety of complex rules. The complexity of these rules generally necessitates reliance on attorneys, and implementing the various procedures involved usually takes a great deal of time; often, several years pass between the time a claim is made and the suit is ultimately argued in court. Due to these delays and the costs involved, the vast majority of all disputes entered for litigated resolution are settled between the parties before the case proceeds to trial. The dissatisfaction engendered by this process has resulted in development of alternatives to litigation, known collectively as *alternative dispute resolution*, or *ADR*.

ADR in itself is not new. Arbitration, a formal alternative, was first applied in the construction industry in the 1870s. Arbitration is not well suited for many types of cases, however, and many new types of ADR have been created as options.

This chapter provides an overview of the civil litigation process, as well as details on a variety of ADR procedures. In fact, there are so many ADR options that it now is possible—and practical—to match a dispute resolution procedure to the nature of the dispute and the attitudes of those involved. Design professionals should be familiar with the various forms of dispute resolution, because construction is so prone to disputes and because so many of them somehow involve engineers and architects. Design professionals can serve the projects for which they are engaged, the clients by whom they are retained, and themselves, by encouraging reliance on techniques that will expedite dispute resolution while minimizing the disruptive emotionalism to which lack of accord often leads.

CIVIL LITIGATION: AN OVERVIEW

Civil litigation is an adversarial process; that is, it pits one party against the other. Those who have not previously been involved in litigation may believe that the decision to sue will result in a battle of sorts and that they will somehow come out holding the defendant's scalp... and wallet. But it seldom works out that way. While a battle may rage, it does so in slow motion and is governed by a precise set of rules that involve innumerable checks and balances. Clearing all the hurdles prior to a court date can take years, and it often is very expensive, in terms of both out-of-pocket expenses for attorneys and experts, and the time that disputants and their staffs must spend. Many lawsuits begun in anger are ended two or three years later through an out-of-court settlement, with the disputants themselves exhausted by the process, united in the belief that litigation is a punishment people must endure for being unable to resolve disputes on their own. For this reason, those who have experienced litigation often are advocates of alternative dispute resolution—ADR—and generally have accepted the dictum that it is best to resolve disputes in an unemotional manner whenever possible.

Since 90 percent or so of all civil lawsuits are settled before trial, the question often arises: Why is litigation used so extensively? There are many reasons, one of the most common being failure to agree to an ADR approach by contract and disputants' subsequent unwillingness to rely on an ADR procedure because it is distrusted or because litigation gives them an advantage they prefer not to lose. In fact, the advantage aspect of litigation can be so pronounced that some parties turn to it almost as a matter of course. For example, certain owners are notoriously slow payers. When they ultimately must be sued for collection purposes by a prime, they will routinely allege that the engineer or architect was somehow negligent and then file a counterclaim. The delays associated with litigation then give such owners use of what should be the design professional's money; resolution of the dispute often involves the design professional accepting a smaller fee, as part of a compromise, two or three years after the project was completed.

Engineers and architects are also sued with some frequency by low-bidding contractors who lose money on a project and then claim the loss was somehow caused by the design professional's negligence. Even though no negligence may have been committed, a massive amount of money will be claimed. A contractor will often call for a jury trial in such cases, realizing that the jury—the trier of fact—is not technically astute, will probably be somewhat confused by all the evidence presented, and will assume that, since huge damages are alleged, some damage must have occurred. In these and other cases, design professionals, as other defendants, are likely to settle the dispute for a payment that is less than what would be required to demonstrate no payment is due at all.

The fact that litigation is subject to abuse does not mean that it is always abused. Nor should it be concluded that owners or contractors are inherently devious or unethical. In fact, many, if not most, of those experienced with construction tend to regard litigation as a last resort. A review of the steps involved, given below, tends to explain why.

Pleadings Stage

Construction disputes between those involved on the same project typically arise when one party claims that the action or inaction of one or more other parties has caused a monetary loss. The party allegedly suffering the loss will often attempt to obtain compensation through informal means, by discussing the issue with those purportedly at fault. When discussions break down, litigation is initiated at the pleadings stage. In some instances, however, no attempt is made to achieve resolution through informal means, and the injured, damaged, or aggrieved party will go directly to the pleadings stage.

The pleadings stage begins when the party who claims to have been damaged—the *plaintiff*—files a *complaint* with the court having jurisdiction. Also known as *statement of claim, petitions,* or *declarations*, complaints identify plaintiffs' allegations against other parties (*defendants*) and the nature of the *relief* sought, usually in the form of monetary compensation for the damages or injury incurred.

Once complaints are filed, defendants receive a *summons and complaint*, usually delivered in person by marshals, sheriffs, constables, or other "process servers." The summons identifies defendants and the actions being taken against them, the name of the court, the name of the plaintiff, and the name and adddress of the plaintiff's legal counsel. Defendants typically are given about twenty days to respond, in order to avoid having a *default judgment* entered against them' that is, failure to respond to the complaint would cause the judge to rule in favor of the plaintiff. Accordingly, defendants usually retain an attorney and make an *appearance before the court*, that is, submit a formal notice that the summons has been received. Defendants and their counsel then meet to determine the best course of action. In doing so, they consider a number of issues, such as the legitimacy of the complaint, its real value, the plaintiff's financial resources and ability to support a lengthy trial, and the reputation of plaintiff's counsel.

In many cases, attorneys for the plaintiff and defendant will at this point attempt to negotiate settlement. If unsuccessful at first, they will likely repeat the process numerous times during the procedure, as positions are altered by time and the course of events. When they are able to negotiate an agreement, they will reduce it to writing, then file the document, which is sometimes referred to as a *stipulation*, with the clerk of the court. The case is then recorded as closed.

As an alternative to, or in conjunction with, efforts to settle the case, defendants can enter a number of motions to test the substance of plaintiffs' claims. One of the most common of these is the *motion to dismiss*, also known as a *demurrer*. Through this procedure, the defendant's attorney argues that the court should dismiss the complaint because the plaintiff does not have a legal right for a favorable judgment, even if all of the plaintiff's allegations were true. Other motions can be made to challenge the court's jurisdiction over the claim or the defendant, or of the formal (as opposed to legal) sufficiency of the claim. For example, if the claim is vague or ambiguous, defendant's counsel may move to require the plaintiff to file a more definitive complaint, or defendant's counsel may move to strike from the complaint parts deemed redundant, superfluous (immaterial), or scandalous.

Among other things, these various motions create an early testing of the complaint and attitudes of the parties toward it. Each gambit changes the scenario, affecting positions relative to settlement.

Once the motions have been made and decisions reached, and assuming the complaint is still "alive," defendants must reveal their position with an *answer* to the complaint. This answer may take the form of a *denial* (an *affirmative defense*), a *counterclaim*, or a combination of the two.

An *affirmative defense* indicates that the plaintiff's allegations are substantially true but that explanatory facts have been omitted. For example, the plaintiff may allege that a structural engineer failed to monitor construction well and, as a result, errors in construction were not revealed in a timely fashion, increasing the cost of correction. The defendant might agree but would inform the court that construction monitoring was limited by virtue of a decision of the owner. A plaintiff's contributory negligence would be another example of facts cited in an affirmative defense, as would a claim that the statute of limitations had expired.

A *counterclaim* is based upon facts that may have been asserted by the defendant, had the defendant wished to initiate a suit. The counterclaim may be based on events that have nothing to do with the event that impelled the plaintiff to sue and could involve damages far in excess of those that the plaintiff seeks to recover. Once filed, the counterclaim becomes a *cross-suit* to which the plaintiff must respond, going through the same steps that the defendant went through in answering the plaintiff's complaint. In this instance, however, the plaintiff's answer to the counterclaim is termed a *reply*.

The complaint, answer, and reply (if the answer is in the form of a counterclaim) comprise the *pleadings*, and, once closed, they identify the only issues that can be raised at trial, with the exception of *amendments to the pleadings*, which can be made within certain preestablished parameters.

After pleadings are closed, either party can move for a *judgment on the pleadings*, whereby the court examines the strength and validity of the various claims.

Pretrial Stage

After pleadings are closed, either side files a *notice of trial*, requesting the clerk of the court to put the suit on either the jury or nonjury calendar. Either side can demand and receive a jury trial; in some cases, a judge will decide on a jury trial even if the disputants prefer otherwise.

Sometime prior to the trial, a judge generally will call for a *pretrial con-*

ference or *hearing*. Attorneys for each side then must appear before the judge in chambers to remedy defective pleadings, eliminate superfluous issues and simplify others, agree which documents are genuine, limit the number of expert witnesses, determine the scope of discovery and decide if a *master* should be appointed to obtain information pertinent to the suit. The judge may also order the attorneys or their clients to make pretrial *admissions*, whereby one side admits to the existence of facts that help the other side or that a point being made by the other side is true. This procedure saves considerable time and cost.

A judge can also order both sides to engage in *settlement negotiations* and report back at a specified time. Although the pretrial conference typically is run on an informal basis, attending it is mandatory.

Discovery The process of *discovery* also occurs before trial, to expedite both trial preparation and the trial itself. It requires mutual disclosure of evidence, thus permitting opposing attorneys to narrow the focus of dispute. It also eliminates the needless time and expense that otherwise would be required to obtain certain facts, and it commits witnesses to that version of facts that they relate under oath before trial. In addition, it creates a better basis for attaining an out-of-court settlement, preserves for use at trial evidence that may otherwise change or disappear, and removes the "game element" (i.e., surprise at trial) from litigation. The three principal vehicles through which discovery is pursued are *subpoenas duces tecum*, interrogatories, and depositions.

Subpoenas Duces Tecum A *subpoena duces tecum* is issued by the clerk of the court (or, in some jurisdictions, by the attorney of record) to compel the other party to provide certain documents. Delivered by a process server, it usually is accompanied by an affidavit that identifies more specifically what is sought and why it is material to the case. The opposing attorney may attempt to reduce the number of documents sought either by reaching an agreement with the other attorney or by seeking a protective order from the court. Expert witnesses also are subject to *subpoenas duces tecum*.

Interrogatories Through an interrogatory, one party issues a list of factual questions that the other party must answer under penalties of perjury.

Depositions A deposition is a form of discovery that gives opposing counsel the right to question the other side's witnesses. Each witness is required to take an oath; a court reporter transcribes everything that is said, and the transcript can be used as evidence in court. Expert witnesses often are deposed and frequently work with counsel in developing questions to ask the other side's expert. In some cases, both sides' expert witnesses attend depositions, one to respond to questions and the other to evaluate answers and suggest additional questions. It is not unusual for an expert's deposition to last a full week.

Motion for Summary Judgment A motion for summary judgment often comprises the final pretrial procedure. It is filed by an attorney to obtain a dismissal of the claim or counterclaim, or both. Unlike a motion to dismiss, it alleges that a case has no merit because the party bringing it cannot prove alleged facts are true.

Trial Stage

As the time for trial arrives, attorneys will answer the *calendar call*, and the suit will be assigned a courtroom. If a jury trial is called for, the opposing attorneys and judge conduct *voir dire*, a procedure whereby prospective jurors are questioned to determine their qualifications. If a juror's answer indicates a prejudice, a relationship with one of the parties, a financial interest in the outcome, or any other situation meriting disqualification, attorneys may *challenge for cause* and state their reasons. Attorneys also are given a certain number of *peremptory challenges*, permitting them to excuse a juror without citing a reason.

Once the jury is impaneled (often with alternates), the trial begins. The plaintiff's attorney typically starts by outlining the facts from the plaintiff's point of view, followed by defendant's counsel who sketches the defendant's case. Then the plaintiff calls witnesses, each being bound by the various *rules of evidence*, including the *parol evidence rule*, the *relevancy rule*, the *hearsay evidence rule*, and the *best evidence rule*.

The *parol evidence rule* bars the admission of any evidence related to understandings that differ from those formally agreed to (as via contract) when the formal agreement was established *after* other understandings were alleged to exist.

The *relevancy rule* holds that the only evidence that affects issues under dispute, as outlined in the pleadings, may be used. It is particularly germane when considering *circumstantial evidence*, which comprises indirect proof or disproof of a fact in question. For example, testimony that an individual was seen walking on a muddy site would be direct evidence that the person was there. Testimony that an individual was seen wearing boots caked with the same kind of mud that was present on the site would comprise circumstantial evidence that the individual had walked on the site. However, even relevant evidence can be excluded by virtue of the *hearsay evidence rule*, which precludes the admission of evidence based on what a witness was told by others, for example, that someone else had seen an individual wearing boots caked with the same kind of mud as that which was present on the site.

The *best evidence* rule holds that the best possible form of evidence must be produced at trial. In other words, if a document is evidence, the original (if available) would be used, rather than a photocopy. Such exhibits are subject to *voir dire*.

It also is a fundamental rule of evidence that lay witnesses may testify only to matters of fact and may not express their conclusions. Conclusions are to be made only by the trier of fact; the judge or the jury. By contrast, expert witnesses are permitted to state their opinions and conclusions, and these opinions and conclusions become evidence. Because the testimony of two opposing experts often will differ, however, it is up to the judge or the jury to determine who is right and who is wrong.

As soon as experts take the stand, they are subject to *voir dire*. The attorney who has engaged the expert will seek to have the court formally recognize the individual as an expert, after the expert has recited pertinent credentials. Opposing counsel may at this time challenge the expert's credentials. This seldom is successful, and, in many instances, both opposing counsel and the judge will *stipulate* that the witness is fit to serve as an expert, making even direct examination of credentials unnecessary. (This can be detrimental in those instances in which the expert's credentials are particularly noteworthy, especially in comparison to the opposing expert's.)

Once a factual or expert witness takes the stand, the party for whom the witness is appearing begins questioning through *direct examination*. Under direct examination, the attorney may not ask *leading questions*, that is, questions that show how the witness is to answer or that suggest the preferred answer.

After direct examination, the opposing attorney engages in *cross-examination*, asking questions about matters brought up during direct examination. The purpose of cross-examination is to test the recollection, knowledge, and credibility of a witness. In practice, however, it is applied to refute or at least to cast doubt upon facts attested to during direct examination, by catching the witness in a contradiction or something that seems like a contradiction (*impeaching* the witness), by casting doubt on the witness's character or capabilities, or by causing the witness to say or do something that causes a jury to dislike the individual.

An attorney will engage in *redirect examination* to correct misinterpretations of answers given during cross-examination or if a witness's testimony seemingly has been impeached. When an expert witness is involved, impeachment will sometimes occur when the expert's testimony at deposition contradicts what is said at trial. In such instances, opposing counsel usually will not give the expert an opportunity to explain why the difference exists. To *rehabilitate* an expert's testimony, the lawyer will ask for an explanation during redirect. If the plaintiff engages in redirect, the defense can *recross*, limiting questions to topics covered in redirect.

The opposing attorney can file a number of motions while witnesses are being questioned, including an *objection* to a question, answer, or both. If the judge considers an objection valid, the jury (if one is being used) will be instructed to ignore whatever was objected to.

Once the plaintiff's witnesses have been fully questioned, plaintiff rests. At this point, the defendant can decide either to present a case or to settle. If it is decided to proceed, one or several strategic motions can be made. One of the most common is a *motion for a directed verdict*. This alleges that the plaintiff has failed to make a case. A directed verdict will be granted when a judge believes that all of the evidence presented, even if true, would result in an unbiased jury finding for the defense. If the motion is granted, the plaintiff may not start another case on the same grounds.

If it appears likely to the plaintiff that a motion for a directed verdict would be granted, plaintiff's counsel would react *before* resting plaintiff's case by making a *motion for a voluntary nonsuit*. This would give the plaintiff the right to begin another action on the same grounds after paying court costs. Presumably, the second action would be stronger than the first, because the various gaps that weakened the initial presentation would be filled in. (A motion for a voluntary nonsuit can be made any time before the judge renders a decision or gives the case to a jury.)

Assuming the plaintiff does not move for a voluntary nonsuit and that a motion for a directed verdict is denied, the defense presents its case, following the same procedures applicable to the plaintiff. After presenting the last witness, the defendant may ask for a motion for directed verdict, whether or not it was sought before. If the motion is denied, the plaintiff may offer evidence to *rebut* what was said by the defendant's witnesses, and the defendant may subsequently introduce evidence relative to that introduced during rebuttal.

Once all of the evidence has been heard and both sides rest, either or both can ask for a motion for a directed verdict. If none is granted, the plaintiff presents a *summation*, recapitulating the claims, commenting on the evidence, and stating the legal principles involved. The defense attorney does likewise, with the plaintiff's attorney being given an opportunity for rebuttal.

After the lawyers have spoken, the judge *charges the jury*, reviewing closing arguments, pointing out the most important issues of law, summarizing the testimony and how the jury should evaluate it. The judge usually points out that whoever brought the claim or counterclaim has the *burden of proof*, with proof being based on a *preponderance of the evidence*. Either or both attorneys may also present *requests to charge*, either orally or in writing, outlining special charges for the judge to consider. Either attorney can also object to the charge as given.

After it is charged, the jury goes into the jury room to discuss the evidence. In some jurisdictions, the jury must reach a unanimous agreement in order to render a verdict. In others, a majority agreement is all that is needed. If an agreement cannot be reached, a *hung jury* results, and the case must be retried.

After the jury delivers its verdict, the attorney for the losing side may make a motion for *judgment notwithstanding the verdict*, or *judgment NOV*. This is essentially the same as asking for a directed verdict, except at this point it would require the judge to reverse the jury's decision. A judge may be more inclined to grant judgment NOV rather than a directed verdict, however, due to the appellate laws. In essence, if the judge errs in granting a judgment NOV, no retrial would be necessary; the jury's verdict would be restored.

If the motion for judgment NOV is denied, the losing party may then make a *motion for a new trial*. Such a motion could be granted for reasons similar to those that would result in a favorable ruling on a motion for judgment NOV or for reasons such as excessive damages being awarded, a procedural error on the court's part, or the opposing party's use of surprise that could not have been guarded against.

If the judge's or jury's verdict is not set aside and no new trial is granted, the judge directs entry of a *final judgment* for the successful party.

Posttrial Stage

A case is not necessarily closed after final judgment is entered. Either party—but usually the party who lost—can enter an appeal to an appellate court, because errors were made at trial, the award was excessive, or for other reasons. The *appellant* must notify the clerk of the court where the original trial was held that an appeal is being filed and must notify the *appellee* as well.

Sometimes an appeal is filed principally to delay payment, in that it bars the appellee from enforcing judgment until the appeal is heard and a verdict is rendered. Accordingly, the appellant must post a bond to cover costs and the final judgment, in case the appeal fails and the appellant's assets are dissipated in the interval.

The appellant's next step is to prepare a *record of appeal*. This identifies exactly why the appeal should be granted, citing whatever precedents may apply. The appellee also files a record of appeal, usually to prove the original verdict should stand. The appellate court consists of several judges, and the court's decision is based on the judges' majority opinion. They may affirm the lower court's ruling, reverse or modify the judgment, or grant a new trial. Once the judges reach their decision, the case is sent back to the trial court for whatever action is necessary.

In some rare instances, the verdict of an appeals court can be appealed but only when a particularly knotty question of law exists. In states that have an intermediate appellate court, an appeal from its decision would be taken to the highest court in the state, usually called the state supreme court or supreme appellate court. And even that court's decision can be appealed, by seeking redress in a federal district court or even the U.S. Supreme Court.

ALTERNATIVE DISPUTE RESOLUTION: AN OVERVIEW

Alternative dispute resolution, or ADR, is a term that characterizes a number of procedures designed to effect or stimulate resolution of disputes, principally, but not exclusively, through agreement of the disputants. The number of ADR procedures and reliance on them have grown substantially in recent years, in response to the ever more costly inadequacies of litigation.

Characterizing ADR

ADR techniques can be categorized as *adversarial* or *conciliatory*, *mandatory* or *voluntary*, *binding* or *nonbinding*.

Adversarial versus Conciliatory An adversarial procedure is one in which each disputant puts a "best foot forward." Plaintiffs attempt to show how major damages have been incurred and why one or more other parties are wholly at fault. Defendants try to show why they are blameless or, as often is the case, why they actually are the ones who have been damaged. After the parties in dispute "go after" each other as adversaries, it is up to a trier of fact, arbitrator, or other *neutral decision maker* (generally referred to as "the neutral") to determine which elements of which argument have merit and, based on those decisions, who gets what.

A conciliatory proceeding is catalytic in nature, designed to lead disputants to a common ground where they can fashion an agreement of their own design. They determine what they will give up—what their compromises will be—in order to settle the matter.

For the most part, conventional arbitration tends to be a purely adversarial ADR procedure, while mediation tends to be purely conciliatory. Most other forms of ADR contain both adversarial and conciliatory aspects. For example, in some, representatives of the disputants engage in adversarial procedures principally for the edification of more senior representatives of their respective organizations, permitting the latter to arrive at a conciliatory resolution on their own. In other words, adversarial proceedings are used only if conciliatory measures prove unfruitful.

Mandatory versus Voluntary A mandatory proceeding is one that disputants are required to engage in, either by a contract between them (*contractually mandated*) or by judicial decree (*court-ordered*). A voluntary proceeding is selected without a requirement to do so, typically after a dispute arises. In some instances, however, disputants may voluntarily petition a judge to mandate a certain type of ADR.

Joinder: Whenever something other than court-ordered ADR is involved, the issue of *joinder* becomes vitally important. It relates to who is required to become involved in a proceeding. When only two parties are in dispute, there is no particular problem. Construction disputes often involve more than two parties, however, and in some cases several are named as defendants because it is impossible to know beforehand who caused a problem. Or party A may file a claim against party B, who in turn files a claim against party C, demanding that C indemnify B for any damages B may have to pay to A. If all those involved do not agree to abide by the dispute resolution method selected, moving forward with ADR can create problems. After A and B resolve their dispute, C may file a claim against one or both, necessitating litigation of the same issues.

For the most part, the issue of joinder can be resolved easily by contract, but only when a project is characterized by careful contract administration. A major construction project can involve fifty contracts or more, leaving significant opportunity for something to "fall through the cracks." In such cases, it is sometimes necessary to use litigation to compel one or more parties to engage in ADR, typically arbitration.

Binding versus Nonbinding In a binding proceeding, the decision made by the presiding officer must be adhered to by the disputants, subject to any appeal that may be available. In a nonbinding proceeding, the neutral does not make a decision but, rather, provides advice or guidance, to help disputants come to an agreement of their own making. When the neutral offers a nonbinding "advisory decision," it is sometimes called "appealable" in that the disputants do not have to live with it; they can continue with or initiate litigation, arbitration, or some other form of binding dispute resolution process. (The word *appeal*, as used in this sense, should not be confused with the formal appeal process associated with litigation.)

Almost any type of nonbinding process can be made binding, simply through mutual agreement of the disputants. Likewise, a procedure that usually is binding can be made nonbinding, for example, voluntary nonbinding arbitration.

ADR Benefits

ADR advocates cite a number of benefits for different processes. The following summary of benefits is general and does not necessarily apply to each procedure.

- *Fairness:* The rigidity of the legal system can affect its fairness as applied to a given dispute. The ability to "custom-tailor" an ADR procedure to the needs of the parties and issues involved can help assure fairness, providing neutrals know what they are doing.
- *Speed:* ADR procedures generally take far less time from beginning to end than litigation, both in terms of hours expended and time elapsed. This aspect of ADR can enhance fairness; over time, witnesses' recollections become less distinct; some may no longer be available.
- *Cost:* By reducing the amount of time required of one's attorneys, less expense is required. (By some estimates, from 50 to 67 percent of all dollars changing hands as a result of civil litigation go into the pockets of those retained to resolve disputes.)
- *Creative Solutions:* ADR awards can be structured in a variety of ways, permitting payment in kind rather than cash (e.g., furnishing products and/or service as partial or full compensation), allowing a defendant to purchase an allegedly damaged home from the plaintiff, and so on.
- *Confidentiality:* Private ADR procedures permit confidentiality, helping to avoid headlines when they are a concern. In situations in which nonbinding ADR, particularly conciliatory ADR, becomes a prelude to an adversarial proceeding, should the ADR process not be successful, confidentiality helps assure that attitudes or possibly even certain information revealed during the ADR procedure will not be used at trial. The confidentiality aspect of ADR can encourage its use.

- *Neutral Selectivity:* Disputants generally have the ability to select the neutral who will "work the case." This permits disputants to select individuals who are familiar with the construction industry and who are generally recognized for fairness, credibility, and intellectual honesty. Mutual regard for a neutral can be of significant value in effecting resolution, particularly in situations in which the ADR outcome is nonbinding.
- *Limitation of Discovery:* In many ADR procedures, discovery can be limited to that extent which disputants agree is fair, reasonable, and sufficient to arrive at a resolution. This minimizes cost and can virtually eliminate the frustration and irritation that can be experienced when opposing counsel use discovery as "fishing expeditions."
- *Maintenance of Goodwill:* Many long-term relationships are marked by occasional disputes. The manner in which a dispute is resolved will determine whether or not the relationship will survive. Adversarial procedures usually test relationships to the utmost and often are destructive. Many forms of ADR permit resolution without adversarial histrionics, thus helping to minimize the impact of a dispute.
- *Less Intimidating Surroundings:* A courtroom is designed to be intimidating, in order to convey the majesty of the law. The forum selected for an ADR procedure can be far more "user-friendly," encouraging direct disputant-to-disputant negotiation.

Attaining an ADR Agreement

Whenever a party believes that ADR, or an attempt to reach a decision via ADR, is preferable, it generally is best to work for contractually mandated ADR. Achieving voluntary ADR early in a dispute can be difficult, especially when emotions are high or when a complainant is looking for enrichment. Voluntary ADR is easier to obtain later, but only because the disputants tend to be drained. After having pursued countless hours of research, motion filing and discovery, and having paid countless thousands of dollars, they often welcome any possibility of "getting this thing over with." By contrast, at the onset of a project, when there are no disputes on the horizon and when all parties are on good terms with one another, there is an excellent likelihood that positive attitudes toward disputes will hold sway.

Reliance on Attorneys

Some people advocate ADR principally from the belief that attorneys are not needed. While this may be true in some instances, those involved in a dispute would probably be ill advised to proceed without an attorney's guidance, at least initially. Is the form of ADR being considered the best one, given the circumstances involved? What are the laws in the state with jurisdiction, should ADR fail? Assuming there is a choice, would it be better to rely on ADR or litigation? If ADR is selected, which points should be emphasized or deemphasized? How much will it cost? These and many other questions should often be addressed to trusted counsel before a decision is made. And particularly when significant sums of money are involved, it may be wise to have an attorney on hand to give advice, if only because the attorney has extensive experience in the matter of disputes and their resolution.

Unless the parties agree otherwise, no form of ADR bans reliance on attorneys, although some may require principals to speak for themselves. Many people experienced in construction disputes will advise that a party should proceed without reliance on an attorney only if an attorney thinks it a reasonable approach.

The Role of the Insurer

The insurers of the defendants involved must agree to the ADR method selected, since they may be required to serve as the principal source of funding. Some insurers are far more likely to agree to a given procedure than others. It is essential to ascertain the insurer's attitude before suggesting a given form of ADR.

SPECIFIC ADR OPTIONS

The ADR options discussed below include:

- informal spontaneous negotiation,
- voluntary prehearing negotiation,
- mandatory pretrial negotiation,
- mandatory binding arbitration,
- voluntary binding arbitration,
- specialized binding arbitration,
- expedited binding arbitration,
- voluntary nonbinding arbitration,
- mandatory nonbinding arbitration,
- court-appointed masters,
- settlement masters,
- early neutral evaluation,
- Michigan mediation,
- mini-trial,
- summary jury trial,
- private litigation,
- mediation,

- mediation/arbitration,
- mediation-then-arbitration,
- med/arb2,
- rent-a-judge, and
- resolution through experts.

Of these, informal spontaneous negotiation and voluntary prehearing negotiation are not commonly thought of as ADR. They are discussed in that context for ease of reference. A concept called "ADR by covenant" also is discussed, although it is a means for effecting ADR, as opposed to an ADR technique, per se.

Informal Spontaneous Negotiation

Informal spontaneous negotiation is the process that occurs when parties to a dispute discuss the circumstances involved in an attempt to arrive at an equitable settlement without further ado. Informal spontaneous negotiation occurs frequently when a dispute still is in the incipient stage. This can occur when a contractor calls an unanticipated or changed condition to the attention of the appropriate party, the cost of doing the work is agreed to, and the work then is authorized. All too frequently, however, the work may not be authorized properly or assumptions may be made as to what is, and is not, an extra. Then, after the work is done, a dispute erupts when the contractor demands payment. In this regard, it is worthwhile to note the dictionary definition of *spontaneous*: "acting or resulting from a natural impulse." Regrettably, the natural impulse in such situations seldom is to negotiate. Owners may suspect that their design professionals or contractors are attempting to "gouge" them for unjustified sums or may simply believe that a quoted price for overall work should be lived up to despite unanticipated new requirements. In some cases, of course, suspicions about providers may be justified, or the failure to provide proper notification may impose difficulties of which those requesting an extra are unaware. In any event, the request for extra payment will result in emotional reactions, and/or the initial denial of an extra will create an emotional response. Due to emotions, attitudes then harden, and the dispute begins to take on personal overtones that can quickly cause a molehill to grow into a mountain. If any party in this type of situation can at this point have the presence of mind to shuck emotional responses and rely on personal diplomacy, the other party or parties often can be encouraged to adopt similar attitudes. Then the dispute can be quickly resolved through personal negotiation, saving all parties considerable time, dollars, and aggravation.

Another effective technique to encourage informal spontaneous negotiation requires at least one party to the dispute to contact a superior with bargaining authorization, encouraging the latter to contact an individual at the same or similar level in the other party's organization, so these individuals can negotiate. This approach is particularly effective, because the negotiators are one step removed from that level where emotions tend to be running most strongly.

As simple as informal spontaneous negotiation may sound—*and actually is, in fact*—the inability of people to employ it is in large measure responsible for making all other types of dispute resolution techniques necessary. Note, however, that little generally stands in the way of one party to a dispute initiating informal spontaneous negotiation procedures after another type of dispute resolution procedure has commenced. In such cases, however, it almost always is advisable to attempt initiation of informal spontaneous negotiation only with advice of counsel.

Voluntary Prehearing Negotiation

Voluntary prehearing negotiation comprises attempts at settlement made by attorneys retained by the parties in dispute. Discussions can begin at any time prior to formal presentations being made, at trial, at an arbitration hearing, or in some other forum. The attitudes of counsel and those retaining them are of vital importance. If emotions are running high, one or more of the disputants may encourage counsel to "go for the jugular."

An experienced lawyer is apt to recognize emotion-charged instructions as hyperbole and abide by them *only* if the attorney believes that doing so is in the best interests of the client. It seldom is, due to the time and cost involved. It always is in the best interest of disputants to seek a fair and reasonable settlement as quickly as possible to minimize the cost of dispute resolution, especially when relatively small sums are involved. The more money at issue and/or the more complicated the case, the more time and money generally will have to be expended to ascertain facts, points of law, and so on. When complicated cases involve relatively small amounts of money, however, it generally is not worthwhile to engage in extensive discovery and research, because the dollars expended in the process can quickly amount to a sum that approaches or exceeds that at issue.

Symbolism is another factor that must be considered relative to voluntary prehearing negotiation. Some attorneys may be averse to beginning the settlement process too early in the dispute for fear it will be interpreted as a sign of weakness. Depending on the nature of opposing counsel, a good-faith attempt to resolve a problem quickly could thus encourage the other side to delay conciliatory behavior from the belief its position is stronger, or perceived as stronger, than it actually is. It always is a wise idea to discuss the full range of options in this respect with an attorney.

The longer a case drags on, the more likely it is that parties to the dispute will prefer negotiation. Discovery gives disputants the ability to evaluate their own and the other side's strengths and weaknesses more acutely, permitting a better evaluation of the likely outcome of an arbitration or trial. In addition, more pressure is being applied to the disputants themselves, as they experience the frustration of delays and the regular monthly arrival of their attorneys' bills.

Some attorneys contend that certain ADR processes, rather than being a substitute for trial, actually become a substitute for voluntary prehearing negotiation and settlement. As noted, 90 percent or more of the civil lawsuits filed in the United States are settled before trial. By contrast, a 1985 study shows that, in each of three recent years, only 45 percent of the construction claims initiated through American Arbitration Association (AAA) arbitration proceedings were settled or withdrawn before hearing.

Mandatory Pretrial Negotiation

The judge to whom a case has been assigned often will call the parties together for a settlement conference. The judge will be somewhat familiar with the case, having read briefs submitted by the attorneys, and may confer with the attorneys alone, the disputants only, or both; expert witnesses may be called as well. The judge may direct the disputants and their attorneys to attempt to arrive at a negotiated settlement in order to expedite resolution without trial, thus to help reduce the clogging of what most likely is an already crowded calendar. If attempts to resolve the issue fail, judges may use their special position in an effort to gain acceptance of specific terms they strongly suggest.

Judicial attempts at pretrial negotiation often occur at the "eleventh hour" before trial, due to demands on judges' time. At that point, disputants have already spent just about everything necessary in order to go to trial. While a trial can be somewhat costly, it is often a relatively small expense when compared to the costs associated with discovery.

The inadequacies associated with mandatory pretrial negotiation have resulted in development of alternative or additive judicially mandated procedures, discussed below.

Mandatory Binding Arbitration

Mandatory binding arbitration is an adversarial procedure through which disputants present their cases to one arbitrator or (in larger cases) a panel of three arbitrators whose decision is binding and court-enforced. Generally speaking, there is no right of appeal, and opportunities for discovery are limited. Arbitration is almost always entered into on a mandatory basis, by virtue of contractual conditions. In most instances, Construction Industry Arbitration Rules of the American Arbitration Association (AAA) are followed.

Arbitration is the oldest type of formal ADR used by the nation's construction industry, having first been introduced into standard-form construction agreements as early as 1871. In the mid-1960s, AAA formed the National Construction Industry Arbitration Committee (NCIAC), comprising representatives of numerous national construction-oriented organizations. The Committee helps AAA develop procedures and panels of arbitrators uniquely suited to the construction industry. More than 30,000 arbitrators, representing all segments of the industry, now serve on the AAA's national construction panel.

In typical application, AAA-administered arbitration begins with either party initiating a claim by filing an AAA demand form or a letter that contains the necessary information. The demand for arbitration describes the dispute sufficiently for the AAA to select appropriate candidate arbitrators. The demand is accompanied by a filing fee, whose amount is based upon the amount in dispute.

The next step usually is a prehearing conference, where attorneys and their clients meet with a senior AAA representative to exchange information or to stipulate uncontested facts, and discuss administrative details, such as the estimated number of hearings required, the timetable, the rate and terms of arbitrator compensation, arrangements for an on-site inspection (if needed), description of claims and counterclaims, and exchange of witness lists.

According to AAA, many cases have been resolved at prehearing conferences, because they give attorneys and disputants an opportunity to "share their perceptions with one another." In these instances, the administrator can help the parties arrive at a solution to their problem, serving somewhat as a mediator.

Following the prehearing conference, if one is held, disputants select an arbitrator or a panel of arbitrators from an annotated list furnished by the AAA. An arbitrator has the authority to consider amendments to the claim or counterclaim; schedule, close, and reopen hearings; grant or deny hearing adjournments; and conduct an arbitration in the absence of a party after due notice. Where authorized to do so by law, the arbitrator may also subpoena witnesses or documents independently or at the request of a party.

Although arbitration can be effective for resolving some construction disputes, it has drawbacks that, in large measure, have stimulated development of other ADR approaches. Generally speaking, the larger and/or more complex the case, the more significant some of these drawbacks become. One of the most significant of these is the absence or reduction of prehearing discovery. This increases the opportunity for surprise at a hearing, and that can work to the severe—and some contend unfair—detriment of one of the parties. In addition, restricted discovery can make it necessary to keep witnesses on the stand far longer, thus making the arbitration more complex and costly.

Another problem associated with arbitration can be caused by the neutrals used. Although arbitrators may be familiar with the industry or specific aspects of it, they may not be familiar with certain points of law and, in particular, certain important precedents that actually may have guided one of the disputants in development of an agreement or certain contractual provisions. This problem is not unique to arbitration; it can affect virtually any type of ADR procedure. While some attorneys claim that litigation can help prevent this problem from arising, the fact is that other types of ADR, or certain types of ADR provisions applicable to arbitration, can be employed. As an example, by using specialized arbitration (see below), parties to a contract can specify who the arbitrator will be in the event arbitration becomes necessary and can select an individual familiar with points of law as well as technical issues.

Neutral selection can proceed far more quickly and cooperatively before disputes arise.

Another significant problem associated with arbitration is the limited ability to appeal, even in cases in which it is generally agreed that the decision rendered is grossly unfair. Courts are reluctant to interfere with arbitration; arbitrators' written decisions tend to be extremely brief and do not need to specify the rationale that led to their decision. Generally speaking, appeal is permitted only when it can be shown that the arbitrator was biased by virtue of an affiliation of some type or when arbitration rules have been seriously breached.

Voluntary Binding Arbitration

Voluntary binding arbitration is identical to mandatory binding arbitration, except disputants agree to mandatory arbitration after a dispute has arisen. This is a somewhat rare occurrence, most commonly being the result of an unjoined party's agreement to joinder, thus precluding the potential for litigation.

Specialized Binding Arbitration

Specialized binding arbitration is essentially custom-developed arbitration designed to meet the specific requirements of those entering into the agreement. A specialized binding arbitration clause may specify the location of hearings, the qualifications of the arbitrator, provisions for prehearing discovery procedures, or expedited hearing schedules. As with other forms of binding arbitration, specialized arbitration can be either mandatory or voluntary.

Given the predominant reliance on the AAA and its Construction Industry Arbitration Rules, it could be said that any form of binding arbitration that deviates from AAA rules is specialized. Note that agencies other than AAA are available to administer or otherwise to facilitate an arbitration of any kind.

Expedited Binding Arbitration

Expedited binding arbitration is not considered specialized, because the AAA has developed specific rules for it, comprising the "Expedited Procedures" division of the Construction Industry Arbitration Rules (Section 54-58). Expedited procedures are applied to any case in which the total claim of any party does not exceed $15,000, exclusive of interest and arbitration costs. If both parties agree, expedited procedures may also be applied to cases in which total claims exceed $15,000.

Under expedited procedures, all notices and announcements are transmitted by the AAA to the parties by telephone and are subsequently confirmed in writing. The AAA also notifies the parties by telephone of the appointment of the arbitrator. The date, time, and place of hearing are set by the arbitrator. The AAA notifies parties of these decisions by telephone seven days before the hearing and follows up with a Formal Notice of Hearing sent to each party.

The hearing and presentation of the parties are generally completed within one day. An additional hearing may be scheduled by the arbitrator, if necessary, to be held within five days. The award is rendered not later than five business days from the close of the hearing, unless otherwise agreed to by the parties.

Expedited procedures can be modified (making them a form of specialized binding arbitration), and these modifications could include (among other things) preselection of the arbitrator, as well as administration by a party other than AAA.

Voluntary Nonbinding Arbitration

Nonbinding arbitration, or advisory arbitration, is a catalytic procedure devised principally to stimulate agreement prior to resorting to a more formal binding process. It is used when a case comes down to a few "threshold issues" which, when resolved, lead to resolution of the entire case. The parties may also use it to see how well their experts do under cross-examination and what impact their testimony may have on the ultimate trier of fact. Likewise, in a situation in which a complicated or novel issue is presented, the point can be briefed and argued to a retired judge whose opinion may be invaluable in putting insight and perspective into the case.

In an advisory arbitration process, counsel are able to tailor their own rules of evidence and manner of presenting the case. They are also able to put "teeth" into the proceedings; for example, if the case is not resolved according to the recommendations and decisions of the neutral, the party who then moves to litigation and does not obtain a more favorable result would be required to pay the opponent's attorney's fees and costs and/or interest on the money in controversy.

Mandatory Nonbinding Arbitration

Mandatory nonbinding arbitration is more commonly known as *court-annexed* or *court-ordered arbitration*—something used in a number of state and federal district courts, usually for cases involving $50,000 or less. Because it is nonbinding, it permits an "appeal" in the form of a traditional trial. Such appeals are discouraged by imposing court costs or legal fees upon the unsuccessful appellant. The cost of court-annexed arbitration generally tends to be nominal, however, because the neutrals are volunteer lawyers, appointed by the court either initially or when disputants fail to agree on one from a list provided by the court.

A report by Benjamin R. Foster published in the September 1985 issue of *Construction Law Advisor* states: "The experience with court-annexed arbitration in Pennsylvania and California has demonstrated that a significant number of cases are resolved in half the time of a traditional procedure and without a new trial as an appeal. Where there have been new trials, the results often are similar to the result the first time."

As with many other procedures, disputants have the ability to make certain modifications, as by making the outcome binding.

Court-Appointed Masters

Court-appointed masters are relied on for many functions; typically, to furnish timely expertise that a judge may be unable to provide. The role of masters can be segregated into three broad categories: fact-finding, case management, and settlement. Of the three, fact-finding is the traditional role under Rule 53 of the Federal Rules of Civil Procedure, "Helping the judge find the facts at the trial stage, or the jury unravel a set of complicated facts." The other two are newer and more ambitious.

Using a master as the developer of a case management plan can be particularly beneficial in large, complex cases. In one of these, for example, a U.S. District Court Judge consolidated a number of asbestos cases in Cleveland. He appointed two special masters who devised a Case Management Plan (CMP) and a Case Evaluation and Apportionment Process (CEAP). The CMP, developed to streamline the pretrial process, embraced new discovery mechanisms that eliminated traditional interrogatories and document production. The CEAP used a computer model to give parties specific information about the settlement value of their cases, based on previously settled cases in the area, creating a solid starting point for negotiations. In just over a year, all of the cases then pending were settled. The cost of the special masters amounted to approximately one hundred dollars per party per case.

The appointment of special masters for purposes of settlement is a newer approach, discussed below.

Settlement Masters

The "settlement master" concept uses procedures somewhat akin to those associated with court-annexed arbitration and mini-trials, discussed below. Within approximately three weeks from the date of a judge's order, each party prepares, simultaneously exchanges with the other parties, and submits to the master a ten-page (or shorter) narrative statement of the facts and issues of the case, with reference to, and copies of, principal exhibits relied on. Then, within two weeks, the parties prepare, exchange, and submit ten-page (or shorter) briefs to the master. A hearing is then held during which each party presents thirty-minute (maximum) oral arguments to the master; afterward, the master proceeds with settlement discussions.

At the hearing and at settlement discussions, each party may be represented by as many as five people. At least one representative must have full settlement authority. No more than three attorneys may represent any given party. The master and the disputants keep all proceedings, documents, and discussions confidential. The court does not receive information except for that submitted by the master, relative to the final result. The costs of the master's service is split evenly between the parties.

Early Neutral Evaluation

Early neutral evaluation, as used by the U.S. District Court for the Northern District of California, is a court-ordered process normally imposed three to four months after the original complaint has been filed. Its purpose is to promote meaningful dialogue about disputes early on and to cause the disputants and their attorneys to confront the issues and assess them in a realistic manner. Achievement of this purpose could overcome some of the significant problems that commonly occur at the inception of litigation, such as shotgun pleadings that can mask, rather than clarify, the real issues in dispute and claims inflation.

Early neutral evaluation relies on a highly experienced, widely respected private attorney as the neutral. At least one week before the evaluation session, each party delivers to the evaluator and to the other party an evaluation statement not exceeding ten double-spaced typewritten pages. The statement identifies legal or factual issues whose resolution could reduce the scope of the dispute or contribute to settlement, as well as the discovery needed to expedite case preparation.

The evaluation session is held at a neutral site. Unless hardships would be created, disputants' attorneys must be accompanied by at least one person with settlement authority. The evaluator presents opening remarks about the session itself and the evaluator's role. Each side then makes a fifteen- to thirty-minute presentation, addressing the issues in dispute. These remarks can be augmented through the presentation of documents. The evaluator can interrupt a party to ask for a clarification during a presentation, but the other disputant may not.

After presentations are made, the evaluator works with the disputants to identify issues about which they agree and may press them to agree on issues about which they are not far apart. Next, with the help of the parties, the evaluator identifies questions whose answers are unknown and issues about which the parties disagree. The evaluator may suggest that the parties jointly discover the truth about unknown facts. The evaluator may probe issues about which the parties disagree to determine how and why each party has established its outlook.

After assessing the attitudes of the disputants, the evaluator gives a candid appraisal of liability and damages, that is, the likelihood of being found liable and an estimate of damages. The evaluator then may provide an informed guess about the costs of taking the case through trial to encourage the disputants to think about settlement. If the parties are inclined to settle, the evaluator can allocate more time for discussion. Otherwise, the evaluator will recommend motions and the appropriate nature and extent of discovery.

Among other benefits claimed for early neutral evaluation, it requires disputants and their attorneys to do their homework early in the dispute, rather than permit delay and then rely on extensive discovery. It also requires counsel and their clients to face the realities of their cases early on and to listen to an effective appraisal from someone in a position to "know the score." It stimulates open communication between disputing parties and can allow some clients the cathartic release of getting the story "off their chest." It may also encourage them to drop unrealistic claims or expectations.

Michigan Mediation

"Michigan Mediation" is a highly successful program used by state courts throughout Michigan and the U.S. District Court for the Eastern and Western Districts of Michigan. Despite its name, Michigan Mediation is not a mediation process. Instead, each disputant draws the name of an attorney from a list furnished by the court. The two attorneys then select a third, who acts as the chair. (Numerous attorneys with considerable experience reportedly volunteer to serve as Michigan Mediation neutrals in order to gain experience in case presentation, because the procedure is used so extensively.)

At least ten business days before the panel hears the case, each side presents pertinent documents, including a summary or brief of factual and legal positions, along with bills, records, expert reports, photographs, and so on. At the hearing itself, each side has no more than thirty minutes to make its presentation, unless an unusual or multiparty case is involved. The panel then has ten working days to arrive at its decision, but—more often than not—it decides "on the spot," indicating who owes how much to whom.

After learning of the decision, the disputants have twenty days in which to accept or reject the decision; failure to respond is tantamount to acceptance. If both parties accept, judgment is entered for the amount indicated. If either side rejects the panel's decision, the case proceeds to trial. If the panel's decision is unanimous (most are), however, rejection of its decision can result in penalties being imposed on the rejecting party, for example, court costs and the other party's attorney's fees for each day of trial, as the court determines. To avoid such a penalty, the plaintiff, if rejecting the panel's decision, would have to be awarded 10 percent more than the panel suggested. A rejecting defendant would have to pay an amount at least 10 percent less than the one recommended.

Michigan Mediation can be voluntary or court-ordered. A court will order it of its own volition or upon the petition of either party. Such requests are most likely to be granted when cases are difficult to evaluate or when emotions play a significant role, making settlement unlikely.

Mini-Trial

The mini-trial is a contractually mandated or voluntary (typically postdispute) form of ADR, characterized by summary presentations by lawyers and experts for each party, presented to top management representatives (with authority to settle) of each party. A mutually selected neutral usually is employed to help establish specific mini-trial procedures and to advise the parties as to the strengths and weaknesses of their respective cases.

Experience to date indicates that best results are obtained in mini-trials of cases involving complex questions of mixed law and fact, that is, where the disagreement centers on how existing law and precedents apply to the facts of a particular case. The American Arbitration Association has developed a set of mini-trial procedures, but the parties involved are at liberty to establish whatever procedures they prefer.

As noted by James F. Henry, president, Center for Public Resources (New York, NY):

> The name [mini-trial] is misleading; it is not a trial at all, but a structured, nonbinding settlement procedure that treats the dispute more as a business problem, rather than a legal problem.... By limiting the scope of both preparation and presentation of a case, the mini-trial procedure forces each party to concentrate on formulating its "best case." Each case is then presented before a neutral advisor and corporate executives of both who, unlike a judge or jury, are already sophisticated about the business that gave rise to the dispute. From the vantage point created by the mini-trial procedure, the businessmen can get, perhaps for the first time, both sides of the dispute and its likely litigated outcome. With the presentations still fresh, the executives meet to discuss settlement. Without the litigation system's restrictions on permissible issues, evidence, and types of relief available, the parties, uniquely aware of their business operations and objectives, are free to create a settlement agreement that works for them. The result: a business-oriented resolution in which both parties may be winners.

> Although parties sometimes choose not to use a third party as neutral advisor, a mutually selected neutral can help the process in several ways.... The neutral's most important function may be to provide, at the parties' request, an advisory opinion of the strengths and weaknesses of each side's case and a prediction of the likely outcome of the dispute at trial. Parties may want to request the neutral's opinion before they begin their settlement talks. Alternatively, they can request the opinion if they reach an impasse in their negotiations. Either way, the purpose of the neutral's opinion is to spur the parties to the settlement. Although the neutral is typically not intended to be a mediator, parties may want to ask the neutral to perform this role.

Summary Jury Trial

The summary jury trial typically is a mandatory (court-ordered) nonbinding ADR technique applied in pending litigation to obtain an advisory opinion from a jury as to the eventual decision in a case. Originated by Federal District Judge Thomas Lambros, Northern District of Ohio, Eastern Division, the technique was first utilized on March 5, 1980. As described by Judge Lambros:

> The summary jury trial is a half-day proceeding in which attorneys for opposing parties are each given one hour to summarize before a six-member jury. Basically, introduction of evidence is limited, and witnesses are excluded from the proceeding. After the evidence has been presented and the judge provides a short explanation of the law, the jury retires and either presents a consensus verdict or, if no consensus can be reached, reveals anonymous individual juror views. The jury's verdict is purely advisory, unless the parties agree to be bound by the verdict. The main purpose of the procedure is to provide parties with an insight into the way a trial jury would view the case without the expenditure of time and money required for a full trial.

As Judge Lambros noted, results of a summary jury trial can be made binding by agreement, subject to a consensus being achieved, or its results can be used as a catalyst for agreement. As with a mini-trial, top management representatives of the disputants, with decision-making authority, should be on hand.

Because the summary jury trial usually is effective only after discovery and other pretrial procedures have been accomplished, it saves only the costs associated with final trial preparation and a trial itself. As such, it usually is applied only to "hard-core" cases that involve wide divisions as to how a jury will evaluate the evidence.

Private Litigation

Private litigation is a generic term that refers to procedures that emulate those used in a court of law but conducted before a retired judge or some other mutually accepted individual with judicial qualifications. Known popularly, and perhaps somewhat misleadingly, as "Rent-a-Judge" (see below), private litigation can be initiated with or without court involvement.

At least six states (California, Nebraska, New York, Oregon, Rhode Island, and Washington) have passed laws permitting the court, on stipulation of all parties, to refer a pending lawsuit to a private neutral. The procedure, known as *trial on order of reference*, has the same force and effect as a trial held in a courtroom before a constitutional judge. The proceeding follows judicial rules of procedure and evidence (unless the parties agree otherwise), results in a record made by a court reporter and findings by a neutral that stand as findings of the court, and creates the ability to appeal from the neutral's decision under rules of appellate procedure, just as from a regular trial court judgment. Disputants may prefer a trial on order of reference because it permits them to select their own neutral, establish their own schedule and proceed quickly, and maintain privacy and confidentiality.

Mediation

Mediation is perhaps the most popular of all the "new" ADR procedures. As a consequence, a number of organizations now provide mediation services. The success of mediation is indicated by a report of the DPIC Companies (Monterey, CA), one of the nation's largest insurers of design professionals. As of June 1988, DPIC Companies reports:

1. An estimated legal expense savings in excess of $10 million occurred on 188 claims.
2. The estimated savings on a per-claim basis exceeds $50,000.
3. The time saved by mediation (i.e., on the anticipated closure date) is estimated as ten months per claim.
4. The timing of the mediation process (i.e., from submission to the provider to the point of resolution) averages from 90 to 120 days.
5. For 188 concluded cases, the average cost of the mediation process itself was $3,000.
6. The claims that DPIC Companies proposes for the mediation process are accepted into the process by all parties at the rate of 80 percent.
7. The claims for which the process has been accepted are successfully resolved at the rate of 74 percent.

Mediation is a voluntary or mandatory nonbinding conciliatory procedure designed as a catalyst for dispute resolution. Beginning in a joint session, the mediator reviews the reasons for the parties to participate and explains the process: the procedures and ground rules covering each party's opportunity to talk, order of presentations, decorum, discussion of unresolved issues, resort to private meetings, and confidentiality of proceedings.

After the preliminaries are concluded, each party describes its understanding of the issues, the facts surrounding the dispute, and what it wants and why. The mediator then gathers facts, clarifies discrepancies, and evaluates relationships and dynamics between the parties and their counsel.

If joint discussions break down, the mediator meets separately with each party, often shuttling back and forth between the two. During each private meeting (caucus), the mediator tries to clarify each party's positions and priorities, loosen rigid stances, and explore trade-offs. If the mediator can identify alternatives, the parties will often move toward acceptable accommodations. Then the mediator makes suggestions about a final settlement, stresses the consequences of no agreement, and formalizes offers.

Even if mediation does not result in an agreement, many issues are disposed of, so resultant litigation or arbitration may address a simplified list of unresolved problems.

Mediation/Arbitration

Mediation/arbitration, also known as med/arb or binding mediation, was developed by ASFE in the mid-1970s. It is unique in that the person who mediates the dispute would also arbitrate any issues not resolved through mediation.

The ASFE med/arb procedure begins when project principals confer to select a neutral before work begins. Since disputes typically arise after the start of construction, it is suggested that the principal parties involved in the selection should be the owner, the prime design professional(s), and the general contractor and/or construction manager. The neutral selected for med/arb should enjoy a reputation for intelligence, integrity, fairness, and intellectual honesty and should be familiar with the construction industry, as well as with both mediation and arbitration procedures. In typical application, the med/arb neutral would be on the scene soon after a dispute was reported or after attempts at informal spontaneous negotiation proved fruitless. The neutral then initiates mediation procedures and is authorized to keep them going until all issues are resolved or the neutral believes full resolution will be impossible. The neutral would then institute arbitration to resolve any outstanding issues. Although mediation as a discrete process already experiences a high success rate, med/arb proponents suggest even more success will be experienced when it is part of med/arb. They base this belief on disputants' realization that the arbitration will be unconventional, given the med/arb neutral's more intimate view of the situation. In conventional arbitration, a neutral only becomes familiar with the parties' adversarial positions.

Med/arb can generate a variety of benefits beyond those associated with other forms of ADR. One of these—now no longer unique to med/arb—is the immediacy of action occasioned by preselection. The mediator can arrive on the scene quickly and obtain information while it is still fresh. If desired, partial discovery can be employed, with the neutral also having the right to conduct interviews of witnesses. If the nonbinding aspect of the procedure fails, the binding aspect can begin immediately and usually can be conducted on an expedited basis.

Critics of mediation/arbitration claim that it is inappropriate for one individual to serve as the neutral in a conciliatory process and then serve as the binding authority in an adversarial procedure. They believe that the neutral becomes biased as a consequence of mediation and thus cannot provide a fair decision through arbitration. But the role of med/arb neutrals is not significantly different from that of judges. During the course of litigation, judges often attempt to facilitate a settlement and, in the process, may perform a variety of mediator functions. If attempts fail, the judges involved may then go on to hear the case without a jury, rendering a final, albeit appealable, decision. In the case of med/arb, however, disputants can select their neutral, relying on a person with the intellectual and emotional qualities required to overcome potential bias problems, assuming bias is a problem. Some say it is not, contending that the predispositions gained through mediation can lead to an arbitrated decision that is very close to what the disputants would have arrived at, had they been able to settle the dispute on their own.

Mediation-Then-Arbitration

Mediation-then-arbitration, or med-then-arb, was developed by the Deep Foundations Construction Industry Roundtable as a modified version of med/arb. The procedure was created specifically to address the criticisms associated with using one neutral for two dissimilar processes. Med-then-arb procedures, known as "The 100-Day Document," are similar to med/arb provisions, except that arbitration, if needed, is conducted under the aegis of a different (but also preselected) neutral, one who is not privy to any of the discussions or materials relating to the mediation process.

Med/Arb2

Med/Arb2 is another new ADR technique being advanced by ASFE. It is almost the same as med-then-arb, with one basic exception. If mediation fails to resolve all issues, either disputant may request the mediator to decide unilaterally all outstanding issues. If the other disputant and the mediator concur, this step is taken and formal arbitration by a separate neutral then becomes necessary. Disputants would stipulate to the mediator's decisions along with those reached of their own accord, all recorded in the same document.

Rent-a-Judge

Although the colloquial "Rent-a-Judge" description is most commonly applied to private litigation, a more literal interpretation is now advisable, given retired judges' steadily growing involvement in private-sector organizations established to facilitate alternative dispute resolution. Through judicial referral or, more commonly, election of the parties to a dispute, these retired judges can be relied on to furnish virtually any type of ADR procedure, including a private litigation that could be (if the parties select) a "bolt-for-bolt" duplicate of conventional litigation, except for the ability to select the presiding officers and expedite proceedings.

In many instances, parties will accord a degree of deference to a retired judge that they would not ordinarily accord to some other neutral. In all cases, however, individual conduct is the key. While it may be easier for retired judges to gain the respect of disputants initially, the significance of their former work will be quickly dissipated if their conduct, comments, or intellectual capabilities do not jibe with "judicial image." Note, too, that the need for expertise in the industry or a specific aspect of it may be more important than expertise in law or dispute resolution in general.

Resolution Through Experts

Reliance on experts can be effective means for resolving disputes that hinge principally on findings of fault, as opposed to matters of law, providing the disputants select knowledgeable experts and direct them to conduct a full investigation and provide impartial results. Agreements relying on experts can be established initially by contract or, later, after a dispute has arisen. Some well-known experts contend that the best approach is for each party to retain its own expert and for the experts to agree then on an overall research program. This approach helps assure that "no stone is left unturned"; it also helps reduce expert costs in that the disputants share the expense of certain research activities, such as testing, that they might otherwise have to bear individually.

At the conclusion of research, the experts meet to discuss their findings and to prepare a report on their general conclusions. These findings then can be used to structure an agreement through which those at fault "make good" through payment of damages, performance of service, and so forth.

An alternative to the forgoing is for each of two parties to select an expert and for the two experts so selected to select a third. It has been said that, in such instances, each of the two disputant-selected experts merely acts as an advocate for the client, while the third does the actual decision making. This does not have to be the case, however, if experts are selected and charged appropriately.

ADR by Covenant

Although not an ADR technique in and of itself, a concept developed by ASFE has particular merit for the construction industry. In essence, an owner would create new covenants for a property, requiring that any subsequent owner of the property would have to rely on a nonbinding ADR procedure in a catalytic effort to resolve a dispute prior to engaging in litigation. The disputes involved would be any between the new owner and any former owner, as well as certain parties who provided services to any former owner, such as surveyors, engineers, architects, and contractors. A covenant could require a specific type of ADR, or it could go so far as to afford selections from a list of ADR techniques.

9

Basic Economics of Professional Practice

Design professionals in independent practice operate in two worlds: that of the professions and that of commerce. Some regard this situation as problematic, creating financial pressures that prejudice professional outlooks. To the extent that this situation does emerge, it is not the fault of business, per se, but rather that of the individuals who manage a firm. In fact, professionalism and the pursuit of profit are not only compatible but also mutually reliant. A professional approach to technical endeavors is essential for long-term business profitability. Profitability is essential for the maintenance of a professional organization. If professional attitudes must be sacrificed in order to keep a practice afloat, chances are that financial management is not being practiced in a professional manner. Financial management provides the means to reduce a number of concerns to a common base—dollars—permitting overall monitoring of many complex, interrelated concerns.

This chapter provides a brief overview of the relationship between professional practice and profits, and it discusses several basic financial concerns, including accounting functions, techniques for identifying overhead, and the various reporting functions necessary to oversee the financial aspects of a practice and the projects it pursues.

THE PURSUIT OF PROFIT

Profit has become a dirty word in some circles, regarded as the motive for unconscionable business practices of all types. But profit is not the problem; people are. Those who adopt an amoral attitude toward business are likely to assume the same barren philosophy toward any activity they pursue, believing the need to fulfill their own needs justifies whatever means are necessary. The real problem in such instances is one of perspective: The desire to accumulate wealth becomes so all-consuming that it obscures other concerns. The profit motive has very little to do with such situations. Most people go into business not so much to seek a profit as to pursue activities they enjoy. Self-fulfillment is the goal; profit is essential to the operation of the business that makes self-fulfillment possible.

The Purpose of Profit

Profit is used for a number of purposes. Two are fundamental. First, it creates the reserves a business needs to survive the unanticipated. Second, it provides the capital a business requires to enhance its operations.

Contingency Reserves We already have discussed contingency reserves to some extent, in terms of those set aside as self-insurance, that is, the amount needed at least to cover the deductible of a professional liability insurance policy. Contingency reserves are needed for many other purposes, however, including sudden price hikes, such as those affecting insurance premiums, an accident, or other loss that is not covered by insurance, or the bankruptcy of a client who owes an appreciable amount for services rendered, among many others. Although some of these losses are partially recapturable through income tax write-offs, full recovery is virtually impossible, and, in any event, tax recovery often is a lengthy process, incapable of satisfying the need of paying ongoing expenses. Of course, the money required can be borrowed, but, to be creditworthy, a business must be run in a profitable manner.

Capital Formation Profit is also the source of the capital needed for investment in one's business, to help improve its ability to achieve business objectives. This may mean the expansion or refurbishing of an existing office, the addition of new equipment, the opening of a branch office, the purchase of another firm, and so on. Although many of the costs associated with such endeavors are paid for through borrowed funds, at least 20 percent—often more—usually must come from the firm itself. In reality, however, 100 percent is derived from profits, because such major investments are regarded as a conversion of assets, typically from cash (a liquid asset) to equipment, a building, or some other form of fixed asset. When this is done through borrowed funds, the interest paid on the principal (the borrowed money) is an expense, because it is part of the cost of doing business. The principal is not an expense, however. In essence, it is an advance based on future profits. The firm obtains this advance from a lender by demonstrating that it should be profitable enough to repay the loan easily. Although the firm pays for the use of the money, it usually derives a larger benefit from being able to convert that money to some other form of asset. This benefit typically takes the form of more business or more productivity and generally will put the firm into an even better position to "retire" its debt.

Other Applications The profits that a firm earns are commonly used for a number of other purposes. One of these is dividends paid to those owning stock in the company, assuming it employs a corporate structure. A corporation is a business entity that exists unto itself and is treated as a separate taxpayer by federal and other taxing authorities. Many design professionals and other businesspeople prefer a corporate structure because it insulates them from certain liabilities a business can accrue. Other business structures, such as partnerships and sole proprietorships, are considered extensions of individuals, making their owners responsible for business debts. Although corporations have owners—those who receive stock in exchange for their investment—the stockholders generally are immune to attachment of their personal assets as a consequence of business activities, except to the extent that they are also officers or directors of the corporation. Larger businesses make their stock available to the general public, and several design professional organizations have "gone public" in order to raise capital for expansion. The vast majority of design professional corporations are "closely held," however, meaning their stock is not publicly traded. Nonetheless, those who have invested their money in the business most certainly merit compensation, since the money could just as easily have been invested elsewhere. The compensation is paid as a dividend (so much per share), which is derived from profits.

Profits also are used to fund retirement programs for employees, through profit-sharing plans. These permit a business to set aside a portion of profits for their employees' future use and thus become an important element of a firm's overall compensation package. Profits may also be used to fund bonuses or other types of employee rewards, to show thanks, recognition, and appreciation when due.

Profit's Impact on Fees

From time to time, a government agency that employs "in-house" design professionals, such as a state's transportation department, will issue a statement suggesting it operates much more cost-effectively than private firms, in part because it does not seek profits. Statements such as this are always unfortunate and almost always untrue. In one incident, for example, ASFE challenged exactly such statements made by a spokesperson for the Virginia Department of Transportation. After several months of analysis, it was found that the state's financial data failed to consider several vital factors, including its cost of professional liability insurance (covered through a self-insurance fund) and the taxes it did not have to pay. In fact, the secretary of the department ultimately admitted to ASFE that the data used for the comparisons were incomplete and that it was impossible for the state to put together the information required for an accurate assessment.

The notion that a public agency can operate at a lower rate because it does not have to earn a profit seems somewhat reasonable, but not when it is recognized that the net (after-tax) profit of many engineering and architectural firms is 5 percent or less; rarely is it in excess of 10 percent. Although some public agencies pay staff lesser amounts than those paid by private firms, state agencies generally offer more in the way of fringe benefits: more paid leave, more generous retirement programs, and so on. Unlike private firms, however, public agencies are not generally in a position to "lay off" personnel when "business" becomes slow. As such, it generally is most cost-effective for public entities (and large private ones as well) to hire sufficient personnel to handle routine work but to rely on outside firms to provide services that demand higher levels of expertise or that are necessary in order to meet occasional peak work loads. In fact, most public agencies operate in precisely this manner, with many relying on private firms for almost all of their design requirements. Those that do not do so assume much higher overhead than private-sector firms, because so much time is wasted.

The Need to Produce Although there are many differences between public-agency design departments and private firms, it is worthwhile to note that they often operate in much the same manner from a financial management point of view, in that they are given a budget to produce a certain amount of work. Any specific project is headed by a project manager who monitors progress using many of the same financial reporting systems employed by firms in private practice. Engineers and architects on the public payroll can receive as much pressure to produce within budget as those in the private sector. In either case, the pressure is applied to achieve programmed results, since failure to do so could start a chain reaction that would affect the organization's ability to fulfill other plans.

In the private sector, the budget for any given project includes an allowance for profit. As such, when someone foolishly compromises professional principles in order to meet the budget, it could be alleged that the compromise was made for the sake of profit. And it is precisely such charges that give "profit" negative connotations. But the real culprit in such instances is not profit; instead, it is poor judgment, often abetted by lax management. In many cases, work is accepted for too low a fee, making it virtually impossible to do a professional job profitably. The result is that corners are cut in order to make a profit or, more commonly, to minimize a loss. Cutting corners only tends to make matters worse, however, and, over the long term, it is far more costly. Shortcuts sow the seeds of errors, omissions, delays, and disputes, all of which, at best, lead to client dissatisfaction and a damaged reputation. In fact, professionals who really understand what profits mean recognize that short-term profits often lead to long-term losses and that short-term losses are sometimes necessary investments to maintain client satisfaction and achieve long-term profitability. Not losing $10,000 worth of time on a major commission is hardly a feat to applaud when the procedures required to achieve the outcome lead the firm to a risk exposure ten or more times greater, or the potential loss of a valued client and a damaged reputation—a design professional's "stock in trade." Of course, it is not very often that one hears that professionalism and integrity were *not* compromised in order to assure profits. Nonetheless, insofar as engineers and architects are concerned, this really is the case.

Profit from a National Perspective

From a national point of view, profit is the lifeblood of the American economy. Businesses of all types employ it to grow and become more competitive, often by modernizing existing facilities and building new ones. The work involved stimulates the buildings sector of the construction industry, as well as companies that fabricate, supply, deliver, and install the new equipment and machinery ordered. The growth that occurs increases employment, increasing the tax base and stimulating consumer demand. Increased tax income allows public agencies to improve and expand the nation's infrastructure, providing even more stimulation to the construction industry. Increased consumer spending creates heightened demand for goods and services of all types, making it worthwhile for business to grow further and become even more competitive. In essence, then, the profit motive is fundamental to the success of the American economy.

ACCOUNTING

Accounting is the process of keeping track of—accounting for—money. It is fundamental to the operation of any organization that handles funds of any type, not only so it can quantify its financial activities, but also for purposes of overall management and planning.

Classifying Accounts

Generally speaking, all of a business organization's monetary transactions can be classified into five major divisions or accounts. These are:

- *Assets*, or all things owned by a firm. Assets are generally categorized as *current* or *fixed*. Current assets include cash or other assets (including accounts receivable) that are readily converted to cash within one year. Liquid assets are those that are "instantly" convertible to cash, comprising cash itself, as well as certain investment instruments, such as certificates of deposit, stocks, and bonds.

    Fixed assets are those with a life that exceeds one year, generally including equipment, furnishings, and real estate. As such, a building

would be considered a fixed asset, even though it probably could be sold in less than one year.
- *Liabilities* comprise anything that a firm owes to another party; that is, claims on the firm's assets. Liabilities also are subcategorized as *current* (those due in one year or less) and *long-term* (those due in more than a year).
- *Net worth* (equity) is the difference between assets and liabilities. For example, if you own your own house (a fixed asset), its current value less the principal amount due on your mortgage (a long-term liability) would equal its net worth.
- *Revenue* is the amount of money earned by a business entity over a given period. The periods typically involved are a month, a quarter, or a year.
- *Expenses* are the amounts that a firm spends during the same period during which revenues are calculated. For purposes of effective reporting, it is desirable to select an accounting method (there are several) that gives a reasonably effective indication of how much it costs to produce the revenue received.

As already noted, money spent by a firm is not necessarily an expense, as when it converts a cash asset into a fixed asset. Note, however, that all fixed assets are assumed to lose value over time, through a process called *depreciation*. Depreciation *is* considered an expense, even though it does not involve the physical spending of money. For example, if a company buys a computer system for $15,000 and depreciates it over five years, it might account for the system's depreciation as a $3,000-per-year or $350-per-month expense. Some refer to depreciation as a "paper expense," because—as a noncash expense—it shows up on a financial report, not in a checkbook.

Guidelines established by Generally Accepted Accounting Standards (GAAS) and Generally Accepted Accounting Principles (GAAP), and through pronouncements issued by the American Institute of Certified Public Accountants (AICPA) and the Financial Accounting Standards Board (FASB) are generally followed in identifying where in these five different classifications an item fits, but the rules are not hard and fast. As a result, questions will sometimes arise as to whether or not something categorized as an expense actually is an expense. In some cases, it may be an expense insofar as a firm is concerned, but not according to the IRS. In such an instance, the IRS will modify the basis on which tax is computed. For example, a firm's records may show that it earned $1 million in revenue for a year and had expenses totalling the same amount, reducing its income tax liability to zero. The IRS, after an audit of the firm's records, may say that $200,000 of the claimed expenses are not expenses for tax purposes and that, as a result, expenses were only $800,000, creating a profit (net income before taxes) of $200,000 and taxes due of $56,000. As a result, assuming the firm does not content the IRS position, government records would show the firm having a net profit after taxes of $144,000, while the firm's books could feasibly show that the company operated at a $56,000 loss.

Charts of Accounts Small firms that experience a limited number of financial transactions each year usually do not require extensive detail on the specific nature of each item. Larger firms do need such specificity because there is so much variability; failure to provide detail could result in chaos.

The highly detailed chart of accounts shown in Figure 9-1 is based on a model developed by the American Consulting Engineers Council (ACEC). As a review indicates, assets alone can be subdivided into almost seventy categories. Many firms use this chart of accounts exactly as presented; others have developed their own. Even those that use the identical chart do not necessarily agree on how specific items should be treated, however, underscoring the fact that accounting is far more judgmental than some may assume. General agreement on the treatment of certain issues is helpful for certain purposes, such as surveys or financial analysis preparatory to a merger or a buy-out. Nonetheless, the key concern is developing an approach suitable for a specific firm and the nature of the reports it prefers. Computerization makes development of reports and the entire accounting function far simpler than reliance on manual methods. Even small firms, including one-person operations, can now rely on a computer for such purposes.

    1000 Series—Assets (All things owned by the firm.)
    1100 Series—Current Assets (Cash or items convertible to cash within one year.)
        1100—Cash (Summary control account for all cash and cash equivalent items.)
        1101—Petty Cash (Amount established for the petty cash fund. Normally this fund is handled on the imprest basis; that is, as funds are used, a reconciliation is prepared and funds are replenished up to the limit of the fund.)
    1102 to 1109—Cash: Bank Account A, B, and so on. (Each bank account—checking and savings—listed separately, except for those listed below.)
        1110—Cash Reserves (Cash reserves held in an account separate from the above.)
        1111—Reserve: Bonus and Incentive Payment (Cash held in an account specifically earmarked for this purpose.)
        1120—Accounts Receivable (Summary control account for all outstanding billings to clients. Record further breakdowns when information is pertinent.)

Figure 9-1: Chart of Accounts. (Based on "ACEC Uniform Chart of Accounts," appearing in ACEC Guidelines to Practice, Volume II, No. 1, Financial Management and Project Control for Consulting Engineers)

1121—Accounts Receivable: Professional Fees (That portion of billings that includes labor, overhead, and profit, when these elements can be distinguished.)

1122—Accounts Receivable: Reimbursables (That portion of billings for reimbursable expenses plus mark-up, if any, when distinguishable.)

1123—Accrued Interest and Dividends Receivable (Interest and dividends due but not yet received.)

1124—Rents Receivable (All rental income not yet received.)

1125—Accounts Receivable: Sale of Capital Assets (Amounts due from sale of capital assets.)

1126—Allowance for Doubtful Accounts (Amounts set aside as reserves for potential bad debts. Reserves can be established against specific accounts, or a general reserve can be established as a percentage of total accounts receivable.)

1127—Salary Advances (All advances against salaries not yet repaid.)

1128—Expense Advances (Advances for travel or other advances given to employees but not yet repaid.)

1129—Accounts Receivable: Other (All other accounts receivable.)

1130—Notes Receivable (Summary control account for recording all outstanding notes from clients and others.)

1131—Notes Receivable: Sale of Capital Assets (All notes received from purchasers in connection with the sale of capital assets.)

1132—Notes Receivable: Officer Loans (The amount of all evidences of indebtedness signed by officers of the firm.)

1133—Notes Receivable: Other (All other notes received for any other purpose.)

1140—Earned Unbilled Revenue (Summary control account for recording the value of work in progress or unbilled accounts receivable, based preferably on selling price, that is, direct labor, overhead reimbursables, and profit representing the amount that will eventually be billed.)

1141—Work In Progress: Not Yet Billed (The amount of labor, overhead, and profit earned but not yet billed.)

1142—Reimbursable Expenses: Not Yet Billed (Unbilled reimbursable expenses plus mark-up, if any.)

1143—Other Earned Unbilled Revenue (Any other unbilled revenue from sources other than projects.)

1150—Investments: Securities (All short-term investments owned by the firm.)

1160—Prepaid Expenses (Summary control account for recording all expenses paid in advance.)

1161—Rent (Prepaid rent.)

Figure 9-1: (Continued)

1162—Taxes (Taxes paid in advance.)

1163—Interest (Interest payments made in advance.)

1164—Insurance (Insurance deposits or other prepaid amounts.)

1170—Cash Deposits (All deposits made for such items as air travel cards.

1180—Other Current Assets (Other current assets not otherwise listed.)

1200 Series—Fixed Assets (Assets with a life of more than one year.)

1200—Office Equipment (Summary control account for all items of office equipment recorded as an asset. Firms generally establish a policy regarding the dollar value of equipment carried as an asset; for example, all equipment costing $200 or more is recorded as an asset and depreciated; anything under $200 is charged directly as an expense when acquired.)

1201—Furniture, Fixtures, Equipment (All items acquired and depreciated over a period of years.)

1202—Accumulated Depreciation (All depreciation charged against the furniture, fixtures, and equipment.)

1203 to 1209—Office Equipment: Other (Any further breakdowns in office furniture and fixtures that may be needed with accompanying account for accumulated depreciation.)

1210—Technical Equipment (Summary control account for all items of technical equipment, such as laboratory equipment.)

1211—Equipment: Instruments, Library (Surveying equipment, instruments, and technical library.)

1212—Accumulated Depreciation (Depreciation charged against technical instruments and library.)

1213 to 1219—Technical Equipment: Other (Any further breakdowns in technical equipment that may be needed with accompanying account for accumulated depreciation. Use additional numbers to segregate technical equipment from library, if necessary.)

1220—Transportation Equipment (Summary control account for all transportation equipment.)

1221—Autos, Trucks, Planes (Capital assets of a transportation nature.)

1222—Accumulated Depreciation (All depreciation charged against transportation equipment.)

1223 to 1229—Transportation Equipment: Other (Any further breakdowns in transportation equipment that may be needed with accompanying account for accumulated depreciation. Use additional numbers to segregate autos, trucks, and planes, if necessary.)

1230—Patents (Summary control account for all patents, including the cost of research, development, and applicable legal fees.)

1231—Patents (All costs associated with acquiring the patent and the value of the patent itself.)

Figure 9-1: (Continued)

1232—Accumulated Amortization of Patents (All accumulated write-downs of the patents.)

1233 to 1239—Other Patents (Any further breakdowns required by individual patents with accompanying amortization account.)

1240—Leasehold Improvements (Summary control account for all types of leasehold improvements.)

1241—Leasehold Improvements (The acquisition cost of all leasehold improvements.)

1242—Accumulated Amortization of Leasehold Improvements (All accumulated write-downs of leasehold improvements, normally written off over the term of the lease.)

1243 to 1249—Leasehold Improvements: Other (Any further breakdowns required by individual leasehold improvement with accompanying amortization account.)

1250—Buildings (Summary control account for recording all buildings owned by the firm.)

1251—Building A (An individual building.)

1252—Accumulated Depreciation (Accumulated depreciation against the building.)

1253 to 1259—Other Buildings (Any additional buildings owned by the firm with accompanying depreciation account.)

1260—Land (Summary control account for all land owned in the name of the firm. Since land is listed separately and not depreciated, an allocation of costs needs to be made between buildings and land at time of acquisition.)

1261—Parcel A (The acquisition cost of an individual parcel of land.)

1262 to 1269—Other Land (All other parcels of land owned by the firm.)

1270—Investments (Long-Term) (All long-term investments, beyond one year, that are owned in the name of the firm.)

1271—Investment #1 (The individual investments owned by the firm. Long-term investments are generally recorded at cost or market price, whichever is lower.)

1272 to 1279—Other Investments (All other investments owned by the firm.)

1280—Other Assets (Summary control account for all other long-term assets.)

1281—Goodwill (The excess paid for a firm beyond its net book value. Goodwill is written off over a period of years not to exceed forty.)

1282—Organizational Costs (The capitalized costs incurred at the time the firm was organized. Organization costs are generally written off over a period of five years.)

1283 to 1289—Other Long-Term Assets (Any other long-term assets not otherwise mentioned.)

Figure 9-1: (Continued)

1300 Series—Deferred Assets (Summary control account for all other items of deferred assets. Payment for some assets is delayed for a period of time, as in the case of an installment contract. In that case, the value of the asset is recorded here.)

2000 Series—Liabilities (Claims against the firm's assets.)

2100 Series—Current Liabilities (Liabilities due within one year.)

    2100—Accrued Salaries and Payroll (Summary control account for all items owed by the firm to employees and officers but not yet paid.)

    2101—Accrued Salaries (Salaries owed but not yet paid.)

2102 to 2109—Other Accrued Salaries and Payroll (Accounts used to break down accrued salaries into finer detail, if necessary, as by officers, engineers, and drafters, for example.)

    2110—Accrued Payroll Overhead (Summary control account for all accrued payroll overhead items.)

    2111—Accrued Bonus and Incentive Payments (Obligations for expected bonus and other incentive payments.)

    2112—Accrued Paid Leaves (Obligations due and not yet paid for leaves with pay.)

    2113—Accrued Employer Salary Taxes and Insurance (The employer's portion of payroll taxes and insurance due but not yet paid. Accounts 2114 to 2118 represent further breakdowns of these obligations.)

    2114—Federal Social Security (FICA) Tax (Employer's portion not yet paid.)

    2115—Federal Unemployment Insurance (The unpaid portion of this tax paid wholly by the employer.)

    2116—State Disability Insurance (The unpaid balance of what is in some states an obligatory insurance.)

    2117—State Unemployment Insurance (Paid by the employer; the unpaid balance.)

    2118—Workers' Compensation Insurance (The unpaid balance of this obligatory insurance.)

    2119—Accrued Employer Contributions and Benefits (Any voluntary benefits owed by the employer but not yet paid. Accounts 2120 to 2122 represent further breakdowns of these obligations.)

    2120—Employees' Group Insurance (Life and Health) (All obligations owing but not yet paid to the group insurance plan.)

    2121—Employees' Retirement: Pension Plan (Amounts due to the retirement plan.)

    2222—Employees' Education (Obligations due for employer-paid education courses.)

2123 to 2129—Other Employer Contributions (Accounts used to record any obligations for employer contributions to any other employee benefit plans not otherwise listed.)

Figure 9-1: (Continued)

2130—Accrued Payroll Deductions (Summary control account for the employee portion of all obligations paid through payroll deductions.)

2131—State Income Tax Withholding (State income taxes withheld from employees but not yet paid.)

2134—Employees' Group Insurance (Life and Health) (Employees' contributions to group insurance withheld but not yet paid.)

2135—Employees' Retirement: Pension Plan (All amounts withheld from employees for contributions to the retirement plan that have not yet been paid.)

2136—Other Payroll Deductions (Employees' contributions to any other benefit plans not mentioned above that were withheld but that have not yet been paid.)

2140—Notes Payable (Short-Term) (Summary control account for any notes of the firm due within one year.)

2141—Note Payable #1 (The individual note and the amount owing.)

2142 to 2149—Other Notes Payable (Any other individual notes in these accounts, if further breakdown is required.)

2150—Accrued Business Taxes Payable (Summary control account for taxes that are owing but not yet paid.)

2151—Federal Income Tax (Deferred) (The amount owed for taxes not yet paid. In many cases, this difference is caused by lesser taxes due under the cash basis than under the accrual basis of accounting. The obligation remains and must be shown, even though the taxes owed under the accrual basis might not be paid for a long time, if ever.)

2152—State Income or Franchise Tax (Any state income or franchise taxes owed but not yet paid.)

2153—City Business or Property Tax (Any city business or property taxes owed.)

2154—Gross Receipts or Sales Tax (Any gross receipts or sales taxes owed but not yet paid. Sales taxes are generally not levied against service in most states but may be collected against anything else the design professional sells, such as reports, plans, and so on.)

2160—Accounts Payable (Summary control account for recording all amounts due to vendors and others.)

2161—Consultant Service Fees (All amounts due to outside consultants for project work.)

2162—Technical Service Fees (Amounts due to outside technical personnel for services performed.)

2163—Rent, Utilities, and Maintenance (Any amounts owing for rent, utilities, and maintenance.)

2164—Office Supplies and Services (Amounts owed to vendors for office supplies and services.)

Figure 9-1: (Continued)

2165—Reproduction and Photo Expenses (Amounts owed to vendors for reproduction and photo expenses.)

2166—Telephone/Telegraph (Amounts owed for telephone, telex, and other types of communications, expenses.)

2167—Miscellaneous Accounts Payable (Any amounts owed to vendors that were not recorded above.)

2170—Accrued Interest and Bank Charges (Summary control account to record interest expense and bank service charges due.)

2171—Interest (Interest expense owed to banks and others on outstanding notes and loans.)

2172—Bank Charges (Any service charges levied by the bank on the firm's account.)

2180—Refundable Income (Summary control account for income items not yet earned.)

2181—Retainers on New Projects (The unexpended portion of prepaid income.)

2182—Deposits Held on Plans (Any deposits that may be returned.)

2183 to 2189—Other Refundable Income (Any other items of prepaid income.)

2190—Dividends Payable (Summary control account to record dividends that have been declared by the board of directors but have not yet been paid.)

2192—Dividends Payable #1 (Record here the dividends payable on one of the classes of stock, such as common stock.)

2192 to 2193—Other Dividends Payable (Use these additional accounts to record dividends declared but not yet paid on other classes of stock.)

2200 and 2300 Series—Long-Term Liabilities (Liabilities due beyond one year.)

2200—Mortgages Payable (Summary control account to record the portion of mortgages that are payable beyond one year.)

2201—Mortgages Payable #1 (The amount of mortgage due on one loan.)

2202 to 2209—Other Mortgages Payable (Accounts used to record the amount of long-term mortgages due on other loans.)

2210—Notes Payable (Long-Term) (Summary control account for the amount of notes due after one year.)

2211—Notes Payable #1 (The amount of the note due on one loan.)

2212 to 2219—Other Notes Payable (Accounts used to record the amount of long-term notes due on other loans.)

2400 Series—Deferred Liabilities (Summary control account to record deferred liabilities, that is, obligations due in the future, such as contingent payments based on earnings of an acquired firm.)

3100 Series—Net Worth (Summary control account to record the net worth or equity accounts. Net worth is the difference between assets and

Figure 9-1: (Continued)

liabilities and represents the ownership interest in the firm. This ownership interest is expressed differently depending on whether the firm is a sole proprietorship, partnership, or corporation.)

IN THE CASE OF AN INDIVIDUAL OWNER:

3100—Capital (The amount of capital originally invested by the owner when the firm was started.)

3200—Withdrawals (Withdrawals made by the owner in lieu of drawing a salary if the owner were an employee.)

3300—Profit and Loss (An account used to close out the net income or loss for the year. It represents the cumulative earnings or losses since the firm began.)

IN THE CASE OF A CORPORATION:

3100—Capital Stock: Common (The original investment of the incorporators when they began, representing their initial purchase of common stock of the firm.)

3200—Capital Stock: Preferred (If any preferred stock has been authorized and issued, the initial purchase is recorded here.)

3300—Retained Earnings (The cumulative earnings or losses since the firm began up to the current year.)

3310—Profit and Loss (Net earnings/loss for the current year.)

3400 to 3900—Paid-In Capital, Treasury Shares, and so on (Accounts used to record additional transactions that commonly occur. For example, paid-in capital is that amount paid for stock in excess of its par value. It is recorded separately from the capital stock account. Treasury shares are stocks that have been repurchased by the corporation and put in the treasury. Treasury shares have not been retired and may be sold again in the future.)

IN THE CASE OF A PARTNERSHIP:

3100—Capital (Summary control account to record the capital investment made by the partners to purchase their partnership interests. The amount changes as a partnership interest increases or decreases due to the entry or exit of partners, and infusion of additional capital by existing partners.)

3101—Capital, Partner A (The individual capital account showing all transactions pertaining to one partner.)

3102 to 3199—Capital, Other Partners (The individual capital accounts of the other partners.)

3200—Drawings (Summary control account to record withdrawals made by all partners.)

Figure 9-1: (Continued)

3201—Drawings, Partner A (The drawing account pertaining to one partner.)

3202 to 3299—Drawings, Other Partners (Individual drawing accounts of the other partners.)

3300—Profit or Loss (Cumulative earnings/losses of the partnership since it began.)

4000 Series—Income (Revenues earned by the firm.)

4100—Professional Fees (Invoiced) (All amounts earned on projects from clients. Fees earned on labor, overhead, and profit should be included when distinguishable in the contract. Otherwise, record all earnings in this account where the elements are not distinguishable, such as in lump sum or percentage of construction projects.)

4200—Reimbursable Expenses: Invoiced (Summary control account to record earnings from reimbursable expenses, including mark-up, if any.)

4201—Purchased Service (Income from outside purchased service, such as consultants.)

4202—Project Reproduction and Printing (Reimbursements for reproduction and printing.)

4203—Long-Distance Telephone and Telegram (Reimbursements for telephone and other communications charges.)

4204—Allocated (In-House) Computer (Portion earned on charges for in-house computer service.)

4205—Project Travel and Subsistence (Reimbursements for travel and related expenses.)

4206—Other Reimbursable Expenses (Reimbursements for any other expenses not listed above.)

4300—Temporary Staff Assignments: Contract Labor (Revenue earned on outside contract labor.)

4400—Interest, Dividends (Income earned from investments, savings accounts, and similar sources.)

4500—Rents (Income earned on rental property held in the name of the firm.)

4600—Sale of Capital Assets (Gain/Loss) (Any earnings or losses incurred as a result of sale of capital equipment owned by the firm.)

4700—Other Income (Any revenues earned from sources not mentioned above.)

4800—Losses on Uncollectable Accounts (Losses incurred as a result of bad debts or work in progress that cannot be invoiced for whatever reason.)

4900—Work in Progress (Professional fees and reimbursable expenses earned but not yet invoiced.)

Figure 9-1: (Continued)

5000 Series—Direct Project Expenses (Cost directly attributable to specific projects.)

    5101—Payroll, Staff (The direct labor cost for design professionals, designers, drafters, and other technical employees, including overtime.)

    5102—Salaries of Officers and Principals (The cost for officers and principals when directly engaged in project work.)

    5200—Direct Project Expenses (Summary control account to record expenses directly attributable to a project.)

    5201—Purchased Service (The expense of purchasing outside services, including consultants. Accounts 5202 to 5206 permit further breakdown of these services, if needed.)

    5202—Site Surveys

    5203—Aerial Photogrammetry

    5204—Subsurface Exploration and Laboratory Testing

    5205—Computer Services

    5206—Other Contracted Services

    5207—Project Reproduction and Printing (All expenses pertaining to reproduction and printing, whether outside services or performed in-house. If in-house facilities are used, calculate rate that would be comparable to outside services.)

    5208—Long-Distance Telephone and Telegram (All telephone, telex, and other communications charges directly applicable to projects.)

    5209—Allocated (In-House) Computer Expense (All charges to projects for use of in-house computer. The rate established should be comparable to charges for equivalent outside services.)

    5210—Project Travel and Subsistence (All charges for travel and subsistence in connection with projects.)

    5211—Other Direct Expenses (All other direct project expenses not listed elsewhere.)

6000 Series—Indirect Costs

    6100—Indirect (Payroll-Connected) Cost (Summary control account for all indirect payroll costs.)

    6101—Bonuses, Incentive Payments (All bonus payments and incentive awards. Extraordinary awards in any particular year should not be used in determining ratios for succeeding years.)

    6102—Vacations (The amount charged to vacations, if there is no accrual for vacations. If vacations are accrued, the amount of accrual is recorded and vacations taken are charged against the accrual.)

    6103—Holidays (Time charged to holidays. If holidays are accrued, the amount of the accrual is recorded and holidays taken are charged against the accrual.)

Figure 9-1: (Continued)

6104—Sick Leave (Time charged to sick leave. Time off for sick leave is normally not accrued in a professional firm because professionals are not likely to use all of their sick leave.)

6105—Other Paid Leave (Any time charged to leave with pay.)

6106—Employer Salary Taxes and Insurance (The employer's portion of payroll taxes and insurance. Accounts 6107 to 6111 permit further breakdowns, if needed.)

6107—Federal Social Security (FICA) Taxes

6108—Federal Unemployment Insurance

6109—State Disability Insurance

6110—State Unemployment Insurance

6111—Workers' Compensation Insurance

6112—Employer Contributions and Benefits (The employer's contribution to employee benefit plans. Accounts 6113 to 6116 permit further breakdowns, if needed.)

6113—Employee Group Insurance (Life and Health)

6114—Employee's Retirement—Pension Plan

6115—Employees' Education

6116—Employees' Recreation

6200 to 6499—Indirect (General and Administrative) Costs

6210—Indirect (Nonproject) Payroll (Administrative, clerical, bookkeeping, and salaries of officers, principals, partners, and associates for time expended on nontechnical work of administration, general supervision, business development, and unassigned (standby) time. Accounts 6211 to 6219 are available for further breakdowns, if needed.)

6211—Payroll—Staff

6212—Salaries of Officers and Principals

6220—Business Taxes (All expenses associated with business taxes. Accounts 6221 to 6229 permit further breakdowns, if necessary.)

6221—Federal Income Tax

6222—State Income or Franchise Tax

6223—City Business or Property Tax

6230—Legal, Accounting (Charges for outside accounting and legal advice, including audits and tax preparation. Accounts 6231 to 6239 permit further breakdowns, if necessary.)

6231—Legal

6232—Accounting

6240—Interest and Bank Charges (All bank charges and interest expense.)

6241—Interest

6242—Bank Charges

Figure 9-1: (Continued)

6250—Rent, Utilities, Maintenance (Costs for office space, including cleaning, repairs, and maintenance.)

6251—Rent

6252—Utilities

6253—Cleaning and Related Expenses: Trash Removal Janitorial Service, and so on

6254—Repairs, Maintenance

6260—Office Supplies and Services (Costs for office supplies and services not charged to projects.)

6261—Clerical Supplies and Stationery

6262—Engineering/Drafting Supplies

6263—Postage and Shipping

6265—Miscellaneous Office Expenses (Include courier services.)

6270—Telephone, Telegraph (All communications costs not charged to projects. Include the cost of the installation and local service as well as any long-distance calls that cannot be charged to projects.)

6280—Professional Activities Expenses (Costs for professional activities for staff members and officers.)

6281—Membership, Dues, Subscriptions

6282—Luncheons and Meetings

6283—Annual Meetings and Conventions

6284—Workshops and Seminars

6285—Professional Registrations and Licenses

6286—Travel and Subsistence (Used for recording this portion of the expense separately, if necessary.)

6290—Auto, Truck, Plane Expenses (Transportation costs not charged to projects. Include actual costs of vehicles owned by the firm, as well as mileage costs for personnel who use their own cars on company business.)

6291—Gas Oil, Lubrication

6292—Garage and Hangar Rent

6293—Maintenance and Repair

6294—Insurance

6300—Business Development, Public Relations (Expenses in connection with acquiring new projects that have not yet been identified. Time for business development is recorded under the appropriate indirect labor account.)

6301—Travel and Subsistence (Travel in connection with following up leads, business lunches with prospective clients, and so on. Firms involved with government contracting should consider combining this account with 6311, because both are considered allowable in overhead if the

Figure 9-1: (Continued)

travel is in connection with prospects of proposal preparation. Entertainment and directory adverstising are specifically unallowed.)

6302—Entertainment (Expenses not associated with a proposal or specific new business activity, such as tickets for sporting events, opera, and so on.)

6303—Professional Directories/Advertising

6310—Precontract Activities (Expenses in connection with proposal preparation.)

6311—Travel and Subsistence (Travel in connection with preaward conferences, making presentations.)

6312—Brochures and Proposals

6320—Administrative Travel and Subsistence (Travel in connection with visits to other firms and firm management.)

6330—Business Insurance (Expenses for liability insurance coverage.)

6331—Professional Liability

6332—Public Liability

6333—General Property

6334—Key Man Life and Disability

6335—Other Types of Insurance (Except Payroll)

6340—Civic Contributions (Charitable contributions and donations.)

6350—Office Furniture, Fixtures, and Equipment (Expenses in connection with leases and rentals.)

6351—Rent

6352—Maintenance and Repair

6360—Technical Instruments and Equipment (Expenses for surveying and other types of technical equipment.)

6361—Rent

6362—Maintenance and Repair

6370—Library and Reference Data (Expenses in connection with the acquisition of catalogs, directories, and reference material.)

6371—Maps, Charts

6372—Books and Texts

6373—Periodicals, Subscriptions (For reference material, as distinguished from professional subscriptions.)

6380—Research and Development (Expenses for research into new technical developments, such as the implementation of CAD systems.)

6390—Miscellaneous Expenses (Expenses not recorded elsewhere.)

6391—Recruiting (Include newspaper ads, agency fees, expenses for interviewing, campus recruiting expenses and literature, moving allowance for new employees, and testing for new employees.)

Figure 9-1: (Continued)

6392—Other

6400—Depreciation (The depreciation of capital equipment over its estimated useful life.)

6401—Fixtures, Equipment, Furniture

6402—Technical Instruments, Equipment

6403—Autos, Trucks, Planes

6404—Amortization of Patents

6405—Amortization of Leasehold Improvements

6406—Buildings

6407—Library and Reference Data

Figure 9-1: (*Continued*)

Accounting Methods

Several accounting methods are available. Cash accounting and accrual accounting are used most extensively. Through cash accounting, income is recorded as a receipt when the cash is received, and an expense is recorded as an expense when the firm pays a bill.

In accrual accounting, money is considered income when it is earned, and an item is considered an expense when it is incurred. For example, consider account no. 1141 in the chart of accounts shown in Figure 9-1, which creates an asset account for work in progress, even though the work has not yet been billed.

Accrual accounting is far more popular than cash accounting, and—for larger firms—is required by the IRS. Although it tends to be more complex than cash accounting, it gives a much better "picture" of a firm's true financial position, because it permits periodic reports to match revenues against the expenses required to produce it. As an example, consider the simplistic report shown for purposes of illustration in Figure 9-2. Through the cash method of accounting, income for the period is shown to be $110,000, because the Project A client has finally paid the last bill for work performed in a prior reporting period, and the Project B client has paid the first bill. Under the expense column, however, direct labor charges associated with five projects are indicated. (Direct labor charges are the payroll expenses of those working on the projects, for the amount of time involved.) No direct labor expense is shown for Project A, because work was completed during an earlier period. The direct labor associated with the other projects greatly exceeds the income shown for them, because no bills have been issued or, if they have been, they have not yet been paid. As a consequence, the firm is showing a $53,000 before-tax profit for the period, but this is extremely misleading; most of the expenses have nothing to do with the income.

	Cash	Accrual
Income		
Project A	$100,000	$ 0
Project B	10,000	30,000
Project C	0	20,000
Project D	0	20,000
Project E	0	5,000
Total	$110,000	$75,000
Expenses		
Project A	$ 0	$ 0
Project B	11,000	11,000
Project C	7,300	7,300
Project D	7,000	7,000
Project E	1,700	1,700
Other	30,000	38,000
Total	$ 57,000	$65,000
Net Profit (Loss)	$ 53,000	$10,000

Figure 9-2: Impact of Cash Accounting versus Accrual Accounting

Now consider the data shown in the accrual column of Figure 9-2, where income includes the billable value of work in progress and expenses incurred, whether or not it has been paid. No income is shown for Project A, because it would have been shown in prior reports, at the time it was earned. The direct labor expenses are identical, because, as payroll costs, they are paid on a regular basis, but "other" expenses are higher because they include items for which costs have been incurred but have not yet been paid. The before-tax profit shown through the accrual method is $10,000, which is a far more accurate representation of the firm's position. As such, accrual accounting gives the firm's management a much better understanding of how much it actually cost the firm to produce its revenue.

ESTABLISHING OVERHEAD

It is usually essential for design professional organizations to know precisely what their overhead rates are, in order to permit formulation of appropriate charges for the work they perform. There are a number of ways of developing this "cost accounting" data, and some are based more on art than science. For example, some firms will indicate that their overhead is far lower than others' by virtue of the calculation method used, even though the amount charged to clients for the same amount of work by the same people would be the same. Design professionals should understand the nuances involved, not only so that they can give proper direction to their own financial personnel, but also to offer explanations to clients who may become confused by the seeming differences in numbers.

Understanding the Variables

The key variables associated with establishing overhead are *direct labor, indirect labor, payroll burden,* and *general and administrative expenses.*

Direct labor comprises billable time spent on projects by direct labor personnel, such as design professionals, designers, specifiers, and drafters. Other feasibly could be considered direct labor personnel, as discussed below.

Indirect labor refers to the hours spent by persons whose time is not normally charged to projects, such as clerical and accounting staff, or the time spent by direct labor personnel on nonproject activities, such as marketing, firm administration, and personnel recruitment.

Payroll burden comprises expenses that the firm covers on behalf of employees, such as payroll taxes, workers' compensation insurance, health insurance, and so on, as well as the value of paid leave. Payroll burden applies to direct and indirect labor alike.

General and administrative expenses are those associated with running a business, such as rent, equipment leases, automobile expenses, income taxes, and professional services.

Applying the Variables

One method for calculating overhead for a coming year would proceed as follows.

Identify Billable versus Nonbillable Hours Staff members would be differentiated in terms of direct labor personnel versus indirect labor personnel. The direct labor hours (net billable hours) worked by direct labor staff then would be identified. This can be done by determining the number of hours for which direct labor personnel are paid, usually (40 hours per week × 52 weeks per year =) 2,080 hours per year, and subtracting from that sum the number of nonbillable hours. Nonbillable hours comprise paid leave for vacations, holidays, and sick leave, as well as time spent on administrative or similar activities. The hours likely to be spent on these latter activities can be estimated based on a review of the firm's personnel records. The total number of nonbillable hours then is calculated and subtracted from the number of paid hours, to derive net billable hours. This procedure is illustrated in Table 9-1, which presumes a hypothetical design professional firm that consists of one principal, two registered architects or engineers, two designers, two drafters, one bookkeeper, and one word processor. Table 9-1 does not consider the hours of the bookkeeper and the word processor because their time would be fully attributable to indirect labor.

TABLE 9-1: Billable versus Nonbillable Hours Determination

	Payroll Burden Hours		Indirect Labor Hours				
Direct Labor	Holidays	Paid Leave	Business Development	Administration	Other	Total Overhead Hours	Total Billable Hours
Principal	80	160	300	300	100	940	1,140
Design Professional A	80	160	150	100	100	590	1,490
Design Professional B	80	120	75	50	100	425	1,655
Designer A	80	120	0	0	100	300	1,780
Designer B	80	120	0	0	100	300	1,780
Drafter A	80	120	0	0	100	300	1,780
Drafter B	80	120	0	0	100	300	1,780

TABLE 9-2: Salary of Direct Labor Personnel for Billable versus Nonbillable (Overhead) Hours

Direct Labor	Annual Salary	Salary per Hour	Billable Hours	Salary for Billable Hours	Overhead Hours	Salary for Overhead Hours
Principal	$ 82,200	$40	1,140	$ 45,600	940	$37,600
Design Professional A	52,000	25	1,490	37,250	590	14,750
Design Professional B	41,600	20	1,655	33,100	425	8,500
Designer A	31,200	15	1,780	26,700	300	4,500
Designer B	27,040	13	1,780	23,140	300	3,900
Drafter A	22,880	11	1,780	19,580	300	3,300
Drafter B	18,720	9	1,780	16,020	300	2,700
Total	$276,640			$201,390		$75,250

Identify Salary for Billable versus Nonbillable Hours Payroll costs in this procedure are considered the salaries received by direct labor personnel, not including any payroll burden items, fringe benefits, and so on. Table 9-2 identifies the annual salary of each of the seven direct labor staff members, their hourly salary (annual salary ÷ 2,080 hours), and that portion of their salary

TABLE 9-3: Establishing Overhead Costs and Rates

Payroll Burden

Holidays	$ 10,640
Paid Leave	18,560
Group Insurance	13,000
Workers' Compensation Insurance	1,000
Education	3,000
Professional Dues and Licenses	1,000
Profit-Sharing/Pension	9,000
Other Benefits	6,000
Payroll Taxes	19,500
Total	$ 81,700

Indirect Labor

Accounting	$ 31,200
Word Processing	22,880
Business Development	17,250
Administration	15,500
Other	13,300
Total	$100,130

General and Administrative

Rent	$ 30,000
Utilities	5,000
Equipment Leases	12,000
Maintenance/Repairs	3,500
Supplies	3,500
Insurance	45,000
Professional Services	4,000
Bank Loans (Interest)	5,000
Other Taxes	30,000
Other	25,000
Total	$163,000

Total Overhead

Payroll Burden	$ 81,700
Indirect Labor	100,130
General and Administrative	163,000
Total	$344,830
Overhead as a Percentage of Direct Labor (Billable Salary)	171.2%
Break-Even Multiplier	2.712

associated with billable and nonbillable hours, respectively. This segregation of billable versus nonbillable hours and salary is important, because the firm's income is derived from the billable hours worked by direct labor personnel. As such, each hour worked must recover the hourly cost of the individual involved, plus a proportionate share of overhead plus profit.

Identify Overhead Costs and Rate   Table 9-3 indicates overhead costs. The listing shown is abbreviated, even for a firm of this size, for purposes of illustration. Both payroll burden and indirect labor computations include the value of salary given for direct labor personnel's nonbillable hours. Failure to include these values would result in their not being recaptured.

As can be seen, the total cost of overhead ($344,830) is 171.2 percent of the salary direct labor personnel receive for their billable time ($201,390). As such, multiplying the value of billable time by 2.712 (the "break-even multiplier") should result in total annual costs being recovered. Total annual costs are the salaries given for billable work ($201,390) plus the cost of overhead ($344,830), or $546,220. As shown in Table 9-4, multiplying billable hours salary ($201,390) by 2.712 yields almost an exact result ($546, 430).

Lowering Overhead

As noted above, different methods are sometimes used to calculate overhead, in order to lower it. Some of the techniques used are discussed below.

Applying Payroll Burden to the Direct Labor Base   One approach to lowering overhead is to apply payroll burden directly to the direct labor base. In the example given above, this would drop overhead from 171 percent to 93 percent, as shown in Table 9-5. (For purposes of illustration, the entire payroll burden has been applied in this manner, not just the payroll burden associated with direct labor personnel.) Although being able to cite a comparatively lower overhead rate may impress a somewhat naive client, the "bottom line"

TABLE 9-4: Verification of Break-Even Multiplier's Accuracy

Direct Labor	Billable Hours	Hourly Salary × 2.712	Total
Principal	1,140	108.50	$123,690
Design Professional A	1,490	67.80	101,022
Design Professional B	1,655	54.25	89,784
Designer A	1,780	40.70	72,446
Designer B	1,780	35.30	62,834
Drafter A	1,780	29.85	53,133
Drafter B	1,780	24.45	43,521
			$546,430

TABLE 9-5: The Effect of Applying Payroll Burden to Direct Labor Salary Rather than Overhead

Direct Labor	
Billable Salary	$201,390
Payroll Burden	81,700
Total	$283,090
Payroll Burden as a Percentage of Direct Labor (Billable Salary)	40.6%
Multiplier	1.406
Overhead (Less Payroll Burden)	
Indirect Labor	$100,130
General and Administrative	163,000
Total	$263,130
Overhead as a Percentage of Direct Labor (Billable Salary Plus Overhead Burden)	92.9%
Break-Even Multiplier	1.929

stays the same. As shown in Table 9-6, the change elevates the hourly salary basis of direct labor personnel.

Applying payroll burden to the direct labor base is not a conventional acounting procedure. Some design professionals frown upon it, regarding it as a commercial strategy that is alien to professional practice. However, there are some clients, particularly in the public sector, who refuse to deal with a firm unless its overhead or multiplier (a factor applied to direct labor cost to recover overhead and profit) is within certain prescribed limits. This is an overly dogmatic outlook that fails to consider that a firm's higher overhead may result from investments in equipment, which boosts productivity, or from higher pay to personnel who are particularly gifted. As a consequence, the overall cost of service might be lower than another firm's with a lower overhead, and/or the higher quality of service might result in construction

TABLE 9-6: The Bottom Line Remains the Same When Payroll Burden Is Applied To Direct Labor Salary, Even Though Overhead Is Lowered

Direct Labor	Billable Hours	Hourly Salary × 1.406	Subtotal	Subtotal × 1.929
Principal	1,140	$56.25	$64,125	$125,697
Design Professional A	1,490	35.15	52,374	101,029
Design Professional B	1,655	28.15	46,588	89,868
Designer A	1,780	21.10	37,558	72,449
Designer B	1,780	18.30	32,574	62,835
Drafter A	1,780	15.50	27,590	53,221
Drafter B	1,780	12.65	22,517	43,435
Total				$546,534

and/or life-cycle savings that greatly exceed the premium paid to derive such savings. If a manipulation of data results in a lower overhead rate in such cases, permitting a firm to work for a client that otherwise would not be interested, it is difficult to criticize the tactic. Do recognize, however, that many such clients have forms that a firm must complete in order to calculate overhead in a uniform manner, making data manipulation very difficult when the client's specifications are adhered to.

Applying Indirect Labor Hours Directly Another technique for lowering overhead is to treat certain aspects of indirect labor as direct labor. For example, clerical staff is often involved in a project, as when they type reports, prepare and issue memoranda, or perform research. Since the time required for performing these tasks can logically be ascribed to the project, the work of clerical staff could be considered direct labor and thus could be included in the workscope and charged to the project. In a similar manner, quality assurance reviews of project-related deliverables could be ascribed to specific projects and included in the workscope, as well as the cost of administering a project, including the time spent discussing projects in staff meetings. The paperwork requirements associated with such efforts are more than they would otherwise be, but the overhead reduction obtained could be a valuable competitive asset. Furthermore, when all is said and done, the time spent by clerical and other personnel on a specific project actually is direct labor. The practice of "lumping" such efforts in the indirect labor category is really a record-keeping shortcut.

The impact of considering clerical time as direct labor is shown in Table 9-7. As can be seen, based on assumptions made, the salary paid for direct labor hours increased by 9.3 percent, while that paid for overhead hours is reduced by 4.8 percent. (In passing, it is worthwhile to point out that the decline in overhead can be spoken of in three ways. One could say that overhead (based on dollars spent) decreased 4.8 percent, that the overhead rate was cut by 13 percent, or that the overhead rate was reduced by 22.2 percent, all depending on the frame of reference assumed. Comparisons such as these point out the art associated with precise numbers and underscore the wisdom of the observation, "There are liars. There are damned liars. And there are statisticians.")

Note that the same approach can be applied to certain equipment, such as computers. The entire cost can be absorbed through overhead, or, alternatively, project-related costs can be allocated as such. Some state that charging computer time against a project can not only enhance profits but also makes the service evident in the workscope or fee estimate, occasioning discussion of the firm's system, how it is employed to enhance productivity, and so on.

Adjusting the Hour Base Yet another technique for lowering overhead is to base calculations on the actual time worked, as opposed to an assumption of a forty-hour week. In many firms, it is not uncommon for senior personnel to

TABLE 9-7: The Effect of Applying Appropriate Clerical Hours to Direct Labor

Direct Labor	Salary for Billable Hours Before	Salary for Billable Hours After	Increase (Decrease)	(%)
Principal	$ 45,600	$ 53,500	$ 8,000	(+17.5%)
Design Professional A	37,250	39,125	1,875	(+5.0%)
Design Professional B	33,100	33,800	700	(+2.1%)
Designer A	26,700	26,700	0	
Designer B	23,140	23,140	0	
Drafter A	19,580	19,580	0	
Drafter B	16,020	16,020	0	
Word Processor	0	8,250	8,250	
	$201,390	$220,215	$18,825	(+9.3%)
Overhead	Before	After	Increase (Decrease)	(%)
Payroll Burden	$ 81,700	$ 83,900	$ 2,200	(+2.7%)*
Indirect Labor	100,130	81,305	(18,825)	(−18.8%)
General and Administrative	163,000	163,000	0	
	$344,830	$328,205	($16,625)	(−4.8%)
Percent of Direct Labor	171.2%	149.0%	(22.2%)	(−13.0%)

*Increase occurs because word processor is included with direct labor personnel, causing paid leave hours—removed from indirect labor—to be ascribed to payroll burden.

spend sixty hours or more each week in pursuit of their responsibilities. By considering this noncompensated additional time in calculations, the hourly salary is reduced, as is the value of direct labor personnel's time ascribed to payroll burden (holidays and paid leave) and indirect labor (business development, administration, etc.). Alternatively, the hourly rates can be kept the same, but the value of noncompensated additional time can be applied as a credit to overhead. The impact of doing this is shown in Table 9-8, in which we can see that the overhead rate is lowered from 171 percent to 139 percent.

Taking the latter approach is not recommended. Although it may help a firm obtain some projects, it severely reduces the inherent contingency that forty-hour-per-week computations establish. Many projects will take longer than anticipated, and, rather than attempting to seek additional compensation for a firm's errors in estimation, additional hours will be logged, if not to obtain profit, then at least to cut losses. Consider, too, that some firms may not include bonuses as part of overhead, intending to withdraw them instead from profits created specifically through otherwise noncompensated hours.

Other Techniques A number of other techniques are available for reducing overhead through statistical manipulation. Do not overlook the potential of reducing overhead by eliminating unnecessary expense. In doing so, however, closely evaluate the benefits to be derived. When extremely difficult straits are

TABLE 9-8: The Effect of Considering Noncompensated Hours of Certain Direct Labor Personnel

	Before	After	Increase (Decrease)	(%)
Direct Labor	$201,390	$201,390	$ 0	
Overhead				
Payroll Burden	$ 81,700	$ 81,700	0	
Indirect Labor	100,130	100,130	0	
General and Administrative	163,000	163,000	0	
Noncompensated Overtime Credit	0	(64,000)	($64,000)	
	$344,830	$280,830	($64,000)	(−18.6%)
Percent of Direct Labor	171.2%	139.4%	(31.8%)	(−18.6%)

experienced, because a client has defaulted on a major fee or business is extremely slow, then virtually anything that can be eliminated usually is. In more conventional circumstances, however, most firms operate in a relatively prudent manner. In other words, it generally is most effective to be diligent with respect to what is being added to overhead. Bear in mind that people tend to miss what they have never had much less than they miss what they used to have. Thus, it may be better to give employees a larger annual bonus than it would be to expand health plan benefits, since the health plan becomes a continuing element of overhead that will rise over time. Likewise, if low overhead is important to your firm, consider the wisdom of making as many costs as possible direct rather than indirect. Although the amount that a client pays as a result may be the same, overhead rates have taken on significant symbolic meaning to some. It is true: All clients should be principally concerned about the quality of service and the reasonableness of fee. Nonetheless, some are far more concerned about extraneous factors. To the extent that being able to quote a lower, yet still fully legitimate, overhead rate gives you a "leg up," so much the better. Experience will determine whether the additional paperwork required (if any) makes the effort worthwhile.

Complying with Client Computational Methods

As previously mentioned, some clients require that all firms interested in working for them calculate their overhead in the same way and provide standard forms for that purpose. The nation's number one "consumer" of design professional services—the federal government—goes a step beyond this approach, because it unilaterally disallows certain types of expenses, including interest paid on loans, client entertainment costs, and charitable contributions, most of which are tax deductible. Making matters even more confusing, rules can vary from federal agency to federal agency—a problem that most of the major design professional associations have been working on for some time, in an attempt to have all agencies use the same procedure. Note, too, that the federal government also reserves the right to audit a firm, to help assure its overhead rates were properly established, and to deny retroactively (or demand repayment of) remuneration if errors allegedly are found. Since the regulations of each agency differ, those who work with several federal departments may find themselves being audited several times each year, once by each federal client. Note, however, that certain projects are not subject to audit, depending on the nature of the payment agreed to, for example, fixed price (lump sum) versus cost-plus.

ESTABLISHING FEES

The fees that a firm establishes for its various commissions are determined through a variety of techniques, all of which usually are based on cost. Additional amounts are allowed for contingencies and profit. In some cases, value-based pricing is used. These factors are discussed below.

Fee Establishment Techniques

Some of the many different techniques that firms apply to establish a fee include: cost plus a fixed fee; cost plus a profit percentage; direct labor times a multiplier (used to account for overhead and profit, which is basically the same as cost plus a profit percentage); time (based on standard rates per hour for a given category of employee) plus materials (expenses); percentage of construction cost; number of sheets (of drawings); fee per square foot (of the type of building involved); and fixed fee (lump sum), among others.

No matter what type of method is employed, the firm must earn more for providing its services than it costs to provide them, and determination of cost is based on the amount of direct labor expended multiplied by an overhead factor. As such, each method is just another way of saying the same thing: Fee equals cost plus the percentage of profit desired. As a consequence, in the fee established for executing a commission should be included the personnel who will be assigned and the time that will be required for them to perform their work. This is far easier said than done.

The Role of Experience

The only effective guide that a firm has in setting fees is experience. Certain types of projects require more time than others; some clients impose more time burdens than others. To the extent that experience can be an effective guide in these respects, so much the better. But many variables must be considered. Due to changes in personnel or raises given, rates may increase during the course of a project. Certain overhead factors also are subject to change, such as the cost of professional liability insurance. One also has to consider the fact that several other design professionals often are involved in a project, and the amount of work each must do and the date of the onset of that work are largely dependent on others abiding by their various schedules and not making any significant errors. The potential for client changes also must be considered, as well as late delivery of shop drawings, construction delays or other problems occasioned by bad weather, inadequate performance, and so on.

Newer firms tend to make more mistakes in their pricing because they do not have as much experience as more established firms. As these younger firms mature, however, they tend to learn some valuable lessons. One of these is the need to keep track of projects closely, not only to help ensure success with each, but also to record experience so it can be referred to later. By reviewing the cost data associated with a project and by determining how changes could or should have been made, a firm is that much better prepared not to make the same mistake twice. Involving project managers in discussions usually is essential, if they are to perform better next time.

Another lesson learned is the need to define workscope as closely as possible, to help prevent a firm from performing without charge the services it did not agree to provide. This underscores the need to have project managers involved in establishing workscopes and budgets, so they are familiar with what has been agreed to and what is anticipated insofar as costs are concerned. For the most part, project managers tend to be reluctant to "change order" a client for each and every adjustment the client wants. However, project managers must be kept apprised of project costs versus budgets over time. While it is important to maintain client goodwill, it likewise is important to achieve a firm's profit objectives. If a project is running particularly well, minor changes can possibly be accommodated without additional fees. Otherwise, additional fees may be necessary.

Contingency Allowances

With respect to the unknowns that may occur, including client requests for additional services, it almost always is advisable to establish contingency allowances. Experience indicates how much of a contingency is appropriate. Contingency allowances can be built into the fee, to account at least in part for errors made in establishing the fee. A special project contingency reserve is also advisable, and the client should be encouraged to establish one to fund the unanticipated, including client-directed changes.

A firm is very seldom given a blank check. Even when a project proceeds on a cost-plus-fixed-fee, cost-plus-fixed-profit, or some similar basis, an "upset" (not-to-exceed) amount usually is established. Should the upset figure allow for contingencies, or should it be based solely on the work agreed to? This is something that a firm and its client should contemplate and resolve during contract formation. In any event, effective record keeping is essential, in order to justify any and all charges and to record experience.

Other Concerns

The method that a firm uses to set its fee will often give rise to other concerns that should be contemplated. If the fee paid by the client will depend on the firm's cost, the contract should permit a firm to increase its rates when its overhead or direct cost increases. When the firm's fee is based on a percentage of construction cost, the contract should require compensation to the firm, should the project not be built. Such contracts might also benefit from incentives that encourage design professionals to achieve lower construction costs. Otherwise, percentage-of-construction-based fees can discourage the additional work needed to lower construction costs, since the additional work would lower the fee.

Lump-sum-fee contracts generally require a provision of some type to address the potential of project abandonment. Any such provision should also indicate how the design professional will be paid for work performed up to the point of abandonment. Similar concerns arise when the client decides to downsize a project whose design fee was based on square footage or a percentage of construction costs.

In fact, most firms will generally employ a variety of procedures in order to calculate the fee. For the most part, however, the procedure used will be of little consequence if each is designed to achieve the same percentage of profit based on time requirements. In some cases, of course, using a given method to accommodate a client preference and "plugging in" a standard rate the client is used to paying (e.g., the standard percentage of construction, the standard rate per square foot or multifamily residential unit, the standard rate per drawing, etc.) may result in more (or less) profit than otherwise. Profit is often a major variable, however, as will be discussed.

Identifying Profit Objectives

A firm will generally establish an overall profit objective for a year, and, for the most part, this objective, stated as a percentage of cost, will be considered in establishing the overall fee for each project. *Considered* does not necessarily mean "applied," however. Some projects may be accepted at less than the targeted rate of profit in order to establish a relationship with a new client. And, even then, the targeted profit for the project, as any other, may not be achieved. Despite occasional shortfalls, however, a firm may still meet its overall profit goal for a year, if not by achieving its percentage goal, then by earning the aggregate amount of profit forecasted, by increasing volume.

The key to a firm's profit performance is the effectiveness of its project managers. Generally, they are responsible for not only the management of each project but also the collection of receivables. Collection of receivables is a particularly important concern, since the firm must make certain payments on a regular basis, for example, salaries, rent, and taxes. If clients are slow to pay, creating a cash flow problem, money must be borrowed in order to meet ongoing expenses. If it is borrowed from a bank or some other lender, interest must be paid. If it is "borrowed" from an internal source of funding, such as reserves, interest that that funding otherwise would have earned is lost. In either case, however, "slow pay" means less profit. In that some clients are known to be slow payers, the cost effects of their slow pay should be considered in establishing the fees charged to them. In a similar manner, the extent to which risks imposed by a project will affect profitability also must be considered, as noted in prior discussion of risk funding. If there is a 5 percent chance that a given type of project will give rise to a claim that will cost the firm $100,000 (including the value of lost time), the firm should want to establish a $100,000 reserve for every twenty such projects that it accepts, that is, $5,000 per project. Failing to do so would result in each such project being underpriced by $5,000.

The overall percentage of profit that a firm seeks to obtain is usually determined by a number of factors, not the least of which are fees charged by principal competitors. Other factors include a firm's expansion plans, new equipment needs, reserve requirements, and experience with regard to slow payers, and so on. Virtually all such approaches are rooted in cost accounting, however. A firm first determines what its costs are likely to be and then establishes fees that include a fixed percentage of estimated cost as its profit. There is another way, called value-based pricing.

Value-Based Pricing

Through value-based pricing, the cost of a product or a service is determined by its value to the purchaser, not by its production cost. Design professionals involved with mechanical and electrical systems should be familiar with this concept. In recent years, manufacturers have developed a wide variety of energy-saving products. Some of these cost less to produce than the standard products they replace. Nonetheless, the more efficient products sell for a higher price, because the savings they create can easily justify the premium charged. In a similar manner, some organizations base their fees on a percentage of savings achieved, just as attorneys base theirs on a certain percentage, as for settling an estate, or flat rates for specific services. The justification for doing this is perhaps best explained by the old joke about a company that purchased a multimillion dollar computer system and was celebrating its acquisition at an open house, at which many important clients were present. When the head of the company went to activate the system, however, nothing happened. He hurriedly called a contractor, who arrived within a half-hour, examined the system, and then moved to one particular area of the mainframe, which he hit with a sledgehammer. The system immediately came on-line. The following day the contractor submitted a bill for $4,050. The company president called the contractor and asked, "What's the $4,050 bill for? All you did was hit the thing with a sledgehammer," to which the contractor replied, "Fifty dollars is the charge for hitting the computer; $4,000 is for knowing where to hit it."

It is somewhat ironic that design professionals rely so extensively on cost-based pricing practices, given the value of their services. Generally speaking, the cost of construction is ten times the cost of design, and the cost of life-cycle operation and maintenance of the designed structure is approximately a hundred times greater than the cost of construction. Nonetheless, the quality of design determines what the other costs will be. This relationship is shown in Figure 9-3. The triangle below the large rectangle represents design fees; the large rectangle represents the cost of construction and life-cycle operation and maintenance. The small dotted-line rectangle inside the larger one represents the cost of design if the fee were to be doubled. The dotted lines at the top of the large rectangle indicate costs plus or minus 10 percent. The quality of services that a design professional team renders can determine whether the top of the large rectangle is at the lower dotted line (−10 percent) or the higher one (+10 percent). If the lower limit is achieved, the savings involved would be worth

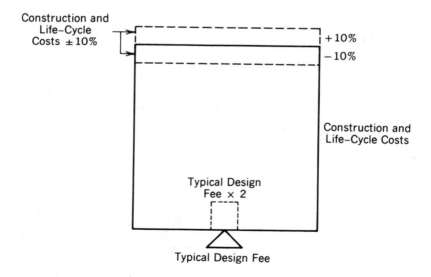

Figure 9-3: The Relationship Between Design Fees and Construction/Life-Cycle Costs

eleven times the typical fee. Were the fee doubled to achieve the savings, the owner would save ten dollars for each additional design fee dollar spent.

Due to the impact of fee competition, design professionals, unlike physicians, attorneys, and other professionals, are generally not in a position to base their fees on value. For the most part, the exceptions are those design professionals (mostly architects) who have gained national or international reputations and whose affiliation with a project will in and of itself create more value to the owner—value that the owner pays for. At this time, however, the cost of high-quality design still is based on cost. From time to time, a firm is able to obtain a value-based fee when it proposes an innovative technique that reduces construction cost and/or construction time, and it receives a percentage-of-savings-based bonus. Certain specific services also may be value-based, among them, value engineering, when a firm receives a percentage of the savings its work creates, or expert witness service, when the expert's credentials will increase the client's odds of winning the case.

The Need for Adequate Fees

No matter how fees are set, adequate fees are essential to professional practice. Inadequate fees create an incentive to cut corners. Cutting corners increases risks. In fact, a substantial portion of the liability problems that confronted design professionals in the 1980s can be attributed to inadequate fees and the shortcuts they made necessary.

FINANCIAL REPORTING

Financial reporting is essential to the effective management of a firm. Due to reliance on computers, reports are now relatively easy to prepare. These reports are not free, however, nor is the time required to review them. Managers must avoid the temptation to obtain more reports than necessary.

For the most part, the larger the firm and the higher an individual's position in it, the more consolidated, or summarized, a report should be. Project managers need to have details about the work they are heading; those who supervise project managers do not need as much detail. Similarly, a branch manager might review summary information for each department, but the regional director may need only summary information for each branch. Because reports will be consolidated, and for other reasons as well, it is vitally important for all persons who provide the information used in reports to report on a consistent basis. Although errors may be caught after the close of a year, permitting appropriate adjustments at that time, the reports issued during a year—usually monthly—are of great consequence. They permit managers to track progress and thus make the adjustments needed to achieve objectives.

Toward this end, many reports compare actual performance to a budget. The budget is established as part of an annual, or more frequent, planning function. In essence, it is up to management to determine a firm's goals and objectives for a year and then to identify how these will be achieved. In doing so, it must establish projected revenue from various sources, as well as the expenses associated with producing the revenue and operating the firm. Failure to plan is almost tantamount to failure to manage. If a firm is not moved in specific directions by management, direction is determined solely by market conditions—the difference between sailing and drifting. If a firm's plans are not materializing, reports will make that fact known, permitting managers to make whatever adjustments may be needed.

Income Statement

An income statement is a summary of revenue, expenses, and net profit for a period, usually a month, as shown in Figure 9-4, for a relatively small firm. Larger organizations might prefer to see revenue itemized in different ways, on a regional basis, by branch office, by departments, and so forth.

What do the data shown in the income statement indicate? In this case, the data indicate that April was perhaps the best month of the year to date, in that revenue was $3,000 ahead of projections, bringing revenue to $1,000 less than projected. For the first four months of the year, direct labor has been maintained at 39 percent of project revenue, which is slightly higher than the 38 percent planned. Overall, overhead has been held to 123 percent of direct labor, compared to the planned rate of 128 percent. Nonetheless, net profit—at 10 percent—is less than the 11 percent called for.

Statement of Financial Position (Balance Sheet)

A statement of financial position or balance sheet is shown in Figure 9-5. It indicates a firm's assets and liabilities at a given point in time (usually at the time the firm's income statement is prepared). Of particular importance are current assets (cash or items convertible to cash within one year) and current liabilities (liabilities due within one year). If there is a trend indicating that liabilities are increasing, chances are collections are slow and a loan will be needed to bridge the gap, so that payroll and other immediate payment needs can be met. More detail usually is provided through two other reports, on cash flow and on accounts receivable.

Cash Flow Report

A cash flow report indicates a firm's cash position, as distinct from the conditions reported through an accrual report. The sample shown in Figure 9-6 indicates at first glance that the firm may be in a somewhat precarious position; at the end of the month it had less than $6,000 cash on hand. However, the

INCOME STATEMENT
APRIL 1991
(In Thousands of Dollars)

	Actual (Period)	Budget (Period)	Difference (Period)	% Difference (Period)	Actual (Year-to-Date)	Budget (Year-to-Date)	Difference (Year-to-Date)	% Difference (Year-to-Date)
REVENUE								
Project Revenue	$ 57.3	$ 54.3	$ 3.0	5.5%	$216.2	$217.2	$ (1.0)	(0.5%)
Other Income	0.2	0.3	(0.1)	7.1%	1.2	1.2	0.0	0.0
Total	$ 57.5	$ 54.6	$ 2.9	4.3%	$217.4	$218.4	$ (1.0)	(0.5%)
EXPENSES								
Direct Labor	$ 22.6	$ 20.4	$ 2.2	10.8%	$ 84.0	$ 81.6	$ 0.7	1.0%
Indirect Labor	8.2	8.5	(0.3)	(3.5%)	33.6	33.3	0.3	(0.9%)
Payroll Burden	7.1	7.0	0.1	1.4%	25.2	27.3	(2.1)	(7.7%)
Rent	2.5	2.5	0.0	0.0	10.0	10.0	0.0	0.0
Utilities	0.4	0.4	0.0	0.0	1.7	1.7	0.0	0.0
Equipment Leases	1.5	1.0	0.5	50.0%	5.0	4.0	1.0	25.0%
Maintenance/ Repairs	2.2	0.3	(0.1)	(33.3%)	1.0	1.2	(0.2)	(16.7%)
Supplies	0.4	0.4	0.1	33.3%	1.3	1.2	0.1	8.3%
Insurance	3.7	3.8	(0.1)	(2.6%)	14.6	15.0	(0.4)	(2.7%)
Professional Services	0.3	0.3	0.0	0.0	1.0	1.0	0.0	0.0
Interest	0.6	0.4	0.2	50.0%	1.7	1.7	0.0	0.0
Other	1.9	2.1	(0.2)	(9.5%)	8.2	8.4	(0.2)	(2.4%)
Total	$ 49.4	$ 47.0	$ 2.4	5.1%	$187.3	$186.4	0.9	0.5%
GROSS PROFIT	$ 8.1	$ 7.6	$ 0.5	6.6%	$ 30.1	$ 32.0	$ (1.9)	(5.9%)
FEDERAL TAXES	(2.5)	(2.2)	0.3	13.6%	(8.5)	(9.0)	(1.9)	(5.9%)
NET PROFIT	5.6	5.4	0.2	3.7%	21.6	23.0	(1.4)	(6.1%)

Figure 9-4: Income Statement

STATEMENT OF FINANCIAL POSITION
AS OF APRIL 30, 1991
(In Thousands of Dollars)

ASSETS			LIABILITIES		
Current Assets			**Current Liabilities**		
Cash	$ 20.2		Accounts Payable	$ 10.4	
Short-Term Investments	6.6		Accrued Expenses	4.2	
Work in Progress	21.1		Deferred Income Taxes	35.5	
Accounts Receivable	45.4		Note Payable	10.0	
Other	0.5		Total Current Liabilities		$ 60.1
Total Current Assets		$ 93.8	**Long-Term Liabilities**		
Fixed Assets			Long-Term Debt	$ 15.0	
Property and Equipment (net of depreciation)	$ 27.5		Total Long-Term Liabilities		$ 15.0
Other Assets	0.8		**Shareholders' Equity**		
Total Fixed Assets		$ 28.3	Capital Stock	$ 25.0	
			Retained Earnings	22.0	
TOTAL ASSETS		$122.1	Total Shareholders' Equity		$ 47.0
			TOTAL LIABILITIES AND SHAREHOLDERS' EQUITY		$122.1

Figure 9-5: *Statement of Financial Position*

CASH FLOW REPORT

Date Prepared: 5/09/91   As of: 4/30/91

	MONTH	NOTES OR COMMENTS
BEGINNING BALANCE		
Checking Account	$ 9,000	
CASH RECEIPTS		
Accounts Receivable		
Professional Services	43,450	$400,600 due 5/15
Project Expenses	4,150	4,800 due 5/15
Capital	0	
Savings	5,000	
Miscellaneous	0	
Loans	0	
TOTAL CASH RECEIPTS	$52,600	
TOTAL CASH AVAILABLE	$61,600	
CASH DISBURSEMENTS		
Payroll	$31,800	
Accounts Payable*	4,000	
Expenses*	14,900	
Capital Expenditures	0	
Loans Paid	0	$10,000 due 5/15
To Savings	5,000	
Stock Repurchase	0	
TOTAL CASH DISBURSEMENTS	$55,700	
ENDING CASH BALANCE	$ 5,900	
NUMBER OF PERSONNEL		
Full-Time	9	
Part-Time	0	
OUTSTANDING LOAN BALANCE	$25,000	
INTEREST RATE ON LOANS		
(Most recent)	9.5% (S-T)/	
	10.0% (L-T)	

*Items requiring a purchase order and recorded in the Invoice Register are considered Accounts Payable. Items not normally obtained with a purchase order and paid directly out of Cash Disbursements are usually considered Expenses. These include weekly tax deposits, insurance premiums, and similar items.

Figure 9-6: Cash Flow Report

report does not indicate how much the firm has in its savings account. Savings data could be indicated as well, and, if so, it probably would be shown that the firm is financially sound. During the month, $5,000 was transferred out of savings to cover expenses, and $5,000 was transferred back to savings.

The report provides a column for notes, to indicate any pertinent information about forecasted future changes in particular. Some firms' financial personnel also prepare cash flow projections for one or several months.

Accounts Receivable Report

An accounts receivable report indicates the status of bills that the firm has issued. The format shown in Figure 9-7, which is for a small firm, reports on a project-by-project basis. In larger firms, such a report would be given to managers who need such precision, such as project or branch managers. Those in higher positions would receive a consolidated report, indicating accounts receivable in general or by branch, division, and so on. Note that the report also indicates work in progress on these projects. The manner of reporting work in progress should be based on how much the work will be billed for, not what it costs to produce. The reporting mechanism involved should track the billing method employed, for example, lump sum versus cost-plus-fixed-fee.

Forecast of Assets and Liabilities

A forecast of current assets and liabilities (Figure 9-8), as a forecast of cash flow, gives senior managers a better understanding of what their immediate cash needs may be. Managers should never have to scurry about to find borrowed funds if it appeared a "crunch" was coming. Banking relationships should be established well in advance, to help assure that lender personnel have a full understanding of the nature of a firm's business and its short-term loan requirements. Many firms operate with a line of credit, meaning that they will be able to rely on borrowed funds automatically, up to a predetermined limit.

PROJECT MANAGEMENT REPORTS

Effective project management is vital to the financial health of a firm. Accordingly, project managers require detailed reports to indicate how well their projects are progressing. Since the fee for most projects is based on a forecast of personnel and time requirements, details should indicate the budget against which actual data are being compared. More than dollar factors are at issue, of course, in that time constraints also must be adhered to.

AGED ACCOUNTS RECEIVABLE/WORK-IN-PROGRESS REPORT
Period Ending 4/30/91

	Total	Work in Progress	Total Accounts Receivable	Current	30 Days	60 Days	90 Days	Past 90 Days
Project A	$20,000	$ 1,500	$18,500	$ 4,300	$ 6,400	$ 7,800		
Project B	12,500	4,500	8,000	4,000	2,000	2,000		
Project C	13,400	6,500	6,900	4,400	2,500			
Project D	8,200	4,600	3,600	3,000	600			
Project E	11,300	2,900	8,400	2,400	3,000	3,000		
Project F	1,100	1,100						
Total	$66,500	$21,100	$45,400	$18,100	$14,500	$12,800		

Figure 9-7: Accounts Receivable/Work-in-Progress Report

FORECAST OF CURRENT ASSETS AND LIABILITIES

Date Prepared: 5/5/91 As of: 5/31/91

CURRENT ASSETS
Cash
    Checking Account $ 6,000
    Savings Account   4,000
    Petty Cash   200 $ 10,200
Accounts Receivable
    Clients   63,600
    Others   0 $ 63,600
Notes Receivable
    Principals   0
    Others   0   0
Work in Progress $ 52,500
    TOTAL CURRENT ASSETS $126,300

CURRENT LIABILITIES
Accounts Payable $ 8,400
Accrued Payroll Taxes   3,200
Notes Payable:

	Total Amount	Current Amount	
Secured Bank	$20,000	$15,000	
Insurance	14,000	10,000	$ 25,000
TOTAL LOANS	$34,000		

Provision for Corporate Taxes $ 44,500
    TOTAL CURRENT LIABILITIES $ 81,100
    CURRENT BALANCE $ 45,200
    CURRENT RATIO   1.6 to 1

Figure 9-8: Forecast of Current Assets and Liabilities

Project Planning

A variety of techniques is available for overall project planning. Ideally, the basic planning should be completed by the time the firm presents its proposal to a client. The method that the client uses to retain a firm is an important factor in this regard, however, and creates somewhat of an irony.

When a commission will be awarded based on a competitive bid basis, a firm's chances of obtaining the project are not nearly as high as they are when QBS is used. At the time technical and fee proposals are developed for a QBS-based project, the firm knows it has almost a 100 percent chance of obtaining the commission, making it worthwhile to invest time in planning. It usually is not feasible to spend as much time when bid-based procurement is involved, but bid-based selection makes precise planning even more essential, since project owners who use this method of procurement will more than likely hold design professionals to whatever their maximum bid might be. Because precise planning cannot usually be performed when bid-based procurement is used, and because of the pressure to keep fees low, all too many bids may be underestimated, making it necessary to cut corners or, at least, to impose change orders for any type of client-directed change, eroding the client-professional relationship.

In any event, planning is necessary at some point before a project has commenced. The techniques most commonly used are critical path method (CPM) or bar charts. Through these or other methods, each task is identified and shown in relationship to all other tasks, indicating the nature of the work involved, when it can be started, and by when it must be ended. These interrelationships are critical, because many tasks cannot begin until others have been completed or until others have arrived at some critical juncture.

Given that overall plans for a project indicate who will be doing what and for how long, plans can be readily "translated" into dollar terms. As such, a review of project "dollar data" can give a project manager a reasonably good idea of the extent to which a project is on or off track. Nonetheless, especially when larger or more complex projects are involved, economic summaries should not be substituted for others, because there is far more to project management than cost control. Cost control is still essential, of course. Without it, a firm cannot hope to achieve its plans.

Project Budget and Reports

A number of different formats can be employed to indicate a project's budget. The approach shown in Figures 9-9 and 9-10 is effective because it identifies a project by phases and the departments involved, and shows how the project budget will evolve over time. By comparing actual time/dollar expenses with those budgeted, any significant variances between forecasts and "actuals" should become readily apparent. Note, however, that this approach does not indicate the direct labor salaries of those performing the work. In many firms, this approach is taken purposely, to prevent individuals' salaries from becoming known. Regrettably, this is a necessary evil, since salaries of persons performing essentially the same work will vary. In such cases, it will generally be up to accounting personnel to provide dollar details to the project manager. In many cases, however, it is far simpler to rely on computerization, through which the individual involved, the individual's supervisor, or accounting personnel enter coded data, for example, project code, department code, phase code, and time. The computer then performs the appropriate calculations and allocations, to indicate where the project is versus where it is supposed to be.

In some cases, a project manager will find that a project is in trouble and, to help improve its status, will work uncompensated additional hours. If an accurate report is to be prepared, this time must be accounted for, so problems can be highlighted and corrections made. In some cases, the client is required

PROJECT NO. 89-18

Date Prepared: 5/01/91   Project Manager: John Doe

BUDGET BY COST FACTORS

	Total Hours	Total Dollars
Direct Labor	800	$11,800
Overhead		20,180
Reimbursables		2,500
Consultants		0
Contingency		3,520
Profit		7,000
Total	800	$45,000
DIRECT LABOR BY PHASES		
Study	60	$ 1,100
Preliminary Design	240	3,600
Final Design	180	2,660
Construction Documents	70	990
Shop Drawing Review	80	1,300
Construction Review	170	2,150
Total	800	$11,800
DIRECT LABOR BY DEPARTMENTS		
Mechanical	309	$ 4,597
Electrical	251	3,731
Plumbing	240	3,472
Total	800	$11,800

Figure 9-9: Project Budget

to pay for the additional time. In others, more typically, the firm does not charge for it. Depending on management's policies, it may be simplest to add the time and its value to direct labor but to *credit* the dollar value of the time to overhead.

Detail Report As the project progresses, detailed reports are furnished to the project manager. The *detail report* shown in Figure 9-11, based on the project shown in Figures 9-9 and 9-10, shows that, after four months, the project is in trouble. The mechanical work has been proceeding well, but the electrical work has been running behind schedule and the plumbing department is only now beginning to catch up.

Project Summary Report A project summary report, one that would be issued to an operations manager or a project manager, is shown in Figure 9-12.

Department	Study		Preliminary		Final		Contract Documents		Shop Drawing Review	Construction Review			Total
	June	July	Aug.	Sept.	Oct.	Nov.	Dec.	Jan.	Feb.	Mar.	Apr.	May	
Mechanical	14 hrs $270	8 hrs $140	60 hrs $900	35 hrs $515	34 hrs $530	24 hrs $380	15 hrs $213	6 hrs $84	43 hrs $675	40 hrs $520	20 hrs $260	10 hrs $110	309 hrs $4,497
Electrical	10 hrs $170	6 hrs $110	50 hrs $770	25 hrs $385	54 hrs $770	24 hrs $360	19 hrs $275	9 hrs $121	14 hrs $250	20 hrs $260	15 hrs $205	5 hrs $55	251 hrs $3,431
Plumbing	14 hrs $270	8 hrs $140	45 hrs $645	25 hrs $385	27 hrs $385	17 hrs $235	15 hrs $213	6 hrs $84	23 hrs $375	20 hrs $260	30 hrs $370	10 hrs $110	240 hrs $3,472
	38 hrs $710	22 hrs $390	155 hrs $2,315	85 hrs $1,285	115 hrs $1,685	65 hrs $975	49 hrs $701	21 hrs $289	80 hrs $1,300	80 hrs $1,040	65 hrs $835	25 hrs $275	800 hrs $11,800
Total	60 hrs $1,100		240 hrs $3,600		180 hrs $2,660		70 hrs $990		80 hrs $1,300	170 hrs $2,150			

Figure 9-10: Project Budget Detail

PROJECT MANAGER'S DETAIL REPORT

Project No.: 89-18
Project Manager: John Doe

Date: 9/31/91

	Current		Project to Date vs. Budget					
	Hours	Dollars	Hours	Budget	Difference	Dollars	Budget	Difference	% Completed	% Spent
Study										
Mechanical			20	22	(2)	$ 370	$ 410	$ (40)	100%	90%
Electrical	14	284	17	16	1	305	280	25	100%	109%
Plumbing			18	22	(4)	350	410	(60)	70%	85%
Prelim. Dsgn.										
Mechanical	52	832	85	95	(10)	1,205	1,415	(210)	90%	85%
Electrical	46	676	83	75	8	1,287	1,155	132	95%	111%
Plumbing			0	70	(70)	0	1,030	1,030	0%	0%
Final Design										
Mechanical										
Electrical										
Plumbing										
Contract Documents										
Mechanical										
Electrical										
Plumbing										
Shop Drawing Review										
Mechanical										
Electrical										
Plumbing										
Construction Review										
Mechanical										
Electrical										
Plumbing										

Figure 9-11: Project Manager's Detail Report

OPERATIONS MANAGER'S SUMMARY REPORT

Operations Manager: Bill Smith
Date: 9/31/91

Project Number	Project Manager	Budget	Activity Status				Billing Status		
			Spent	Remaining	% Completed	% Spent	Billed	Received	AR
88-31	Doe, J.	$ 95,000	$104,000	$ 0	98%	109%	$ 90,000	$ 70,000	$20,000
88-43	Doe, J.	115,000	90,000	25,000	90%	78%	80,000	80,000	0
89-18	Doe, J.	11,800	3,517	8,283	26%	30%	1,000	0	1,000
88-24	Jones, R.	55,000	56,000	0	99%	102%	53,500	47,050	6,450
88-52	Jones, R.	210,000	165,000	45,000	84%	79%	142,700	110,000	32,700
89-20	Green, A.	85,000	5,000	80,000	10%	6%	5,000	5,000	0

Figure 9-12: Operations Manager's Summary Report

In reviewing the work of John Doe, certain patterns of behavior are suggested, based on assumptions that the operations manager would have to verify through discussion with John Doe. It appears as though Project 88–43 is moving along very well, and the client, judging by the paid-up account, is pleased with the work. Project 88–31 is running over budget, is still not complete, and the client still owes $20,000. Project 89–18 does not seem to be in as serious a condition as the project manager's detail report (Figure 9-11) indicates, but the astute operations manager would be extremely concerned that it is running over budget and that only 8 percent of the fee has been billed at a time when more than 25 percent of the work has been completed. The client's slow payments are yet another worry. It can also be seen that the other two projects are far larger and that perhaps Mr. Doe is not paying as much attention as he should to the smaller one. If the operations manager is aware of certain liability loss prevention lessons, he will recognize at once that it could be that the small project that will lead to major problems, especially if slow performance leads to pressure being exercised by the prime design professional.

OTHER REPORTS AND CONCERNS

Many other reports are commonly prepared by design professional firms, much as for any other type of business endeavor. A number of ratios will also be established, comparing, as examples, actual overhead rates to those predicted, debt to equity, and so on. Also, for ease of review, many of the reports will be prepared graphically, using pie charts, bar charts, and others. Graphics can be particularly helpful in reviewing certain situations, because they can indicate discrepancies in a much more dramatic fashion than numbers alone can.

Those involved in the management of the firm or its projects should be ultimately familiar with all of the economic factors associated with their work, comprising those factors they work with directly, those their subordinates work with, and those their superiors use to judge performance. In some cases, individuals will try to indicate that a situation is better than it actually is by intentionally misallocating hours spent or by not reporting noncompensated overtime hours. This usually is a serious mistake, because it permits problems to go undetected. As a consequence, an individual will not receive the additional education or assistance required, because it would appear none is needed. A firm that encourages and rewards excellence generally wants its personnel to have the support needed to achieve excellence. If it does not, attitudes need adjustment. If it does but is not perceived in this manner, then its human resources management needs fine tuning. In a professional organization, virtually all personnel should recognize that no one is perfect. Effective financial management provides a means to identify the improvement needed and the wherewithal to support improvement.

10

Professional Human Resources Management

An engineering or architectural firm is a collection of people. The effectiveness with which these people are selected and managed determines the effectiveness of the firm. If individuals are not carefully evaluated before they are hired, their ability to contribute to the firm will depend more on luck than on intent. If they have little desire to achieve excellence, the firm will be forced to accept mediocrity as its ultimate product. If they are given no incentive to act in a professional manner, they probably will not. What do you want your firm to be? How do you want it perceived? The attitudes and actions of those who comprise the firm are determining factors.

This chapter provides a brief overview of some of the many issues associated with human resources management. Professional instincts alone are not enough. The proper performance of many tasks must be learned. As with so many other responsibilities, gaining knowledge and striving to apply it in the best possible way comprise the professional approach. It takes more time, of course, but the investment required can be justified on the bottom line, through increased productivity, fewer errors, and reduced turnover, one of the most costly concerns of all. When employees leave, completing their assignments becomes an immediate concern. If the work has not been done well, client relations and quality control problems may emerge. Then comes the expense of locating replacements, advertising dollars and placement fees, and the time required to interview and perform reference checks. Once new employees are retained, orientation and training are required, as well as additional supervision as they work their way up the learning curve.

The amount of money saved simply by minimizing turnover can be substantial. The amount of money earned by having a work force that is pleased to be associated with your firm and anxious to do well for it can be even more significant. For many, however, the most important benefit of professional human resources management is the sense of contribution and belonging that can be derived when those who comprise the firm proceed as a team toward the common goal of professional fulfillment.

HIRING

The process of hiring new employees and orienting them to the firm, their responsibilities, and preferred attitudes are among the most important of all human resources management tasks. Each new employee will be subject to a number of first impressions about a variety of issues, functions, and people. Each of these first impressions will be lasting. By making every reasonable effort to hire and orient new employees effectively, a firm can retain personnel longer and derive more productivity from them.

The hiring process frequently begins with identification of the specific technical competence required by the position and then advertising or using other means to find people who possess the desired skills. As commonsensical as this approach may seem and as widespread as its application may be, it is not effective. It ignores a number of key concerns and thus explains why some firms experience much higher turnover rates than others.

Preparing a Job Description

The first step in the hiring process should be identification of the tasks that the new employee will be expected to perform. Task identification can be accomplished by developing a job description, a procedure that can be approached in a number of ways. One is to prepare a listing of tasks and to indicate lines of authority and responsibility. Here is an example:

> Obtain input from a project manager as to the project involved, budget, schedule, and other key factors
> Prepare conceptual designs, for review with supervisor and project manager.
> Develop design for review by peer, supervisor, and project manager. Supervise drafters responsible for preparing final drawings.
> Review final drawings with peer, supervisor, and project manager.
> Prepare specifications in conjunction with specifiers.
> Review similar work performed by peers.
> Participate in meetings of professional societies, with emphasis on one.
> Participate in in-house "brown bag" loss prevention seminars.
> Develop at least one client-oriented magazine article per year.

The foregoing requires much more detail to be an effective job description. Nonetheless, it is sufficient to illustrate a key point: The description does not indicate the amount of time associated with each task. By doing so—by expressing the time requirements of each task as a percentage of overall work hours, it is far simpler to identify the particular strengths that a candidate should possess. For example, if proposal writing is a major aspect of a position, those considered should not only write reasonably well but also should understand something about tort law, contract law, and a number of related issues. If only 5 percent of the employee's time will be spent developing proposals, the need for relevant skills is not as strong; time can be made available to help the employee gain the knowledge required. If proposal development will take up half the employee's time, however, having relevant skills is vitally important. In a similar manner, if a candidate has particular strengths in the area of proposal development and clearly enjoys the work, the individual may quickly lose interest in a position that does not rely extensively on these attributes. By carefully identifying the tasks and attributes needed for the performance of each task associated with a given position, and the relative importance of each task in terms of time requirements, a firm can develop a highly effective means of establishing its actual needs and can screen candidates accordingly.

Given the importance of comprehensive job descriptions, they should be developed and reviewed on a team basis. This approach not only helps assure accuracy but also involves others in the hiring process. Those performing the same work as the one who will be hired should be asked to review a draft job description. Have all tasks been identified? Are the time proportions reasonable? If the position is new, inquire among others who may be affected, such as those who may be giving some of their tasks to the new employee.

Identifying Personality Requirements

Recognize that, in all cases, you are hiring a person, not a machine. The individual has a personality, and it should mesh as effectively as possible with coworkers' personalities. Will coworkers be happy with the new employee? Will the new employee be happy with coworkers? One way of answering such questions is to make an assessment of desirable personality traits. If some are wanted to offset certain negative attributes of coworkers, some consideration should be given to either encouraging those coworkers to change their ways or to changing coworkers. In other words, having to hire someone gives you an opportunity to review your work force and how it functions in depth. Do not let the opportunity pass by. No opportunity to improve should be overlooked.

Obtaining Candidates

Once you identify the type of person you need to hire and the skills that the person should possess, you are ready to encourage people to apply.

Promoting from Within In larger organizations, it usually is appropriate to inform employees that a position is available. This can be done through two methods—informing supervisors as a group and encouraging them to make recommendations, or letting the opening be known to employees directly.

When a firm has effective human resources management practices in effect, supervisors will already know who is best qualified for the position. In a well-run organization, personnel are evaluated and counseled on a regular basis. They are given opportunities to express their professional aspirations and are given the support needed to work toward realization of those aspirations. By the same token, advising all employees directly that a position is available can encourage all interested to speak up so that the firm has an opportunity to become a staff member's vehicle to the future.

In firms in which human resources management is carried out on a "hit-or-miss" basis, advancement opportunities can create problems. In some instances, supervisors may be reluctant to recommend someone qualified, since doing so may force them to go through a replacement cycle. Supervisors need to realize that not recommending a qualified staff member can result in that staff member leaving in any event, due to dissatisfaction with limited advancement opportunities. These departures can be particularly troublesome, since the people who leave will not be available to orient, train, and counsel their replacements. In addition, those who leave often tell others about their dissatisfaction with a firm, just as they can "talk up" their new firm to former coworkers and possibly encourage more to leave.

When all staff members are informed of an opening, some will become dissatisfied when they do not obtain the position, and some will be even more dissatisfied because they believe they are more qualified than the person selected. The pitfalls involved can be avoided, but only through effective human resources management. Individuals need to know why they were not selected, what other opportunities (near-term and long-term) the company foresees for them, and the specific attributes they will have to gain in order to capitalize on opportunities. If a person wants to advance but perceives that there is no room for advancement with your firm, it would be pointless for the individual to stay with your firm.

If you recognize the potential problems beforehand, you can begin to take action now to help prevent problems from materializing. Do not merely hope that there will be no problems or that, should problems occur, they will go away of their own accord. Know the people involved; communicate with them to maintain their satisfaction with the organization. It is a fundamental rule of organizational behavior that, if one member of the organization has a problem, the organization as a whole has a problem.

Seeking Outside Applications Many larger firms obtain new employees by recruiting them from college campuses. Some prefer to work with inexperienced people because neophytes do not have to overcome what their employers consider bad habits gained through experience. Smaller firms generally cannot afford the time required for comprehensive training, however, and thus they prefer to hire those with experience. Since this often means attracting an employee from a competitor, problems can occur.

Employee Piracy Recruiting people who are now employed by another firm is not a new problem. In fact, it was one of the concerns that resulted in creation of the ancient guilds. Merchants agreed not only on overall employee policies but also on salary scales, to prevent one master from enticing another's employee by offering more money. Such tactics cannot be pursued in the United States. Accordingly, you must deal with the problem as a professional, by having a professional attitude and following professional procedures.

As a professional, you must realize that few relationships are lifelong. Sooner or later, someone in your employ is likely to leave. It is also likely that those who leave will obtain employment with a similar firm, where they will perform similar functions.

If employees are satisfied in every respect with the work they are doing, their remuneration, and the overall work environment, chances are they will not leave, unless it is to establish their own firm. It is doubtful that you will be able to satisfy anyone in every respect, however. As such, an employee's move to a competitor does not mean that employee piracy has occurred. Hopefully, your competitors will see it in the same way, if one of their employees joins you.

To avoid accusations of employee piracy and the activities that may justify them, you and any agents you employ should adhere to certain professional procedures. Do not contact potential candidates during business hours to ask if they would be interested in a position; discourage those in your employ from doing so. It is particularly unfair to have new employees contact their former coworkers to "lay it on thick" about how wonderful the new firm is, even if it is the truth. Such activity can be destructive to the other firm and could feasibly be grounds for an unfair competition claim (an attempt to put the competitor out of business). This does not mean that former coworkers must stop seeing one another or that a new employee who is truly thrilled about working for your firm has to plead the Fifth Amendment when asked, "How's the new firm?" If, as a result of contact such as this, several people inquire about working for you, so be it. The "other guy" obviously has a problem and needs to make some changes. The question revolves around who initiates the contact and what the response should be.

Recruitment Techniques In selecting recruitment methods, it is not enough to ask, "Which technique will bring in the most candidates?" You should be concerned about the quality of candidates, particularly in terms of the length of time they may stay with the firm. The Stanford Research Institute performed a major study in this regard, using a sample of 38,000 engineers and scientists from twenty-five organizations located in five widely separated geographic areas. As shown by the results presented in Table 10-1, there is a definite correlation between recruitment method and turnover. As indicated,

TABLE 10-1: Professional Employee Retention as a Function of Recruitment Method[1]

Recruitment Method	% of Current Professional Work Force Recruited[2]	% of Terminated Work Force Recruited[3]	Ratio of Current % to Terminated %	Ratio of Current Employees to Terminated Employees
Employee Referral	51.1	28.6	1.79:1	13.19:1
Aware of Company's Work	17.0	12.1	1.40:1	10.36:1
Newspaper Advertising	10.6	19.1	0.55:1	4.09:1
Placement Service	5.1	14.3	0.36:1	2.63:1
College Recruitment	4.8	10.2	0.47:1	3.48:1
Perceived Advancement Opportunity	2.8	3.9	0.72:1	5.31:1
Individual Recruitment	2.8	0.7	4.00:1	28.33:1
Magazine Advertising	2.2	4.4	0.50:1	3.72:1
Journal Advertising	1.8	1.9	0.95:1	6.88:1
Other	1.9	4.9	0.39:1	2.90:1

[1] Howell, Richard P., "Engineers on the Move," *Engineering Manpower Bulletin No. 11*, Engineering Manpower Commission, New York, NY, June 1968.
[2] N = 3,045
[3] N = 413

employee referral was the most common source of personnel, accounting for 51.1 percent of the sample. This method also enjoys the second-highest current employee-to-terminated-employee ratio. Individual recruitment enjoys the highest such ratio, but the technique was used sparingly, accounting for less than 3 percent of the sample. The second most popular recruitment method can be attributed principally to effective public relations and promotional activity, resulting in individuals seeking out the company because they were aware of its activities. This recruitment channel also enjoys a strong current-to-terminated work force ratio.

Although the research data are of interest, the key concern is your own experience and that of your firm. Some people have found personnel recruiters who have been excellent for them or have had excellent success with some other means. In this regard, it may be worthwhile for your firm to analyze its own records to determine techniques that it has found to be most successful.

EMPLOYEE REFERRAL: If you will be encouraging friends to join your firm, or if you will be suggesting to subordinates that they recommend the firm to their friends, recognize that calls should be placed to prospects' homes, not their offices. Some firms provide incentives for making such calls, as by providing a payment when friends join the firm and another payment after the new employees remain with the firm for six months or so. This approach is considered unprofessional by many engineers and architects, and—regardless of perceptions—it can encourage unfortunate statements or activities for the sake of monetary gain. Ideally, your firm should be operated in such a manner that employees believe they are doing friends a favor by making opportunities known to them.

RECRUITMENT FIRMS: On the opposite end of the recruitment spectrum are "headhunters"—employment agencies or executive search firms that charge a fee for their service. Regrettably, some of these outfits have little understanding of what it means to be professional. They use underhanded, if not illegal, means of identifying persons a firm has on staff and then contact these people during business hours with offers of far more pay, more responsibility, and so on. If they are successful in luring someone away, they in some cases will contact the firm from which they have "stolen" an employee to offer their services in finding a replacement. Through actions such as these, some firms have besmirched the reputation of an industry that includes many fine, reputable businesses.

If you have decided to consider a search firm, first determine how much the service will cost. Irrespective of what you may be told, almost all of their fees are negotiable. Nonetheless, several thousand dollars still may be required.

If you do decide to use a recruiter, select a firm (or firms) in much the same manner as you would retain a design professional firm. Ask several to identify their experience in your field and to provide the names of five firms served within the past twelve months. Contact principals of these firms to determine how satisfied they were with the recruitment agency's services. Did they screen candidates according to the firm's requirements? How long did the employee stay with the firm? Also ask those with whom you speak about some of the other search organizations you may have heard about, to learn what they know. Be wary when a recruitment firm representative asks something along the lines of, "What firms would you most like to see people from?" or, "What firms are most similar to your own?" Either question translates into, "What firms do you want me to lure someone from?" Identify any firms from which you will not accept people, because you enjoy too close a working relationship with them or for whatever other reasons. And be very circumspect when you are told that you will have to offer far more money than you plan to in order to bring in a qualified individual. Frequently, employment firms say this, because it makes their job much easier. Recognize that it can require you to raise your compensation scales all the way down the line.

OTHER METHODS: Between the two extremes of relying on your own personnel for referrals and relying on employment agencies, your own advertising can be used. It is easiest to advertise locally, of course, because those being hired would not have to move, may already know some of their future coworkers, are familiar with local codes and standards, and so forth. If the local market is active, however, it may only offer "slim pickings," unless you are offering a career move people cannot obtain from their current employers. The alternative is to offer a significantly higher salary, but an "above-market price" approach may erode your competitive position or create animosity among existing personnel. If the amount you offer is just slightly more than average but is still enough to attract candidates, those interested may be willing to leave you for just a few dollars more, too, unless you can provide something over and above salary.

Some firms advertise nationwide through trade magazines or society journals. They may also place advertisements in newspapers that serve areas that are economically depressed at the time. For example, in the late 1980s, oil-producing areas of the United States were hard hit by the decline of oil prices, and their economies stagnated. Many design professionals were out of work and eagerly responded to ads for out-of-town employment.

Culling

The process of culling is the same as short-listing—something that occurs when your firm is being considered for a project. Resumes are reviewed to determine who seems most qualified for the position. Technical knowledge and experience are key concerns, of course, but, as has been mentioned already, they are not be-alls and end-alls. Gifted designers who cannot get along with their coworkers may create more trouble than they are worth. The process of culling is performed solely to identify who seems to have what it takes. In this regard, it may be appropriate to advise candidates about the information you want to see. For example, you could ask applicants to submit a list of the projects they have worked on and to identify their role in each.

Interviews

Interviews can be very telling, but realize that human beings—alone among the animal kingdom—have the ability to present themselves as something they are not. Some people like to think they can gauge others by their facial expressions or the firmness of their handshakes, but such beliefs are poppycock. An effective proposal is no more an indication of a firm's design capabilities than an effective interview is of an individual's work capabilities. This is not to say that the interview is superfluous. It is extremely important. Nonetheless, you must always be on the lookout for charlatans.

Effective interviewing is an acquired skill. The goal is to ask leading questions that encourage an applicant to talk and to pose probing follow-up questions that can yield information about attitudes and preferences. The most frequent mistake made by unskilled interviewers is monopolizing the conversation. Skilled interviewers ask brief questions and spend most of their time listening and taking notes.

Another common mistake is making the desired answer clear via the question posed, for example, "Can you handle a lot of deadlines?" The answer almost unquestionably will be, "Yes." Instead, consider questions such as, "What types of deadline pressures have you faced? How did you handle them? How often did you face them? What were the results?"

Be careful, too, about overselling the company or overselling the position. First learn more about the candidate, and then provide factual information. It does no one any good to hire a person who will leave the firm in six months.

In all cases, know who will be interviewing a candidate (often more than one person), and identify the types of information that should be learned. And do whatever is necessary to gain good interviewing skills. Seminars are given on this subject, books, audiotapes, and videotapes about it are available, and it is frequently the subject of magazine and newsletter articles. Those responsible for interviewing should absolutely gain more knowledge if they have had relatively little training, especially so because candidates may have obtained training on how to excel at interviews.

When interviewers are not highly skilled at the craft, they should work with a list of questions, so that all-important issues will be covered. The answers given should be written down, because it is essential to corroborate them.

Review candidates' prior work experience. If there are gaps in employment history, determine why they occurred and what the applicants were doing in the interim. Why did they leave previous positions? "A personality conflict with my supervisor" seldom is a good response. What type of conflict? What type of person was the supervisor, and why did the conflict occur? It could be that the personality of the candidate's former supervisor is similar to that of the person who supervises the position you are trying to fill.

Ask what candidates liked most about their former position and what they liked least. If they left or are planning to leave due to limited advancement potential, what kind of opportunities can you offer? In fact, will the position being offered provide more of what a candidate seeks or less?

Be aware that various statutes forbid you from asking certain types of questions, such as those relating to race, religion, marital status, and so on. *All questions must be job related*. This does not mean you cannot ask questions that will yield the information you need. For example, asking a woman if she has any children is taboo. However, you can ask either men or women if there are any circumstances that would require more than occasional use of company phones for personal business; any circumstances that would require them to take time off for reasons unrelated to health, and so forth.

Depending on state statutes, it may be permissible to require any prospective employee to undergo a polygraph examination, at least to verify the accuracy of claims made in the resume. (There are also firms available who do nothing except verify claims on resumes, checking with colleges and universities, former employees, and the like.) It may also be permissible to have prospective employees undergo drug screening, if you believe that is important.

Reference Checking

The interview process should give you an idea of who the individual is or, at least, seems to be. The next step is to check with former employers. This usually is done by telephone, but eyeball-to-eyeball discussion usually is more effective.

The typical approach involved is first to verify important dates, responsibilities assigned, and salary history. This can be done through the personnel department of a large organization. However, whether the former employer was large or small, it is essential to speak with the individual who actually supervised the prospective employee. The leading question usually asked is, "Would you hire this person again?" If the person with whom you are speaking hesitates before responding, a problem of some type usually is indicated.

A difficult situation may arise when a competitor's employee responds to an advertisement you have placed or has otherwise learned about the opening and asks for a position. Such job seekers generally do not want their employers to know about their search for another position. By the same token, however, it is necessary for you to speak with the employer, unless you already know the candidate well enough to proceed without asking questions. If this familiarity has not been gained through a work situation, however, you probably do not have enough information. The manner in which individuals present themselves at something such as an association activity is not necessarily representative of their conduct at work. In fact, some people may be having problems on the job because they are devoting too much time to association involvement.

If candidates do not want you to speak with their current supervisors, the matter could be laid to rest there. If they are willing to take employment at a position where no inquiries will be made about their current position, they are naive. Of course, such individuals do not want to lose the positions they now have and thus may request that you contact their employers only if they have a very good chance of getting the job. This is a reasonable request. Before contact is made, however, it would be wise to check with the CEO of your firm, to inquire about any relationships the CEO or other top officers may have with their counterparts at the candidate's current firm. If such contacts exist, it may be best to initiate discussion at those levels, to keep everything "above board."

In these and many other situations that you are bound to encounter during your professional career, there are no easy answers. Your best guidance is likely to come from the Golden Rule: Do unto others as you would have them do unto you. Recognize, too, that at least some value can be derived from losing employees, by conducting an exit interview to learn about the mistakes they believe were made and errors that encouraged them to look elsewhere. By taking action when appropriate, you can reduce the likelihood of such losses in the future.

ORIENTATION AND TRAINING

Once you have found the individual whose capabilities best square with the job's requirements, orientation and training are needed. A planned program is essential.

Program Planning

In planning the program, work with the new employee's supervisors, peers, and subordinates to determine the most important orientation and training needs from their respective points of view. This information can be obtained at the same time the job description is being developed, and much of it will relate to understanding the job itself—who reports to whom and who is responsible for what.

Of particular concern are not only those tasks associated with the work itself but also the various quality management mechanisms the firm has in place, something with which new employees may not be familiar. On the other hand, they may be very familiar with them and may even be able to offer suggestions for improvement based on their prior experience. Accordingly, when new employees are being introduced to various procedures, should be stress that new insights are welcomed.

Loss Prevention Issues

Depending on the level of responsibility given to new employees, it may be appropriate to test their knowledge of professional liability loss prevention. ASFE has material that can help in this regard, at least by serving as models for development of firm-specific items. For example, a firm could develop a hypothetical case history that demonstrates how problems occurred. Employees would be instructed to read the case and then respond to questions created to determine if they identified the problems and, if so, the preventive or mitigating measures that could have been applied.

Mentors

Also, depending on the level of responsibility given to new employees, it may be advisable to assign mentors to them. A mentor usually is a senior member of the firm who works with employees to answer questions about their position in the firm, relationships with coworkers, and similar issues. Mentors should not supplant the authority of the new employee's supervisors; rather, they should provide guidance on attitudes and discuss the pros and cons of alternative action. In all cases, mentors owe their principal allegiance to the firm and those who comprise it. It is their responsibility to help their "charges" do their best for their own benefit and for the firm's. Some employees will invariably attempt to establish a triangular relationship whereby the mentor is played off against a supervisor. An effective mentor recognizes this and prevents it from happening, to help assure employees learn how to handle relationships on their own, without the luxury of excuses for failure or not trying.

PERSONNEL POLICIES

Every firm has personnel policies, but few commit them all to writing, as they should. Some do not even prepare a manual of the basic policies and procedures that affect all employees or those within a certain class.

Committing policies and procedures to writing is hardly an academic exercise, nor is a firm's small size ever an excuse for not performing the task. Professional management requires that those managed have an understanding of where and how they fit in, what the company expects of them, and what they should expect of the company. It also requires ongoing review and fine tuning of policies and procedures, which cannot be done if they are committed to memory only. For example, hiring procedures should be written down, along with the questions that should be asked during an interview. This not only permits upgrading but also helps assure that the valuable lessons of experience are not lost simply because different people perform a function. It can also provide valuable documentation in the event a disgruntled job seeker alleges that discriminatory questions were asked.

Management determines the health of a firm, and human resources management is a key aspect of overall management. It is somewhat easy to put off the work of having to commit procedures to writing because there is no client-imposed deadline to meet. *Professionals do not wait to the last minute to do anything.* They are motivated by the desire to perform well, and this desire carries over even to work they do not like to do.

Having a comprehensive written employee policies manual helps all employees better understand the firm and mutual firm-employee responsibilities. It also helps assure even-handed application of policies, so there is no favoritism. It gives prospective employees an opportunity to evaluate key concerns before they decide to join a firm and possibly make a mistake that both they and the firm will regret. The process of developing the manual helps to ensure that all policies will be considered and evaluated closely.

The specific policies adopted by a firm should be not only fair and reasonable, but also competitive. A firm cannot hope to attract and retain good workers if its pay, benefits, or opportunities for advancement are significantly below those offered by competitors.

Manual Content

In reviewing the suggested coverage of a policy manual, recognize that each subject should be written to answer the questions "Who?" "What?" "Why?" "When?" and "Where?" as applicable. "Why?" is generally the most important and frequently the one most often overlooked. Typical content might include the following.

1. *Notice of Intent* In some instances it has been successfully alleged by former employees that they were unjustly dismissed because the cause for their dismissal was not covered in the policy manual. It is thus essential to assure the firm has as free a hand as possible and that the policy manual does not create precisely the type of problems it is designed to prevent. Wording such as the following may be advisable:

    Notice

    This policy manual has been prepared to present information about most policies in force at the time the manual was presented. It is expressly understood that, absent specific employment contracts to the contrary, continuing employment of any individual is a matter of mutual consent. Any employee may leave at any time; the firm may dismiss an employee at any time, for any reason the party initiating the decision believes appropriate at the time. The company reserves the right to change policy at any time and will make a reasonable effort to notify all employees of any such change as soon as possible after it has been adopted.

2. *Introduction* This section should briefly describe the purpose of the manual, which usually is to acquaint all personnel with policies of the firm and to help assure their consistent, uniform application.
3. *History of the Firm* All employees should have some familiarity with the background of a firm, at least so they feel part of a continuing tradition. This section could also include discussion of the firm's philosophy.

4. *Organization and Services* This section can discuss the manner in which the firm is organized and the various services provided. Charts that show responsibilities and reporting relationships can be helpful.
5. *Discrimination* It is prudent to make clear that the company does not discriminate based on race, religion, country of natural origin, sex, or age and that it is against company policy for any employee to do so.
6. *Sexual Harassment* This section should explain what sexual harassment is and that it is cause for immediate dismissal.
7. *Confidentiality* Certain documents, such as personnel records and client lists, should be regarded as confidential and not for voluntary disclosure to any other party except with the express permission of the firm. The manual should identify what the confidential documents are. This can be of value, should an employee seek to benefit by using that confidential information at another firm.
8. *Personal Assistance* The company should have some type of employment assistance program (EAP) in effect to help employees who run into a problem, such as dependency on drugs or alcohol. Assistance should be made available on a confidential basis.
9. *Drug and Alcohol Testing* If a company has programs of drug and/or alcohol testing, reasons for it should be made clear. It should also be carried out on a nondiscriminatory basis.
10. *Workday/Workweek/Pay Period* Explain when the workday begins and ends, the number of days in a workweek, and the pay period. As necessary, provide separate discussion for salaried employees versus those who receive an hourly wage. Does the firm prefer that employees eat lunch outside the office? Does it not care? This information should be expressed.
11. *Apparel and Appearance* Some firms prefer to indicate the type of clothing that people should wear, depending on their work for the firm. Others indicate what is not acceptable, and some just don't care. Attitudes should be stated.
12. *Use of Company Phone* The policy usually is to minimize personal calls, either outgoing or incoming.
13. *Use of Other Company Equipment* Policies affecting employee use of other company equipment should be made clear. This would include computers, copying machines, and so on, as well as automobiles.
14. *Leave* Identify how the company's leave policies operate, what they encompass, which leave is paid and which unpaid, and so on. This section should discuss vacations, holidays, sick leave, personal leave, leave for jury duty and military reserve service, extended leave (as for pregnancy or illness), and so on. Policies should also be developed as to employees' accumulation of vacation time and other paid time off from one year to the next. Some firms require employees to take their vacations or simply lose the benefit.

15. *Insurance Programs* Explain what is offered, where more information can be obtained about the programs, what the various sharing formulas are (e.g., 100-percent paid employee coverage but no contribution for family coverage), and so on.
16. *Retirement Programs* These are handled much as insurance programs. In most instances, manuals provide a summary of the benefits, with more comprehensive discussion being available from another source, such as the director of personnel or the CEO.
17. *Education Benefits and Programs* Explain the company's attitude toward continuing education and the types of additional education individuals should obtain. Indicate how one goes about learning more about educational programs available, whether or not time off is given to attend them, who pays, and the extent to which certain types of educational attainment, such as earning a license, is met with increased responsibilities, more pay, and so on. In the case of certain in-house programs, such as those relating to professional liability loss prevention, indicate whether attendance is voluntary or mandatory.
18. *Raises and Promotions* This section should indicate the frequency of salary reviews, should identify the persons who conduct them, and should discuss key factors considered in granting salary increases. The same should apply to promotions, as well as lateral mobility. Almost all employees want to know what they have to do to get ahead. A professional organization makes this known and follows through on its promises.
19. *Bonuses* If the firm gives bonuses, employees should be told how they are calculated. It generally is wise to indicate that a bonus is not automatic, nor can one be guaranteed from year to year, when, in fact, one cannot be guaranteed.
20. *Moonlighting* A company usually cannot forbid all moonlighting, but it may be able to forbid moonlighting that would expose the company to liability. This latter result could occur if employees do on a moonlighting basis the type of work they perform for the company, and it could somehow be construed that the ability to moonlight is a company benefit, thus making the company liable for any negligent acts of moonlighting employees.
21. *Disciplinary Action* Most firms want the right to dismiss anyone at any time, for no stated cause. To do this without legal repercussion, it generally is necessary to make such a policy known, both in an initial advisory notice and in a disciplinary action section. This does not obviate the need to spell out the specific types of actions that could result in immediate dismissal for cause, such as lying to one's superior or too many unexcused absences. It may also be appropriate to have policies relative to a "monitoring period" that would be tantamount to a probationary period, as well as suspension.

22. *Termination Procedures*   This section would indicate who owes what to whom in the event of termination, in terms of unused vacation time, educational loans given, and so on. It usually is wise to conduct exit interviews with employees who are leaving, to learn how the company may be able to improve itself.

A number of other matters should also be discussed, and, in all such cases, some type of legal review may be beneficial, given the growing amount of litigation between employers and employees or former employees. In this respect, it is commonly advised that employees should sign a receipt when they are given a copy of the policy manual and that later, for example two weeks, they should be required to sign a statement indicating that they have read the manual, understand its implications, and agree to abide by it.

It may also be appropriate to develop procedures manuals. Some procedures apply universally, while others apply only to certain categories of employees. For example, proper procedures relative to answering the firm's telephone should apply to all employees; procedures that drafters or others should use when they have questions should also be indicated. Information about documentation should be given, and so on. Reviewing this information with new employees is important. General reviews are sometimes called for, too, as through staff meetings or "brown bag" luncheon seminars.

PERSON-TO-PERSON RELATIONSHIPS

After new employees have spent several months or so with a firm, a great deal should have been learned about their personalities, attitudes, and work habits. By evaluating the various pluses and minuses, it should be possible to develop a program—educational and otherwise—to help specific employees advance; to do their work more effectively; and to prepare for additional tasks, to benefit the employees and the firm.

Supervisor-to-Employee Communications

Every six months or even more frequently, supervisors should meet privately with each person they manage to discuss progress, specific attitudes, work habits or skills that need improvement, techniques for achieving improvements, and general likes and dislikes. The likes and dislikes portion of the discussion should be candid, and supervisors must be willing to learn, not just listen. The purpose is to help the firm improve by improving its personnel. Such discussions are often conducted as part of salary reviews. If these are to be meaningful, however, employees need guidance. They should be given specific objectives to accomplish, and, when possible, they should be given ideas on how to achieve these objectives. The more specific a manager can be, the better. It will then be up to the manager to monitor employee progress, in order to determine the extent to which guidance is being followed.

Note that effective managers generally do not make up lists with the goal of saving observations for revelation once a quarter or once every six months. If an employee is observed doing something that should not be done or doing something in less than a fully effective manner, guidance should be given immediately, *in private*. Never "dress down" an employee in front of others. This is humiliating and will make working for you a distasteful experience. And, even in private, do not lose your equanimity. State your anger; do not act it out. Although the message may be the same, there is a world of difference between, "You stupid jerk. I saw you. . . ." and "I saw you doing something that bothered me."

Because recognition is something that most people crave, be generous—and sincere—in your praise. If a subordinate has done particularly well on a task, let the individual know. This type of behavior is sometimes referred to as "giving warm fuzzies." People appreciate it and generally will work harder to earn more of it. Let them know, too, that they are appreciated in general, not just for specific actions. A remark such as "It's gocd to have you with us on this project team" can have a highly positive influence on attitudes from the outset.

As effective as praise may be, recognize that additional techniques are appropriate and should be fashioned to meet individual needs. Do not assume that every employee is eager to take on each assignment and has a strong desire to perform it well and turn in error-free work on schedule. The context that surrounds an assignment can be vitally important. Some employees respond best when they are "under the gun"; others go to pieces under such pressure. Similarly, some employees like the challenge of a complex assignment, while others far prefer a routine assignment. By understanding employees, you can make far wiser assignments, based on more than familiarity with the type of work involved. Similarly, you can present assignments more effectively. In the case of a project with a long lead time, for example, it may be effective to say, "This is due in three weeks, but I think you can handle it in two. That gives us time for a strong internal peer review, and we'll be able to put you on another assignment we have coming in."

Much has been made of motivation over the years. In fact, however, no one can motivate anyone else to do anything. Motivation is internal; it comes from within. Good managers are leaders who can say certain things that touch a common chord within all under their supervision; they can also phrase their thoughts to harmonize with a specific person's outlook.

Development of Career Ladders

As important as effective supervisor-subordinate communication may be, it is virtually impossible to derive maximum performance from people who have been promoted to positions they do not enjoy. While a "do it for the firm" pep talk may be of value for a brief period of time, the constant drudgery of doing something one doesn't like will ultimately take its toll. If personnel are to be groomed for tasks they will assume in the future, the work assigned them should somehow jibe with their individual capabilities and preferences. Personnel on the way up should be given opportunities to partake in a variety of different tasks, to determine which are most harmonious with their make-up. Some people who want to become CEO of a firm may say "forget it" once they realize what the job really entails. Likewise, some dedicated engineers or architects who get a taste of management responsibilities may find they like these even more than those associated with design work.

Every design professional firm with more than just two or three employees should establish career ladders for each category of staff. Each ladder should be indeterminate, in the sense that educational and experiential opportunities should create diversity, so employees can obtain a better sense of how they would like to contribute to the firm—and their own development—in the future. In the case of young designers, for example, career ladders should create an opportunity to sit in on sales calls, contract negotiations, site visits, and preparation of marketing plans, among other activities. Such involvements create a much better understanding of the firm and the activities in which it engages. It also gives people an opportunity to determine what they do, and do not, like to do. Thereafter, based on what the firm sees as its own long-term needs, individual courses of advancement can be laid out for given individuals. This does not necessarily mean that any one person will automatically advance to a given position. However, by having people headed in that direction, the firm can help assure itself of continuity and of having a group of dedicated people willing to do well that which they may not like to do, with the understanding that it is necessary in order to achieve overall career goals.

Generally speaking, a company is best served when its personnel are well served. Individual needs go far beyond salary and benefits. By taking individual attributes into consideration, assignments are more effective—something that benefits the company as much as it benefits its employees, now and in the future.

TEAM BUILDING

Although there is a strong need to recognize each employee as an individual, there also is a need to create a feeling of team—an "esprit de corps." In fact, every employee of a firm is important to the firm. Each has a job to do, and sub-par performance can have a seriously negative impact. All employees must realize this. Merely telling them is not enough. After a period of time, they will learn how to hide some of their failures or failings. Too often the result is mediocrity, and frequently this will be accepted as the best a firm can do.

Every firm should work to develop comradeship among its employees, so each feels part of a team and recognizes that individual failure hurts the team. Each employee should also realize that fellow workers, including supervisors and subordinates, are willing to be of assistance. All should want to seek assistance when needed because they recognize that not doing so would be unfair to others, would hurt the team and, ultimately, themselves.

Creating a sense of team effort imposes a considerable challenge on management. Several procedures for doing it have already been alluded to, albeit for accomplishing different objectives. Directing employees into other activities helps them learn more about the infrastructure of a firm, giving them more identity by explaining how their work fits in with others'. It also underscores how others are dependent on their work and how and when it is completed. Having a get-together for all those in a project likewise creates more sense of team involvement, as do occasional but regular firmwide meetings. Every person on staff somehow contributes. Being kept aware of how the company is doing thus illustrates how a contribution or extra effort is paying off.

Several times each year a company should sponsor team-building events, such as picnics in the summer or a Christmas party. Activities such as these create bonds of friendship. They can also help create a more comprehensive image of some people. For example, supervisors are seen in one role day after day. They may seem far different at a social gathering or with a spouse present. Having a softball team, basketball team, or some other company-sponsored program can also be of value (but be sure to consider various legal niceties. Unless certain i's are dotted and t's crossed, an injury while playing softball could make a person eligible for workers' compensation benefits).

In some firms, bonuses are used as incentives for effective work. However, to maximize impact, bonuses are given when the project has been completed, not at the end of the year. Several firms make more use of this approach by making bonuses dependent on team performance, rather than individual effort. High levels of individual effort and cooperative effort both are encouraged.

An effective team-building technique used by one architectural firm is particularly noteworthy. After a project has been built, the firm shuts down for a day and takes *all* employees on an inspection tour of the new building, to help assure all realize how important the firm's work is, what it leads to, and that the work is really a result of what the firm does, not just what designers do. In other words, the firm's final product is a representation of assignments of all people in the firm, and seeing the completed work helps them recognize this fact. Exactly the same type of technique can be used for virtually any type of firm participating in the design of any type of structure.

In essence, every member of a firm should be able to develop a sense of accomplishment from work of the firm, and each should understand how he or she has contributed to the accomplishment. By encouraging formation of this type of attitude and by otherwise encouraging employees to feel part of a team, individuals can come to understand that the best approach to personal advancement comes from working for team—firm—success.

OWNERSHIP OPPORTUNITIES

Some people have an entrepreneurial nature, and some do not. Those who do often tend to be the most dedicated of a firm's staff. If this dedication is not met with an appropriate response by the firm, however, some of the best employees may leave, often to start competing firms.

Just as firms need effective policies to help employees advance, so, too, do they need policies that give partial ownership of a firm to those who help make the firm more valuable. In smaller firms, such programs may come under the heading of "ownership transition." They help assure there is a market for the firm as the owners—often the founders—prepare to phase out. But the ownership plan must be reasonable. If employees must purchase stock each year, it would be somewhat unreasonable to expect them to work as hard as possible if doing so simply makes the stock that much more expensive. As such, employees should be given the wherewithal to purchase the stock, and this wherewithal also should come from the fruits of an employee's labor, as from a bonus, stock option plan, or some other means.

Some firms' owners are reluctant to share ownership, and, in reality, little can be done about it unless their attitudes change. Upon their retirement, the value of their firms will be far smaller than otherwise, in part because there will be few dynamic individuals inside the firm or, even if there are, their ability to purchase will be limited. While a merger with, or buy-out by, another firm is possible, the value to be obtained will be far less than otherwise, since there will be no vested interests.

Employee Stock Ownership Plans (ESOPs) can be of value in helping employees (typically of larger firms) derive some sense of firm ownership. If employees are unable to vote their stock, however, the plan has little meaning. In short, unless a firm develops a means for its most dynamic employees to have a say in the firm's development and to obtain rewards for efforts, the most dynamic employees may leave.

In some cases, of course, employees will leave in any event, simply because they want to be on their own. In many areas, the genesis of twenty firms or so can often be traced to just one or two progenitors. The way in which a firm handles such development says a great deal about its professionalism. Some regard a leave-taking to go out on one's own as an act of betrayal, forgetting that, in most instances, this was exactly how their own firms were founded. By contrast, other firms recognize the natural phenomena involved and do far more than give their best wishes to those leaving. Often they will offer assistance and may even become direct backers by providing "venture capital." They recognize that there may be opportunities to send certain work to the new firm and also realize that, in several years, the new firm may be one they could merge with or buy out, for everyone's direct benefit. This is not to say that having a new competitor is something a firm should look forward to. Every effort should be made to keep "the best and the brightest" and to offer them appropriate rewards and challenges.

TERMINATION

Whenever an employee terminates voluntarily or must be discharged for reasons other than moral turpitude, it is effective to conduct an exit interview. The purpose is to obtain information directly or indirectly that will benefit the firm. In the case of an individual who decides to work elsewhere, determine why. Is it merely the offer of better pay, or is there also more opportunity to advance or take on more challenging work? Are the firm's pay scales too low? Were an individual's capabilities improperly assessed in assigning work? Was there too much pressure or not enough? Was too little done to create a sense of loyalty to the firm? In the case of someone who leaves not too long after joining the firm, were the wrong questions asked during the interview, or were responsibilities not discussed effectively?

Whenever an employee terminates or is discharged, a failure has occurred, and it bears analysis. In some cases, of course, the employee could simply be lured away by false promises. But even this situation should not be ignored. If you believe your firm is a good one, you should let employees know, and you should also inform them that other firms will try to lure them away. Have you given them guidance on how to evaluate offers from other firms? Have you established enough of a positive image so that someone who is offered a position with another firm speaks with you about it, knowing that your professionalism will lead to honest comments? The latter situation is the best to be in, of course, because it indicates that your human resources management is where it should be. And, if you can realistically promise employees that, within a reasonably short period of time, the gains sought through employment with another firm can be gained at yours, chances are they will stay.

POSTTERMINATION

Firms should keep track of former employees who separated themselves under amicable or, at least, nonhostile circumstances. They should be sent annual holiday greetings, if not birthday greetings, too, and these should be signed by an individual's past supervisor or the CEO. The former employee should also be sent the firm's newsletter, news releases, new brochures, and so on. By taking this approach, former employees are treated as alumni of a firm, helping to maintain goodwill and open lines of communication. It also establishes an avenue through which employees can return if they so choose. And, should an opportunity arise, it makes it possible for a firm to get back in touch, to bring aboard an individual who has gained good experience with others but who still retains many of the basics your firm imparted in the past.

THE LAW

Many different legal rulings affect the way in which you treat employees. Various techniques can be used to help assure you comply with the law, not only to protect yourself from government suits, but also from suits filed by former employees. For the most part, the laws are not unreasonable. They have come to pass in large part because some employers have been unreasonable. While compliance with some regulations may increase paperwork, the paperwork generally is necessary to protect the firm. Many of the potential problems can be solved simply by having a comprehensive, clearly written employee policies manual and—most important—by implementing policies in a consistent manner. The regulations that apply should not be allowed to create an "us-versus-them" attitude among employees and management. The precepts of effective personnel management can be implemented fully while abiding by the various laws and affording appropriate protection to the firm. However, in all cases, the firm should be fully familiar with all of the various laws, regulations, and court interpretations that apply. This, too, is a matter of professionalism.

As a general rule, document each employee, for purposes of legal defense (should it come to that) and for purposes of effective administration. A personnel folder should be prepared for each employee as soon as the individual is hired. The individual's employment application and resume should be included in the file. Any notes prepared by supervisors should be placed in the file, too, along with records indicating the projects to which the person has been assigned, educational attainments, memoranda or letters summarizing points made during a salary review, letters of commendation, and so on.

In the case of an employee whose actions are subject to reprimand, it generally is advisable to state the reprimand in writing and give a copy to the employee. If the action involved cannot be tolerated, the memo or letter should say so. It should also advise that any repetition will subject the employee to dismissal. This is done not only to provide guidance to the employee but also, in the event dismissal becomes necessary, to defend an unjustified unemployment compensation request.

Recognize, too, that a company can be held liable for the actions or inactions of its employees and that personnel files can provide documentation that can possibly mitigate certain problems. For example, an employee's driving record should be checked as thoroughly as possible before permission is given for the individual to drive a company car or to drive a personal vehicle on the company's behalf. Should the person cause an accident, and should it be revealed that the employee actually had a poor driving record, your files would at least demonstrate that you made an honest effort to ascertain the truth; that employee dishonesty (e.g., failing to reveal a prior name or residency) prevented you from learning the truth.

Note that personnel files can be of more value than addressing concerns rooted in legal issues. By keeping experience and educational records up to date, perhaps by transferring them to a computerized database, it should be that much simpler for the firm to put together effective teams, for example—those with specific experience with religious facilities, certain types of medical buildings, clean rooms, and so on.

11

Professional Services Marketing and Business Planning

Business planning is a fundamentally essential business activity for which marketing serves as a foundation. Design professionals do not generally recognize what marketing actually entails, however. Many use the term as a synonym for promotion, as in the phrase "marketing one's services." Marketing—*real marketing*—is far different. It comprises a variety of tasks designed to analyze a firm's traditional markets (the sources of its commissions) and to identify those that the firm needs to enter or penetrate further in order to achieve the business plan's objectives. These objectives are established by the firm's principals for attainment of their personal and professional goals.

This chapter provides some generalized details about basic marketing and business planning tasks. Many firms ignore these tasks altogether. Others approach them in a somewhat haphazard manner, developing business strategies by intuition rather than by research and analysis. Although some of those who proceed intuitively might prosper in the long term, many do not. And those who do often are victimized by market forces that could have been predicted but were not, forcing the rapid acquisition or disposal of space, equipment, and personnel, expenditure of reserves, and loss of profit.

Effective marketing and business planning do not preclude the potential for unanticipated events, but they do make the unanticipated far less likely. And, should the unanticipated occur, planning can be of substantial value in fashioning a response that stems from contemplation rather than panic.

Marketing and business planning, more so than almost any other business tasks, give management the ability to shape an organization's future. This ability is particularly important for engineers and architects, because it gives them the wherewithal to achieve their personal and professional goals.

Creating an organization to achieve personal and professional goals is an aspect of practice that distinguishes professional service firms from commercial enterprises.

Growth is a common goal of organizations that exist solely for commercial purposes. But growth for the sake of growth can lead to a barren business existence. By increasing revenues year after year and by acquiring steadily more assets—there can never be an ultimate—the goal is always beyond reach. While growth is often important for engineering and architectural firms, it should be an objective, not a goal—something that is necessary to permit the realization of the principals' professional ambitions. This is why the professional goals of those who lead a design firm are—or should be—the foundation of its marketing and business planning. When there is no genuine marketing, however, it is far less likely that personal goals will be achieved or even that time will be taken to assess what these personal goals are. This is why so many architectural and engineering firms have become purely commercial enterprises. By failing to identify and to strive for professional goals, the principals of these firms have adopted commercial goals by default. But the acquisition of wealth has never been the goal of true professionals. It has been a concomitant—something that occurs almost of itself as a by-product of their pursuit for professional excellence. For this reason, many of the nation's most successful design professionals epitomize professionalism, in their technical pursuits as well as in their business activities. And nowhere is professionalism more important than in the conduct of the marketing and business planning function, where professional attitude and approach can mean the difference between an organization that charts its course to a definable future and one that simply weighs anchor and hopes for the best.

MARKETING AND BUSINESS PLANNING: AN OVERVIEW

Because marketing and business planning involve such a variety of tasks, it is worthwhile to review a synopsis of the activities involved, to create an understanding of the forest before examining the trees.

For firms that have not previously established a marketing effort, work begins by analyzing commissions awarded over the past five years. For planning purposes, it is essential to segment these commissions based upon certain salient attributes, in order to establish *marketing units*, or MUs. Using a broadbrush approach, MUs can be described in terms of general marketing segments, such as overall building types (e.g., commercial, residential, industrial, and educational), general project types (e.g., highways, bridges, tunnels, or wastewater treatment plants), or by some other means that is in keeping with a firm's overall practice.

Once general or primary marketing segments have been identified, it is necessary to describe each more precisely. (The larger and more diverse a firm's practice, the more precision usually is required.) Common MUs thus may be commercial/office building, commercial/mixed use, commercial/ retail, and so on, or, more precisely commercial/office building/speculative/ high-rise/more than 1 million square feet.

Once MU segmentation has been completed, past projects are grouped into the various MUs for analysis. Through this procedure, the firm can learn important information: which MUs have been generating the most revenue, which are the most profitable, which are subject to the most or fewest problems, and so on. And, by arranging the MUs in a chronological pattern, it is possible to identify trends, that is, the kinds of projects that the firm has been obtaining over the years or those that are increasing or decreasing in frequency and profitability.

At the same time that a firm's experience with MUs is being analyzed, the firm can review MU developments in general to determine how the firm has been responding to market conditions. For example, if the number of schools being built has been increasing at the rate of 10 percent per year for the past five years, while the number of school commissions being awarded to the firm has been decreasing by 5 percent per year, the firm's market share has been eroding. Why? Determining the cause gives the firm an ability to determine what it needs to do in order to generate more business from the MU, assuming this suits its purpose.

Analysis of the past leads to an analysis of the future. Through literature search, data review, interviews, and other techniques, the firm can gain insight into what will probably be occurring in those MUs in which it is active and among those organizations and individuals that comprise its clientele. At this point, if they have not done so before, the leaders of the firm need to assess their professional and personal goals. What do they want to be doing five years hence? How big a firm do they want? With what types of commissions and clients do they want to be associated? These and innumerable other questions must be answered, so that the planning function—and thus the firm—can be given direction. The next question that must be answered is basic: How do we get there from here? With this in mind, each MU in which the firm currently is active is reviewed to determine its direction, the degree to which increased or decreased penetration will help support attainment of the firm's goals, and the degree of difficulty associated with obtaining continued commissions from it. Next, the firm analyzes MUs associated with those in which it already is active, to determine the opportunities they present, and, finally, other MUs that hold particular promise but with which the firm now has no relationships.

The next step involves analysis of all that has gone before, and, at this point, the marketing and business planning functions begin to merge, as alternative future scenerios are presented and discussed. Discussion results in the formulation of a plan with respect to the MUs on which the firm will rely to achieve its objectives, the degree of activity in each, and the methods through which the firm will achieve its objectives. Five-year plans are the most common, with the first year being highly detailed, the second somewhat less detailed, and so on.

The business plan should be reviewed and finalized approximately six months before the first "future year" begins. Thereafter, on a monthly basis, those who lead the business planning function receive the various reports necessary to track progress and to determine the extent to which plans are being realized. Then, nine months or so prior to the start of "future year two," the plan can be analyzed once again to determine the cause for any differences between "actuals" and projections and to assess changed external conditions that make plan modifications necessary. Principals of the firm then make the planning changes needed, provide the appropriate amount of additional detail to each of the remaining four years, convert "future year two" to "future year one," three to two, and so on, and add a new "future year five." This procedure helps the firm continually gear its efforts toward identifiable business objectives, using the best possible knowledge of its own capabilities and how they should be applied to take advantage of market conditions that research shows are emerging. The process also helps assure that the work being pursued and obtained is that which is most satisfying for the people who comprise the firm. This helps assure the type of professional response necessary for long-term success and the fewest professional problems.

Those associated with smaller firms may regard some of these tasks as monumental and unnecessary until they achieve a larger size. Abandon such notions. The extent of work involved is directly proportional to the size of a firm, and, to a very real extent, the value derived is inversely proportional to size, since smaller firms are far more affected by growth than larger ones.

The most difficult aspect of marketing and business planning is overcoming inertia. Once the basic work has been completed and once the analysis of the past has been accomplished, it is relatively simple to keep up to date, just by keeping track of data the firm should have available to it in any event. These are tasks that professional business managers pursue as part of their everyday routine. If design professionals charged with the operation of their firms are unwilling to undertake those responsibilities in a professional manner, their professional ambitions will be extremely difficult to realize.

MARKET SEGMENTATION

A firm's overall market consists of all those persons and organizations who have a need for the services the firm provides and the ability to pay for them. In order to perform marketing and business planning, it is necessary to divide the overall marketing into discrete, manageable elements, called *marketing units* (*MUs*). This is accomplished through the process of market segmentation. Several criteria can be used; the basic one usually is project type. Thus, for a firm involved in building design, the market could be segmented according to basic socioeconomic subdivisions: commercial, residential, industrial, and institutional. Each of these *prime segments* can be made more specific through subdivision, or the prime segments themselves can be more narrowly defined.

For example, the institutional segment could be restated as health care, educational, governmental, correctional, and so on. Each prime segment is made more specific through reference to project type, such as health care/hospitals, health care/nursing homes, and health care/emergency facilities. In a similar manner, the commercial sector could be segmented as commercial/retail/enclosed malls, commercial/retail/strip shopping centers, and commercial/retail stores. Alternatively the commercial segment could be restated as more narrowly defined prime segments, e.g., office, retail, and "other commercial."

A firm that is engaged principally in public-sector work may be more inclined to use the specific client as a basic market segment, (e.g., First County, City A, State 1). Prime segments then could be subsegmented by the specific agency involved and, to the extent that it makes sense, by the type of project, such as water treatment facility or waste disposal facility.

Firms that seek work from both the private sector and public sector could include such a differentiation as a prime segmentation variable. This would come under the heading of "Type of Owner." In this regard, one could amplify "Private Sector" by applying variables such as "owner/occupier" or "speculative developer."

It may be important for some practices to differentiate in terms of client type, that is, "direct" where the owner is the client or "interprofessional" where a firm subcontracts with another, for example, interpro/civil or interpro/architect.

Type of service could be a major consideration in some instances, permitting segmentation based on factors such as design-phase-related, construction-phase-related, forensic, and economic studies. Likewise, it may be appropriate to differentiate based on new construction or existing construction, with the latter possibly being subsegmented by factors such as conversion, rehabilitation, expansion, or adaptive reuse.

Project location may be an important factor for some firms. For most, it is not, because most consist of one office only, with the majority of their work coming from the basic *geographical marketing area*, or *GMA*. When there are somewhat frequent exceptions, however, it may be appropriate to identify projects as GMA or "non-GMA," with most of the projects fitting into the latter category likely to consist of work for GMA clients who are expanding into other areas.

For firms with highly specialized services and little competition, a GMA may be regional, national, or even international. Firms with branch offices often regard each branch as having a GMA of its own, but they will often have a special "national accounts" activity located in their headquarters operation.

As should be evident, there are no hard-and-fast rules when it comes to establishing the variables by which MUs will be defined. In each case, the nature of the variables used and the extent to which they are applied should be based on needs of the firm. The needs will become clear through the process of internal marketing research.

INTERNAL MARKETING RESEARCH

Internal marketing research involves a review of company records to determine historical trends. Also known as organizational research, it requires project-by-project analysis to identify key variables.

Project Analysis

Project analysis can require some time if it has not been done before. Once the initial work has been completed, however, keeping records up to date is relatively easy and can be handled through software.

The work begins by gathering records on all projects with which the firm has been involved for the prior five years. Basic information then is extracted and recorded. The "Project Analysis Form" shown in Figure 11-1 indicates an approach to recording information, using variables pertinent to the firm involved. Factors to be recorded are the following.

1. *Year*: A firm can use either its fiscal year or a calendar year to indicate when the work was obtained (when the contract was signed). Calendar year is suggested, because the definition of a firm's fiscal year is subject to change.
2. *MU*: The variables used to identify the MU would be those that make the most sense for the firm. It is easiest to employ a coding system, whereby MFR/C would designate a multifamily residential con-

PROJECT ANALYSIS FORM

1. Year _____
2. MU _____
3. New or existing _____N _____E
4. Services Provided _____
_____
_____
5. Owner _____
6. Client _____
7. Location of Project/Client _____
8. Size of Project _____
9. Fee _____
10. Percent Profit (Loss) _____%
11. Problems _____
_____
_____
_____

Figure 11-1: Typical Project Analysis Form

dominium project, and C/MU/O&R would indicate a commercial (C) mixed-use (MU) building with office and retail space (O&R).
3. *New or Existing*: Indicate if the project involved new construction or existing construction.
4. *Services Provided*: A code can be used to designate basic services. For a structural engineering firm, A might symbolize "preliminary evaluation," B could stand for "design and specification," and C could mean "construction monitoring." For a large multidisciplinary firm, A could indicate "preliminary site analysis," B might mean "economic evaluation," C "HVAC design," D "structural design," and so forth.
5. *Owner*: If a firm is small, it would make sense to indicate an owner by name. Otherwise, type of owner could be indicated, for example, Pub/HA for "public sector/housing authority" or Pri/Spec for "private sector/speculative developer."
6. *Client*: For firms that work extensively on both a prime and interprofessional basis, it may be sufficient to designate the client by "owner" or "other design professional." The latter could be stated more specifically as "architect," "civil engineer," or some other design professional. The client could also be identified by name.
7. *Location of Project/Client*: This can be stated in general terms, with four basic options: GMA Project/GMA Client, GMA Project/non-GMA Client, Non-GMA Project/GMA Client, Non-GMA Project/Non-GMA Client. More specificity could be added, by referring to specific jurisdictions, on a state and/or local basis.
8. *Size of Project*: Here one can use a letter or numerical code to indicate overall project size, based on factors pertinent to it. On some, "A" could designate "less than $1 million," whereas on others it might mean "less than 25,000 square feet." In either case, however, it would mean "small," from the standpoint of the firm involved.
9. *Fee*: Once again, a letter or numeral could be used to designate a range, for example, "less than $10,000," "$10,000–$19,999," and so on. Alternatively, actual gross billings (with or without "pass-throughs") could be indicated.
10. *Percent Profit (Loss)*: This information should be available through project management or other records. Otherwise, it would be necessary to compute the information, based on direct labor and overhead factors. Guessing is not advised.
11. *Problems*: Typical problems include slow payments and claims, among many others. Where these are known, they should be indicated. If they occur frequently enough, a coding method could be used.

Realize that the form shown in Figure 11-1 is just an example. The format used by a given firm should reflect the variables it most often encounters and/

or it most wants to record. In some cases, it may be prudent also to identify the manager of each project, so that later, when annual trend data are being developed, the average or median profitability of projects managed by a given individual can be assessed. Similarly, it may be appropriate to indicate how the firm obtained the project and, when a referral is involved, to indicate the person from whom the referral came.

Two warnings are appropriate in the case of firms that have not previously recorded project information. First, do not become so enthused by the process and its benefits that too much information is recorded. When this happens, the extent and complexity of the data can become overwhelming. Second, do not become so intimidated by the task that data are overgeneralized and thus almost meaningless.

Annual Marketing Unit Analysis

Figure 11-2 illustrates an "Annual MU Analysis Form," used to record information about each MU on a year-to-year basis. The type of detail the form is designed to obtain indicates why computerization can be of value to the internal marketing process. Again, recognize that the form shown is designed to meet the specific needs of a specific firm. The information being developed consists of the following elements.

1. *Year*: This would be either the fiscal year or calendar year, whichever was selected in developing project analysis data.
2. *MU*: The overall marketing unit involved. Through data manipulation, one can develop several reports on different aspects of an MU. For example, reports could be developed separately for Residential/Multifamily/High-Rise/Condo, Residential/Multifamily/Mid-Rise/Condo, and so on, each to the fourth variable (condo as shown, as well as rental). Data could then be developed to the third variable (ignoring the fourth), then to the second, and then for the entire residential segment. The principal purpose in doing so would be to identify any significant differences. For example, if it is shown that condominium projects tend to be far more profitable than rental projects, the difference would be significant and would have a major impact on the marketing plan.
3. *Total MU Projects*: It is worthwhile to know how many projects each MU has generated, in comparison to other MUs. If a firm relies heavily on an MU that subsequent external marketing research indicates is in trouble, a rapid change might be called for.
4. *Total Fees*: Total fees represent billings (including pass-throughs) or fees for all projects, data that can be used in conjunction with total MU project information.
5. *Average Fee/Project*: This is derived simply by dividing the answer to

ANNUAL MU ANALYSIS FORM

1. Year _____
2. MU _____
3. Total MU Projects _____
4. Total Fees $_____
5. Avg. Fee/Project $_____
6. Avg. Percent Profit/Project _____%
7. New Projects
   Total Projects _____
   Total Fees _____
   Avg. Fee/Project _____
   Avg. Percent Profit _____%
8. Existing Projects
   Total Projects _____
   Total Fees _____
   Avg. Fee/Project _____
   Avg. Percent Profit _____%
9. Services
   Type   No. of Projects (% of Total)   Avg. Project Profitability
   _____   _____   _____
   _____   _____   _____
   _____   _____   _____
   _____   _____   _____

10. Service Combination
    Type   No. of Projects (% of Total)   Avg. Project Profitability
    _____   _____   _____
    _____   _____   _____
    _____   _____   _____
    _____   _____   _____

11. *Owner*   No. of Projects (% of Total)   Avg. Project Profitability
    _____   _____   _____
    _____   _____   _____
    _____   _____   _____
    _____   _____   _____

12. *Client*   No. of Projects (% of Total)   Avg. Project Profitability
    _____   _____   _____
    _____   _____   _____
    _____   _____   _____
    _____   _____   _____

13. *Location of*
    Project/Client No. of Projects (% of Total)   Avg. Project Profitability
    _____   _____   _____
    _____   _____   _____
    _____   _____   _____
    _____   _____   _____

Figure 11-2: Annual MU Analysis Form

14. *Size of Project*   No. of Projects (% of Total)   Avg. Project Profitability
    _____   _____   _____
    _____   _____   _____
    _____   _____   _____

15. *Size of Fee*   No. of Projects (% of Total)   Avg. Project Profitability
    _____   _____   _____
    _____   _____   _____
    _____   _____   _____

16. Problems _____
    _____

Figure 11-2: (Continued)

item 3 into the response for item 4. Some may find a median to be a more valuable number or may prefer to identify both the median and the average.

6. *Average Percent Profit/Project*: Again, a median may be a more valuable indicator, or both median and average data could be obtained.
7. *New Projects*: As shown, the goal here is to identify new construction's share of the MU in terms of projects and billings, as well as its relationship to the overall MU relative to average fee and average profit.
8. *Existing Projects*: The same type of information would be identified here as for "New Projects," for the same reason.
9. *Services*: In this response, each basic service is reviewed in terms of the number of MU projects for which it was applied, the percentage of total MU projects involved, and the average profitability of those projects.
10. *Service Combinations*: This is basically the same as item 9, except combinations of services are analyzed. In this way, it may be shown that, when both new design and construction monitoring are provided together, the average percentage of profit for the projects involved is much higher than that associated with either service alone.
11. *Other*: For items 11 through 16. In each case, the objective is identifying the number of projects for a given variable (each type of owner, each type of client, and so on), the percentage of overall MU projects, and the average profitability of the project involved. The purpose of item 16 is to identify a specific type of problem and the types of projects on which they most frequently occur.

Reviewing data drawn from past projects by using an approach similar to what has been described creates insights that otherwise are not possible. What has happened in one given year does not necessarily hold true for other years, however. For this reason, it is important to conduct trend analysis also.

Trend Analysis

Trend analysis involves collection of annual MU data for at least three years; five years is far preferable. Trend analysis is generally pursued as a two-part process. First, trends with respect to any given MU are analyzed. Second, trends are reviewed with respect to interrelationships among MUs.

Trends Within an MU   Trends within a given MU are analyzed by using data shown on Annual MU Analysis Forms. For example, the MU Trend Analysis Form shown in Figure 11-3 would be used to compile five-year data from items 3 through 8 of Annual MU Analysis Forms completed for each of the four prior years and the current year (with projections being used to suggest full-year data for the current year). This approach allows you to see how business has been increasing or decreasing year after year, changes in the average (or median) fee for each project, and—assuming it would be appropriate—the number of projects involving new versus existing construction (and the percentage of total MU fees each involved), average fees for new versus existing construction, and so on. Additional forms could be prepared to analyze trends associated with item 9 through 16, assuming the effort would be worthwhile.

Graphic representations of data are beneficial, because graphics make trends far more evident. Figure 11-4 illustrates a hypothetical situation.

Trend analysis is a valuable exercise. If the total number of MU projects varies considerably from year to year, why? The answer may be found by considering more variables. Is the firm dependent on an interprofessional client getting the business? Have losses or gains been due solely to the effectiveness of the client, or have there been other factors present affecting owner's decisions, such as demand for certain types of space, availability of money, or taxes?

Is there variability in profits? If so, has this been due to changes in service combinations? Types of clients? Location of projects? Could new competition in a given MU be responsible? If so, and if profitability must be cut in order to compensate, has there been any increase in the frequency of certain problems?

What about problems? Are certain types associated more with one type of client than with another? More with one type of project than with another?

By reviewing data in this manner, it should be possible to obtain an intimate understanding of each MU.

Overall Trends   Once those who direct the marketing research effort have analyzed each MU for trends, the MUs can be put together to indicate how each affects overall firm activity, as shown in Figure 11-5. There, three factors are considered for each MU: the percentage of overall firm projects it accounted for for each of five years, the percent of overall fees, and the percent of overall

MU TREND ANALYSIS FORM

MU _____

Total Projects

19__	19__	19__	19__	19__

Total Fees

19__	19__	19__	19__	19__

Average Fee and Profitability Per Project

	19__	19__	19__	19__	19__
Fee	$	$	$	$	$
Profit	%	%	%	%	%

New Projects

	19__	19__	19__	19__	19__
Total Projects (%)	%	%	%	%	%
Total Fees (%)	$ %	$ %	$ %	$ %	$ %
Avg. Fee/Project	$	$	$	$	$
Avg. % Profit	%	%	%	%	%

Existing Projects

	19__	19__	19__	19__	19__
Total Projects (%)	%	%	%	%	%
Total Fees (%)	$ %	$ %	$ %	$ %	$ %
Avg. Fee/Project	$	$	$	$	$
Avg. % Profit	%	%	%	%	%

Figure 11-3: MU Trend Analysis Form

profits or, as may be more appropriate, percent of overall profits before bonuses. Additional factors can be analyzed, of course, and additional analysis will lead to more questions. The value of addressing additional questions cannot be overemphasized, because this is the process through which those in charge of the firm's business planning functions come to learn more about key factors associated with the firm's work. For example, if a certain client or type of client is associated with high profits in one MU but not another, why? If existing construction creates problems in one type of MU but not another, why?

External Trend Analysis As valuable as trend analysis can be, it does not answer all of the questions about past performance, because it cannot give full

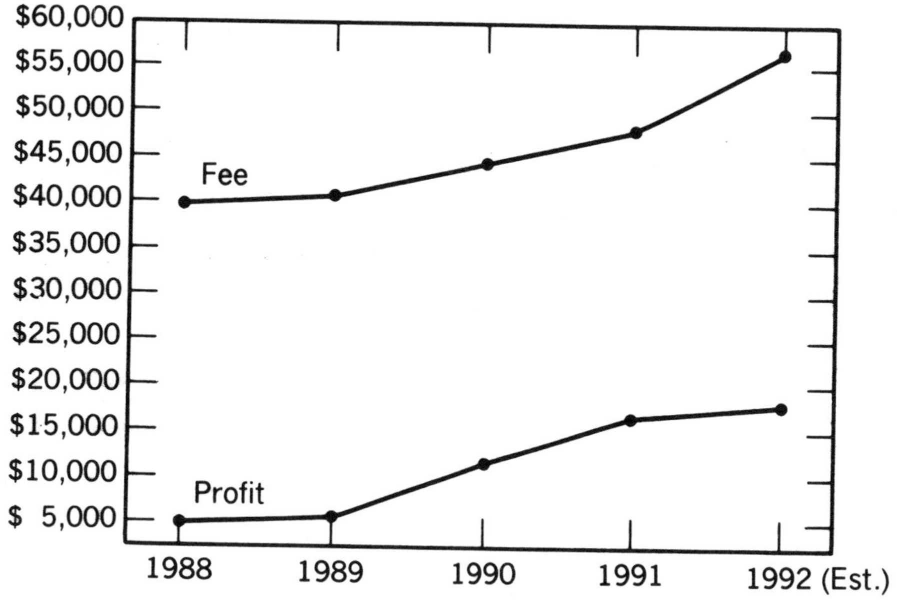

Figure 11-4: Graphic Presentation of MU Trend Analysis Data

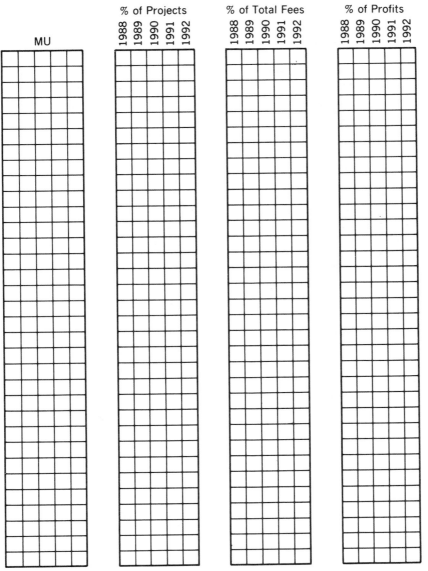

Figure 11-5: Overall Trend Analysis Form

consideration to external influences. It is unwise to assume that a given MU's reduced contribution to overall projects and fees occurred because of reduced activity in that MU, when actual activity in the MU is unknown. It is essential to determine how effectively a firm has been responding to opportunities created by MU activity. Several approaches are available to make such assessments. One is to review requests for proposals (RFPs) or quotations (RFQs) received for projects that the firm has not obtained. These will indicate which MUs have been most active. Determination of who obtained the work will indicate who the competition is for certain types of commissions, and it may reveal what your firm has to do in order to increase its ability to obtain such work in the future. If your firm does not keep solicitations on file, and if it has no other means for identifying projects in which it expressed interest, a system for recording relevant data should be developed for future use.

Literature search comprises another technique for external trend analysis. Many daily newspapers publish a special year-end issue that, among other things, reviews the area's economic trends, new buildings or public works opened, and so on. These can provide excellent insights, as noted in the following section. The same applies to trade publications relating to your clients and the design professions, association journals, and national business publications. By becoming more familiar with the issues covered in these and other sources, you can gain a solid understanding of how major national issues affect the area and clients your firm serves and, in turn, how they affect your firm.

In sum, internal trend analysis can lead to an intimate understanding of exactly what has been happening with a firm's business sources. To understand *why* it has happened, knowledge of external marketing factors is also necessary. Familiarity with external factors is particularly important in determining what is likely to occur in the future.

EXTERNAL MARKETING RESEARCH

External marketing research comprises the identification and analysis of factors that will affect a design professional's business in the future. It is inexact at best, since it relies on predictions. Major influences tend to evolve, however, so the astute design professional does not have to be taken by surprise. Through the use of effective planning, key issues can be identified well in advance, and predictions can be tested before a firm is committed to a given path.

Sources of Assistance

As with internal marketing research, external marketing research is greatly simplified once the function has been established in the firm. For example, those who are responsible for following up on leads, responding to RFPs, making cold calls, and so on can keep track of their activities to facilitate ongoing MU trend analysis. The same individuals can meet on a monthly basis for general discussion, to determine how emerging national or statewide trends in areas such as tax reform, housing, and environmental protection may affect the firm's clientele. Where the answers are unknown, appropriate personnel can be encouraged to discuss the issues with clients. By becoming more attuned to external development, a firm can move swiftly to anticipate needs and thus be in an excellent position to take advantage of them when they materialize.

Many sources are available to help with external marketing research efforts. One of these is a marketing consultant: an experienced individual or firm retained to perform all the research or selected elements of it. Those engaged should work closely with someone from the firm, to assure the two are "in sync," particularly when the assignment goes beyond an analysis of external marketing factors.

Another source is someone in the firm who is given the assignment. Although this approach might work out well, group discussion of central issues usually results in a more balanced appraisal of the various factors to be considered. As such, the best source often is a group of individuals from the firm, possibly working under the general guidance of a marketing consultant.

Many publications can be studied for information. The best place to start often is the local public library. The librarian may be able to provide some excellent assistance, at least to the extent that certain reference works are made available. These will include directories of periodicals and listings of virtually every magazine and newsletter published in the United States. Each is publishing articles at this moment that are of interest and concern to the future of any design professional's practice: articles about new technology affecting your way of doing business, as well as your clients' and prospective clients'. Statistical information is available from the U.S. government, the world's largest publisher, particularly from the Census Bureau. A call to the closest Census Bureau office should yield assistance. For the most part, however, much of the Census Bureau's information will be used and digested by others, and published in magazines and newspapers on a regular basis.

Reporting services can provide some excellent information. These include McGraw-Hill's *Dodge Reports*, which can provide insights about the type of construction ongoing in any given area. McGraw-Hill and others also publish annual forecasts of construction activity.

General business publications can provide a wealth of information. These include *Forbes, Fortune, Business Week,* and *Inc.,* among many others. Small business management newsletters, which often include tax tips, can be helpful. Knowing that the nation's tax laws have a profound impact on what people and businesses do with their money can result in early warnings on a number of important marketing trends. The *Wall Street Journal* is virtually "must" reading, in part because so many of its articles relate to major corporate executives' interpretation of market developments.

Reading local newspapers as well as the major daily in the state (if it is not the local paper) can be of value. In many respects, it will indicate how national trends are being realized in a firm's GMA. Local magazines can be helpful, too, along with local periodicals describing new construction projects. Local governments' legislative dockets, which commonly include requests for zoning variances, can also be helpful sources of information.

Perhaps one of the most valuable sources of all comprises the conversations one can enter into as a member of various local associations. Design professionals have a long history of being willing to share information, and much can be learned during hospitality periods and from informed speakers. Others who can provide insights include clients, manufacturers' representatives, utility representatives, insurance agents, contractors, lenders, government officials, and others who are in a position to learn about important developments and trends, and who may have performed some marketing research of their own.

Doing the Homework

Those with responsibility for external marketing research should be "up to speed" before they ask "outsiders" for information or develop market projections. In other words, they should first read periodicals and books to develop ideas on what might happen in the future, what the major events are likely to be, and how they could affect others.

The need for knowing your GMA cannot be overstressed. When your GMA is a local area, it is advisable to spend several days driving through it to examine conditions and to form opinions. Government zoning and planning officials should be able to help furnish maps that indicate the types of structures that already exist in certain areas, zoning in those and other areas, and so forth. Interviews of certain individuals could also be helpful, but, as will be discussed, they should be saved for later, after initial projections have been formed. In this way, those conducting the interviews are in a better position to test their assumptions and to share information, so that the interview will not be a wholly one-sided process. Generally speaking, people are far more willing to volunteer time for an interview if they believe they will gain something from the process, too. When they do, they are more apt to serve in the same capacity in the future.

Creating the Initial Projections

Initial projections can best be developed through a series of meetings, presided over by one individual. The person chosen should *not* be the firm's CEO if the CEO will intimidate others into silence.

The individual selected—the chairperson—must prepare for each meeting by creating an agenda that will encourage discussion. There is no "best approach" in this regard, because the issues involved are highly interactive; events affecting one MU will likely affect others. The agenda could move from

MU to MU, or it could focus on major factors influencing MUs, such as the economy, taxes, or prison overcrowding, and discuss how each will affect all MUs.

In many cases, it is easiest to begin by discussing MUs in which the firm is active and consider their future in light of changes occurring locally, statewide, regionally, and nationally. The objective is to determine how traditional MU commissions will be affected in the future and how predicted changes may create new opportunities that could be realized by providing nontraditional services. Participants should also consider how owners and other MU decision makers will be responding to the opportunities or pitfalls these changes may create.

In discussing the fate of MUs in which the firm already is active, a number of other MUs will likely be discussed. Take notes about these other MUs, because they could represent markets for future penetration.

As an example of issues that could be considered, assume that yours is a mechanical/electrical consulting engineering firm that has been extensively involved in the commercial office building market. A boom was experienced in the early to mid-1980s, but this has led to overbuilding. This situation, along with tax law changes, suggest that the MU will lose vitality. Does the firm simply write off this MU? What about the existing office MU? What are its needs? Will older buildings be renovated or expanded? Will their owners want to add telecommunications capabilities to help the buildings compete? What about existing mechanical systems? Do they need updating? If they are oil-fired, do they possibly run afoul of any new antipollution laws on the books? Will some of them be changing hands in the near future, and, if so, will there be a need for comprehensive investigations of the mechanical and electrical systems? If such is forecasted to be the case, then there will be a need for related services the firm is not now in a position to provide, including structural review and roofing review. It would also be advisable to determine if asbestos has been installed in the building.

In short, before writing off an MU with which the firm has extensive experience, the planning team should consider alternative nontraditional services it can provide. In this respect, it should also consider single-source "packaged" services that the firm could put together by engaging others, such as structural engineers, for each engagement, or that could be otherwise provided.

The specific means of bringing the new services to the MU should not be a major discussion item at this point, because it could lead to digressions. For example, participants could enthuse about a new subsidiary firm offering full service for existing buildings and the benefits of serving as a prime consultant. This type of reaction could become a major focal point, obscuring numerous other opportunities, as well as the pitfalls associated with the "new venture." The chairperson needs to keep discussion on track. Discussion of the MU would be concluded by noting that the prospects for new development look bleak but a significant opportunity is presented by meeting the needs of existing building owners. The chairperson would then itemize the nature of the needs and the approximate potential overall.

A related consideration may be the future of the developers who have fled the speculative office building market. Some may opt to develop other buildings in non-GMA areas. Others may look for different types of development, as in the retail sector. A consulting mechanical and electrical (M&E) engineering firm's ability to take advantage of predicted retail MU growth might be limited to architects' capabilities. Nonetheless, it should not be an issue at this time. The goal should be identifying the opportunities. Worry about taking advantage of them later.

Once the firm has considered its traditional markets, it should consider nontraditional MUs. One of these might be the industrial sector. If few major factories are being built, are any existing plants being upgraded? How many factories exist in your firm's GMA? What types of products do they fabricate? Is there a need to expand these facilities? What about their pollution controls? Lighting? Materials handling systems? Is there any likelihood that the factory's owners will be considering robotics or direct digital control? In discussing such issues, one must consider the cost of electricity, assuming the plant relies on it. Where the costs are expected to increase within the next few years, energy conservation may be an important issue. The same might be said of the existing office building market, if the issue was not previously discussed. If not, then the prior summary would be expanded, and possibly discussion would have to be reopened. This underscores the importance of the chairperson's role. Brainstorming sessions tend to be highly dynamic in nature, and the chairperson should take advantage of new ideas or perspectives as they arise. Digressions into other areas of concern—for example, how a firm enters a new market or a new aspect of an existing MU—can thus be destructive because of the loose structure that would prevail.

Third-Party Analysis

Once initial market projections are roughed out, it would be wise to subject them to third-party analysis, especially through interviews with key individuals. Those in the higher echelons of a firm usually are in the best position to arrange these interviews, as through luncheon meetings. If they are not intimately familiar with the initial projections and the thinking behind them, however, someone who has this familiarity should be taken along.

The principal purpose of these meetings is to test the validity of certain assumptions. In doing so, the interviewee must be encouraged to talk. Note, however, that meetings such as these can serve purposes that go far beyond marketing research alone: The people who represent a firm, and thus the firm itself, will be making an impression.

Typical candidates for interviews would include the following.

Owners/Developers: What plans do they have for the immediate and long-term future? Will existing buildings be sold? Expanded? Renovated?

Modernized? What types of changes do they forecast for various "downtown" areas? The suburbs? The exurbs? What do they most like and dislike about dealing with design professionals?

Government Officials: What are general plans for existing and new public facilities, such as water and waste treatment plants, streets and highways, bridges and tunnels, airports, and so on? What will be the most significant differences between the area as it now exists and as it will be ten years from now? What types of facilities will be needed more? Less?

Interprofessional Clients: What are they doing about the future? What are their clients doing? What opportunities do they foresee, and what changes are needed to take advantage of them? Will there be more emphasis on quality of design or cost of design? How are they dealing with professional practice and liability concerns? What do they see happening to them in the future?

Contractors: What MUs do they sense will be the most promising for the future? What new technologies will be applied more frequently? How do they rate the plans and specifications produced by different firms, including the one asking the questions?

Others: Numerous others can be interviewed, and each should be able to provide valuable guidance from his or her own perspective. These include, in no particular order, association executives, bankers, insurers, utility representatives, manufacturer's representatives, property managers, and union officials, among others. It may also be possible to arrange for a discussion of the future as a program of a local chapter of a national association of design professionals, such as ACEC or AIA. The more input that is obtained, the more accurate the projections are likely to be.

FINALIZING FORECASTS

Each of the MUs identified initially can now be reexamined in light of comments derived through interviews or other means. In adding more detail, it would also be worthwhile to note any constraints that may have been mentioned by persons interviewed or that participants may be aware of. It is not necessary to discuss them in any depth at this point. They should merely be listed for discussion later.

Comparing Trends and Forecasts

At this point, you have developed two types of "intelligence." You have an intimate knowledge of the MUs in which your firm has been active, how they relate historically to the firm's business, and what the trends have been. You also have knowledge of what these MUs and others are likely to do in the future. By analyzing how your firm has responded to MU changes in the past, you thus are in a position to evaluate how it will react to them in the future, assuming no changes are made to the firm itself.

Many firms base their business planning solely on trend analysis. In other words, when certain trends are evident, these trends are projected to determine what the future holds in store. For example, if the number of projects a firm has garnered from the commercial/retail/strip center/renovation MU has been increasing by 10 percent per year for the past five years, it is assumed that the increase will continue. It is dangerous to base projections on this approach alone, because it ignores marketing forecasts. It also overlooks changes in the competition, such as new firms arriving on the scene. Nonetheless, trend projection is valuable, as a comparison with market intelligence. For example, if trends show that 60 percent of your firm's business will come from the commercial/retail/strip center/renovation MU in three years, while market research suggests that this MU will start drying up next year, planning for significant change should begin now.

A solid understanding of how trends and forecasts interact can be established by preparing a projection of the firm's future business activity on an MU-by-MU basis based on existing trends. Then, discuss each MU trend projection based on the results of marketing research. This approach will also lead to discussion of MUs in which the firm is not now active. As an example, consider the case in which trend projections indicate that business obtained from one of the firm's principal MUs should be on the rise but external market research shows it will be declining. Are the market forecasts reasonable? Did most people agree with them? If so, what are some of the other aspects of the MU that may be worthy of consideration, to take advantage of experience and expertise?

What about trend projections that shown the firm will obtain steadily less business from an MU that is predicted to grow strongly in the years ahead? Assuming people are confident in the forecasts, why is the firm losing ground? Evidently, the competition can offer something it cannot. Is this specialized expertise? Higher quality? Lower fees?

Where trend projections and forecasts agree, the firm probably should pursue market share increases, to help it maintain market share in the event more competitors enter the market. Where trend projections are characterized by peaks and valleys, compared to forecasts that are relatively stable, reasons for the historical fluctuations should be determined. Has the cause been failure to control marketing activities?

The end result of this activity should be estimates of future firm business in each MU in which it is now active, based upon marketing intelligence and trends. If a firm is strong in a weak market, the business ebb associated with the MU involved will be slower than otherwise. If a firm is weak in a weak market, MU business is likely to dry up entirely, assuming there has not been a change to the firm: no new services added, no new personnel added, and no new promotional techniques used.

Where it is indicated that the firm is weak in a strong market, the firm needs to consider increasing its market share. Where research predicts strong growth for MUs in which the firm is not active, consider the potential of getting involved.

At this point, it would still be premature to determine specifically what the firm's business plans should be. First it is necessary to consider goals of the firm's principals.

GOAL SETTING

If they have not as yet done so, principals of the firm must consider their professional and personal goals with respect to the firm. As already discussed, making money generally is not a firm's purpose. Money matters are important, of course, because the firm must make money to continue its existence. If it is well managed, however, making money should not be difficult. It will be far more difficult to make money by offering services that not only are profitable but that also satisfy the personal needs of those who lead the firm. A perfect match is doubtful; some compromises will be necessary. Nor can one hope to achieve personal professional goals overnight. It is something toward which the firm must strive, highlighting the importance of having an effective long-term business plan.

Identifying Individual Goals

One way of addressing goals is to encourage each person whose goals will be considered to prepare a list. (At the direction of the principals, more than the principals' goals could be considered.) Some of the goals that may be identified are discussed below.

Interprofessionals often express the desire to be treated more as professionals than as subcontractors. Translated into marketing terms, this goal would encourage pursuit of more business where the firm serves as the prime. For mechanical and electrical engineers, it could mean expanding services in the existing building market by offering more in the field of energy management, systems enhancements, or lighting consultation. For structurals, it could mean offering specialties in the field of roofing inspection and consultation. In other words, those who want more involvement as the prime should determine the types of services they could offer to gain prime consultant status. This can be done in MUs in which they now are active, as well as those in which they could become active.

Another common goal is the desire to become involved in major projects, those of "true significance." In this case, business planning should consider a staircase approach, that is, a method by which the firm can gradually step up to continually more significant commissions. Gradualism must be the key, because it is necessary to gain experience. Without experience, it will be difficult for the firm to obtain commissions, and, if it does obtain them, lack of experience could pose a liability threat, particularly if the firm uses reduced fees as a means to attract clients. In determining the nature of the staircase needed, planners can be either general or specific. In other words, they can examine MUs in which they are active to determine the direction needed in order to gain the larger projects, or they can consider specific clients and owners and the techniques that can be used to "reach" them.

Some engineers and architects may feel that the nature of their existing practice is not demanding enough in terms of the quality produced. In this case, they may want to look for work in MUs where quality is of vital importance (historic restoration, for example), or they may want to seek out clients who place a high premium on quality. The same might apply to those who feel the need to be more innovative.

For some who have been bruised by the spectre of professional liability claims and losses, a shift to MUs or services that pose less liability problems may be in order. This may mean becoming involved in more analysis of existing construction or forensic engineering, or it could mean switching to certain types of projects or clients that traditionally involve less risk exposure.

Some professionals may feel a need to offer more to benefit society. In this case, more effort could be made to welcome projects that are designed principally to meet societal needs. Alternatively, the firm could stress growth and expansion of staff, to give certain key individuals more time for public service, through involvement in public commissions of one type or another. Such a marketing strategy could also benefit those who want to spend more time teaching.

Expansion can also help those who want to become more involved in design and less involved in management. Alternatively, such individuals could assess their desire to stay with the firm. In fact, a number of senior managers have left larger firms to enter "solo" practice. They are better off from the point of view of personal satisfaction and generally feel no monetary bind. The firm may be better off, too, by finding a replacement who regards the management tasks involved as more of a personal challenge and who derives more satisfaction from performing them well.

The list of personal goals is endless, but almost all can be shaped into marketing or business objectives. However, do not simply assume that marketing is the only, or even the best, answer. Consider, too, the many non-marketing methods availble to satisfy personal goals. As examples, many of the "hassles" associated with being a subcontracting design professional can be eased through use of multiple or separate contracts. The desire to produce better quality can often be met through stressing the benefits of quality to existing clients and pointing out that the fee premium paid for higher quality is relatively small. The liability problem can certainly be eased through careful project selection, but this is only one of numerous other approaches available, including better delineation of workscopes and more effective contract formation. And, for those who want to become involved in education, it should be remembered that a firm's existing staff almost always will be better served through educational experiences that can be offered by senior members of the firm. None of this is to say that one should look to meet personal goals solely through alternative means. However, the discussion of goals should reveal that many changes can be made to the status quo to help principals and others derive more satisfaction from their work, and that these avenues should be pursued just as vigorously as those associated with marketing.

Establishment of the Firm's Professional Goals

A firm's professional goals are those that satisfy most of its principals' professional goals. Various methods can be used to convert individuals' goals into the firm's goals. Those preparing their lists of goals could submit them to one party, who would then prepare a list of all goals expressed. In some cases, of course, those expressing their goals may be reluctant to disclose them to others. As such, a trusted outside party, such as the firm's attorney, would be required to prepare the overall list. Alternatively, each person preparing a list could have it typed; all lists then would be given to one typist, who would then prepare one overall list, without knowing whose list was whose.

The overall list of goals could be distributed to all parties involved at a meeting, and then—to the extent necessary—several could be combined into one when they are essentially the same but stated differently. Once the unified list has been created, each participant could be asked to rank each goal in terms of its importance on a personal level, using a scale of from 10 (extremely important) to 1 (of no importance). This, too, could be done in a confidential manner. By adding the number of points each goal is awarded, it would be possible to rank-order goals that are of most importance to the firm.

Evaluating MUs in Terms of Professional Goals

The various MUs in which the firm is involved or could be involved can be evaluated in terms of the degree to which each will satisfy one or more of the firm's professional goals, either immediately or in stepping-stone fashion. At this point, then, those involved in the marketing and business planning function have almost everything they need to begin a comprehensive discussion of all the MUs involved. This discussion would focus on MUs' importance to the firm, given historical trends and trend projections, marketing research and forecasts, and the firm's professional goals. The next question to be answered is, "How do we get there from here?" This requires a review of optional business and marketing strategies.

SELECTING FUTURE OPTIONS

Any change from the existing marketing strategy engenders at least some risk, because it amounts to moving into unfamiliar territory. The first step, therefore, must be to identify what the options are for entering or increasing penetration in any given MU and then to consider the drawbacks and benefits of each.

Options in General

Many options are available for entering or increasing the penetration of any given MU.

The first and easiest is *creating more awareness of existing services* through business development activities. This creates no particular risk, except to the extent that it may result in the firm offering its traditional services to nontraditional clients.

Another approach, which can be considered half-marketing and half-promotion, is *repackaging existing services*. This usually takes the form of *niche marketing*, that is, creation of a unique niche within a large marketplace. For example, consider a consulting structural engineering firm that, as do most others in the area, offers a variety of roof-related services. It is recognized that most of the buying decisions relative to existing roofs—and there are far more existing roofs than new ones—are made by property managers. So as to make their services unique from those offered by competing firms, the structural engineer establishes a separate subdivision: "The Roofing Professionals." The company has its own brochures, business cards, and so on, but, in reality, these trappings and the name itself comprise the only factors that make the company any different from what it always has been. In essence, the firm has put old wine into a new bottle in order to attain more awareness for itself and a larger share of an existing market. The principal risk, of course, is dealing with a new type of client—one that may not be used to dealing with professional organizations.

Adding new services often is required to make the most of opportunities with existing or new clients in existing MUs and to penetrate new MUs, preferably with existing clients. Several strategies are available for adding services. These include the following.

- *Staffing up*, provided that at least one member of the existing staff is knowledgeable enough to run the new service section. Significant problems could otherwise emerge.

- *Merger* or *buy-out* represents a strategy for obtaining a firm that already offers the services a firm needs to add. There is an advantage to this approach in that the firm being acquired is already established in certain MUs. Ideally, all of the individuals associated with the acquired firm will remain with it, along with all existing clients. This gives the acquiring firm a relatively stable base to build on, the availability to offer new but "time-tested" services to existing clients, and the ability to offer certain services of the acquiring firm to the acquired's existing clientele. There generally is not much difference between a buy-out and a merger, except

to the extent that the principals of the merged firm become principals of the "new" firm created through merger. In a buy-out, the acquired firm basically becomes a division (if that) of the acquiring firm. In either case, the acquiring or merged new firm has the ability also to *repackage the additional new services.*
- An *affiliation* can be established whereby two (or more) existing firms agree to pool their resources in establishing a new one. This can be done through establishment of a new corporation or partnership, by joint venturing, or by other means.

Other techniques also are available for adding new services, but most are simply variations on a theme.

Another strategy that must be considered is GMA expansion, in which one offers repackaged, existing, or new services in different areas. Branching is most conveniently accomplished when a new field office is necessary to serve the needs of a GMA client operating in a non-GMA area.

Evaluating the Options

Once the various options for each of the prospective changes has been identified, the pros and cons need to be discussed. For some, the principal risk is a small outlay for new graphics and other business and promotional materials, and for dealing with new clients. Others involve major commitments of capital, manpower, and time, and may impose certain additional risks. For example, adding services may create competition with certain major clients, as when a consulting structural or mechanical and electrical firm adds architectural services, or when a geotechnical engineer adds roofing review and consultation. One also must consider the risks inherent in the work and its insurability, as well as the nature of the clientele and what has to be done in order to secure engagements.

In some instances, an option will be seen as a natural "fit" with ongoing developments with a firm and the GMA. Others may entail more risk and may be used if an opportunity is simply too strong to be passed up or if conditions affecting other MUs deteriorate too badly.

PUTTING IT ALL TOGETHER

At this point, everything discussed can be put together in the form of a business plan, which shows the direction in which the firm should be moving over the next five years. Each MU in which the firm now is involved can be discussed and projected, in terms of the growth or downsizing likely, based on factors such as the degree to which the firm wants to pursue it and how much business it will offer. Next, the firm can look to other MUs, including those that may be new aspects of existing MUs. Long-term approaches will be appropriate where it is necessary to gain experience or to obtain capital investment. In some cases, however, immediate pursuit of the MU may be feasible, as when niche marketing can be applied.

Developing the Plan

Plan development begins with review of a draft plan, development of alternatives, and more discussion, until a final plan emerges.

The hypothetical prefinal business plan shown in Figure 11-6 illustrates an approach created for a small consulting structural engineering firm. As can be seen, its three most substantial MUs—commercial/office/new, commercial/office/existing, and commercial/hotel/new—all are expected to shrivel. Modest pick-up is anticipated five years from now, but the further in the future the estimate, the less accurate it is likely to be. The retail sector is expected to take off, due to the forecast of more consumer spending (tax law impact) and the maturation of a major new suburban area.

Two options are presented for implementation in one year. The first (A) is the opening of a branch office to obtain local work in a growing suburban area. There will be a need for new schools and probably a publicly owned hospital, plus other health care facilities. The branch would generate about $85,000 in additional fees. Option B is the formation of a new roofing engineering firm to take advantage of a growing opportunity. Two years out, a third option (C) is presented, namely, the purchase or start-up of an engineering laboratory, and extension of construction monitoring services to other firms and public agencies.

Assuming predictions related to the current organization are reasonable, it is apparent that the firm must do something in order to maintain its workforce. Certain projects slated for the coming year have already been abandoned. Next year's billings are expected to be 25 percent below the current year's. The option of opening a branch—something that one of the firm's architectural clients wants to do—is attractive. The structural firm's principal responsibility would be financial, keeping time requirements to a minimum, and the income will permit maintenance of staff. But the forecast for the branch's future is not particularly heartening, in part because the newly developing area is expected to attract competition. Establishing an office there could feasibly give the firm a "leg up," but some of the newer firms may be "hungrier"; principals may be willing to work longer hours for less remuneration. In addition, there is always the potential for a dispute with the architect over how the money is being spent. The architect, as a client of the structural firm, is in a position to exert leverage.

Option B is one already mentioned—putting old wine in a new bottle—to begin penetration of the growing—and aging—existing construction market. If Option A were not pursued, chances are more promotional effort could be put into Option B, to generate far more business volume (and profit) than indicated. It also would be reasonable to conclude that contacts made through roofing services would lead to more potential for other

	Current Year							
	1988	1989	1990	1991	1992	1993	1994	1995
New—Interpro								
Com/Office	150	190	140	100	50	40	30	50
Com/Ret1	60	25	45	55	90	100	110	120
Com/Ho-Mo	90	110	80	0	25	30	30	35
Inst/Edo	0	15	0	15[A50]	15[A60]	0[A20]	0[A0]	50[A50]
Inst/Hosp.-Med. Ofc.	0	45	0	0[A50]	0[A50]	0[A30]	0[A0]	0[A0]
Inst/Nsg. Home	25	30	40	60	80	50	40	0
Inst/Parkg.	30	0	25	40	30	15	15	20
Inst/Relig.	20	0	25	0	0	0	0	0
MFR/Hi-Rise	40	50	50	0	50	65	65	75
Existing—Interpro								
Com/Ho-Mo	0	0	0	0	40	0	0	50
Com/Office	110	130	165	60	50	40	50	50
Com/Ret.	85	70	80	125	160	100	50	60
MFR/Convert	0	0	0	0	60	65	65	65
Existing—Prime								
Forensic	15	15	15	15	15	20	20	20
Roofing	20	17	20	20[B60]	25[B90]	25[B100]	25[B125]	30[B150]
Paving	0	0	0	0	0[C30]	0[C40]	0[C40]	0[C45]
Lab	0	0	0	0	0[C100]	0[C110]	0[C110]	0[C50]
Const. Mon.-Other	0	0	0	0	0	0[C30]	0[C40]	0[C50]
	645	697	645	490	690	550	500	625
				[A575]	[A785]	[A600]	[A500]	[A625]
				[B530]	[B755]	[B620]	[B600]	[B745]
				[A,B615]	[C820]	[C730]	[C690]	[C835]
					[A,B850]	[A,B675]	[A,B600]	[A,B745]
					[A,C915]	[A,C780]	[A,C690]	[A,C835]
					[B,C885]	[B,C805]	[B,C790]	[B,C955]

Options
A = Open branch in suburbs
B = Form roofing consultancy
C = Purchase CMT lab

Figure 11-6: *Business Plan (with Alternatives for) Small Structural Engineering Firm*

services, including those associated with new commercial and multifamily residential building design.

The potential for purchasing or opening a small construction materials testing laboratory in two years is strong, if the roofing business goes well, since the two tend to fit together. In addition, having its own laboratory would enhance the firm's construction monitoring capabilities and permit it to branch out more.

Finalizing the Plan

After the prefinal plan has been reviewed, the final plan can be created. In the case of our structural engineering firm, Option A would be eliminated; Option B would be substituted with projected gross billings of $120,000 instead of $60,000, because it will be established immediately. This urgency is justified, because principals of the firm have already seen several projects head for abandonment. And more emphasis on the existing construction market appears wise, given uncertainty about the direction of new construction and the principals' desire for more prime consultant work. Option C is left in place, because no decision has to be made at this time. Nonetheless, principals of the firm will keep their ears open for news about any laboratories that may be interested in selling, including those with which it regularly works.

Once the plan has been finalized in general, it must be finalized in particular. This means creation of month-by-month budgets to consider both historic revenue and expenses, as well as any new expenses that may emerge, as for marketing and promotion. Plans need to indicate the impact of income and expense projections on assets and liabilities, on a monthly or some other basis, since a build-up of reserves may be required to make acquisitions of an ongoing company, new equipment, or some other major items.

In addition to preparing budgets based on the financial impact of the plan, it would also be appropriate to prepare schedules, indicating certain major tasks that must be performed by a given date, such as securing a bank loan, hiring and training additional personnel, developing new letterhead, and so on.

Given the schedule imposed by the plan, it may also be appropriate to identify the specific business development activities that may be necessary, the budgets involved, and those who will be given responsibility.

By providing as much detail as possible, the firm can help assure that its plans will be realized. And, if they are not realized, the firm will be in a position to know why not and the measures needed to get back on track.

Unquestionably, all of this work together comprises a major undertaking, but only if it has not been done before. Once done, staying current is relatively simple and requires a much lower level of effort. To put the work off is to cross one's fingers and hope for the best. This is not professional management. Failure to provide professional management can make attainment of professional goals impossible.

Contingency Planning

Before putting the plan "to bed," one other element of concern should be addressed: contingencies. What if one or several MUs collapse? What if one or several take off to a far greater extent than anyone could possibly have predicted? Generally speaking, it is sufficient for a group to discuss and arrive at consensus decisions as to what the most valuable strategies would be to help ensure that the firm does not falter or fail (to avoid the bottom of the downside), or that it does not grow so quickly that it gets into trouble, or that it fails to grow quickly enough to take advantage of the opportunities. The decisions reached with regard to these can be fashioned into a memo that could be attached to the plan, so that, should one of the contingencies emerge, the thinking done during a calm period can be applied.

KEEPING THE PLAN CURRENT

One person in the firm should be considered the director of planning, so that people will know to whom they should bring market intelligence. This would be information about new trends, legal decisions, zoning changes, and so on that people learn about from clients, manufacturer's representatives, newspaper items, trade magazines, and any other sources. *All* members of a firm's staff should become attuned to the fact that the better prepared the firm is to meet the future, the better off everyone in the firm will be.

Internal marketing information can be obtained on a regular basis, simply by completing a Project Analysis Form for each project. Until such time as actual gross billings and profit are known, projections can be used. External marketing information can be collected for review.

For the most part, plan reviews should be conducted at least every six months. At that time, all "knowns" can be updated and external factors can be discussed. Plans can be adjusted accordingly. When major changes are imminent, as is the case with the hypothetical structural engineering firm, more frequent meetings would be prudent. At least once a year, the plan should be updated so that it continually projects five years into the future.

In analyzing the actual changes that occur compared to the forecasted changes, it is important to ascertain the causes for any significant variances. If predicted slumps or booms have not occurred, why not? Were certain factors not considered? If this is the case, what were the factors and how will considering them affect other MUs? How will developments in the MU in question affect other MUs?

If the company does better than expected in a given MU or does not achieve what was predicted, why? Is the cause external or internal? Has business development been as effective as hoped or not as effective? All these and similar questions must be answered to help assure a complete grasp of the factors that influence a firm's future direction.

12

Professional Business Development

Design professionals for many years relied almost solely on referrals from clients and colleagues—"word of mouth"—to develop business. As discussed by Greenwood, reliance on peers was so important for this purpose that it encouraged strict adherence to professional norms of behavior. Although referrals still constitute an important source of business, firms generally resort to far more aggressive methods. As discussed in this chapter, these methods can be segregated into two broad categories: *marketing communications*, which is applied to create more awareness of a firm and thus to attract business to it, and *direct selling*, which comprises a range of activities pursued to identify and create opportunities for specific projects.

Although a variety of marketing communications and direct selling tools are used, a firm's most important business development asset is the type of positive reputation that can be gained only by achieving client satisfaction. Client satisfaction is created by producing high-quality work on time and within budget. While technical competence is important in this regard, effective management is equally essential. If the workscope is inadequate, if an insufficient fee has been allocated, if scheduling is improper, or if there is any other breakdown in management functions, the mechanism needed to satisfy clients will not be in place. Is it any wonder, then, that surveys show consistently that the most profitable design professional firms in the United States are those that achieve a high level of client satisfaction?

Some firms still believe that it is unprofessional to rely on anything other than reputation. But times have changed. In fact, pursued effectively, business development techniques can actually enhance a firm's professional image, by making people more aware of what it is to be professional. Marketing communications is particularly important in this respect, because it can help create far more awareness of, and appreciation for, the vitally important contributions that engineers and architects make to society itself. Consulting engineers, in particular, have been remiss in this regard. Although they perform far more design work than architects, relatively few people know who they are or what they do. Since the business development activities of individual firms contribute to a reversal of this situation, it can be said that such activities can benefit the profession itself. Those who design America merit recognition for their accomplishments. For this recognition to be forthcoming, however, business development activities must be conducted in a professional manner, with full awareness of what must be done to maintain an appropriate image and to achieve the business objectives necessary to accomplish professional goals.

MARKETING COMMUNICATIONS

Marketing communications comprises a variety of techniques that firms use to make themselves known to, and noticed by, those who are in a position to use their services or to recommend their services to others. It is essential for those involved in marketing communications to understand the process of image development, since it is an image of the firm that will be communicated.

Image Development

Within the context of marketing communications, an image is a mind's-eye picture of someone or something, often based on some fragmentary fact fleshed out by imagination. Just about everyone forms images of what someone or something is "really like," and those images stand as reality until more details are provided. How often have you heard "You don't look anything like I imagined you" from someone who has talked with you previously only by telephone.

The process of image formation is illustrated in Figure 12-1. Output from a source is picked up by one of our sensory receptors as perceived input. The word *perceived* is important, because what we perceive the output to be may not be what it actually is. Sometimes we mistake a truck's backfire for a gunshot, just as we can sometimes confuse a complimentary remark with sarcasm.

Input is immediately subject to prejudice, a combination of memory and attitude toward it. Cologne worn by an associate may remind you of someone else and momentarily affect your mood. The clothing worn by another party, or the tone of voice or some other sensory input, may have a similar effect, subtly causing or evoking an attitude.

Context is an all-important concern, too, in that the inputs received and attitudes toward them will be evaluated relative to the context being considered. An individual who is being interviewed will likely form an image of

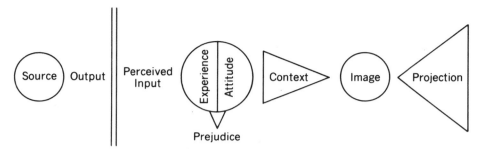

Figure 12-1: The Process of Image Formation

the interviewer as an employer, using inputs derived through the interview as cues to indicate how the person may behave as a boss.

The end result of the process is an image of the individual—a picture of what that person is really like. The image itself can then be treated as perceived input when it is *projected* to create the image of an organization. For this reason, some people will harbor a negative attitude toward a company of 5,000 employees, simply because they were treated unfortunately by one employee.

It is essential to recognize two key aspects of image development. First, even though the image that a person forms may be wholly inaccurate, it will nonetheless be believed—it will be accepted as reality—until additional facts (inputs) provide additional information. Second, image is *always* being created, as long as there are outputs and people to receive them. In other words, you are creating an image of yourself, whether you mean to or not, and, more than likely, you are also creating an image of your firm, whether you mean to or not. This being the case, it obviously is in your own and your firm's best interests to effect as much *image management* as possible, in order to improve the chances that the images being created are what you and your firm want them to be.

Image Management

In establishing what you want your image to be, bear in mind that no individual or organization has just one image, because contexts are variable. For example, any given man may have an image as a brother, father, son, neighbor, coworker, and employee. Each relationship creates a context that determines how image-forming information will be applied. Thus, job candidates who enter your office for an interview may appraise what they see to learn more about your firm as a prospective employer. More "employer image" information is obtained through the interview process. Prospective clients may appraise the same facilities and personnel, but the information they derive will be used to form an image of the firm as a provider of services. In either case, of course, the image actually formed will depend on the prejudices of the individuals involved. Some may like what they see and hear and form a positive image. Others may form a negative image from the identical inputs. As such, while it may be possible to manage the outputs that lead to image formation for a given context, it is impossible to control image, because it is impossible to control how outputs are interpreted.

Given the manner in which images are formed, it is relatively easy to manage outputs in an attempt to create an image that makes the firm look better (in a given context) than it really is. *Do not do it*, if only for practical reasons. Creating a false image in order to attract people or projects to a firm will result in problems. The closer they get, the more their exposure to image-forming information and the more accurate the image will become. A significant disparity between what was portrayed to exist and what actually exists will result in an image of dishonesty.

For positive results in the long term, image management should be applied to establish an image of what really exists. If something about your firm might create a negative image, *change it; strive to be what you want your image to be.* If you want prospective employees to think your firm cares for its newest employees, care for your newest employees and have them speak briefly with prospects during a tour of your facilities. If you want your firm to be regarded as an organization that achieves technical excellence, try to do just that: Enter design competitions, and let your clients how you have entered.

It is true that everything done to create a specific image is subject to misinterpretation. But failure to attempt image management does not conceal image; it merely makes misinterpretation that much more likely. Besides, when image management activities are pursued effectively, in a professional manner, the majority of the public's image of you and your firm will be exactly what you want it to be.

Establishing a Marketing Communications Plan

A marketing communications plan is formulated to implement the business plan. The business plan should be examined to identify the MUs that are important to the firm. Each MU consists of people who are in a position to make or influence decisions about using the services of your firm. These people can be grouped into *publics* who have something in common, permitting you to target your marketing communications specifically to each, so what you have to say will be of interest and is less likely to be misinterpreted.

Consider the situation of Able Geo, a hypothetical geotechnical engineering firm that decides to enter the hazardous materials field. Research has indicated that many buildings in its GMA store heating oil in underground tanks that might be leaking. Due to a new state law, the commercial/office building MU could become an important new source of business for Able Geo. As such, the MU's decision makers must be told about the law and its potentially costly consequences. They should be encouraged to determine if their tanks are leaking, and they should be informed that Able Geo has the capabilities needed to perform testing and to design remedial measures.

In establishing a marketing communication plan for this MU, Able Geo must first identify the MU's publics. The decision makers in this case are the buildings' owners, but, in many instances, their decisions are merely rubber-stamp approvals of recommendations submitted by their property managers. Therefore, property managers also comprise an important public, but not the only one. Others are in a position to make referrals and recommendations, including public officials, oil suppliers, mechanical contractors, and real estate agents.

Once Able Geo identifies all the MU's key publics, the firm must decide what each public must do to help Able Geo achieve its objective, that is, penetration of the commercial/office building MU by offering hazardous materials engineering services. Obviously, it wants owners and property managers to become aware of the need for storage tank evaluations and to turn to Able Geo for assistance. It wants real estate agents to become aware of this need as part of preliminary site assessments and to mention Able Geo as a source of professional assistance. It also wants mechanical contractors and oil suppliers to mention the need and to recommend Able Geo as an organization that can help.

In essence, for this and virtually all other MUs, it is essential to determine who makes the decisions, who influences those decisions, who is called on to make recommendations, and who would alert someone to the need for concern. Figure 12-2 indicates the communications channels that then would be applied. As indicated, information sources send their messages to decision influencers who, in turn, send their messages to the decision makers. Accordingly, the firm communicates with the information sources, decision influencers, and decision makers. The information being provided by Able Geo in this instance is, "Leaking underground storage tanks could be a serious, costly problem. Check them now and take remedial measures if necessary. Contact Able Geo for more information and assistance."

In all cases, it is essential to recognize that a one-time effort is not enough. The message must be repeated to all important publics on a continuing basis, so that the name of the firm and its message will be seen and heard by important publics on a regular basis, to achieve what is known as "the serendipity factor." Ideally, you would somehow call a person's attention to your capabilities just before or just after that person recognized a need for those capabilities—serendipity. When your message is given just once, serendipity seldom occurs. However, if you regularly call attention to your firm's capabilities among a wide range of persons who might be in a position some day to need or recommend your firm, serendipity is far more likely to occur.

Note that the same message usually cannot be given to each public a firm has to consider. Since most people are self-interested, at least in part, it often is appropriate to tailor messages to the self-interests of each public. Going back to our hypothetical example, why would real estate agents want to inform people about potentially leaking underground storage tanks? Among other

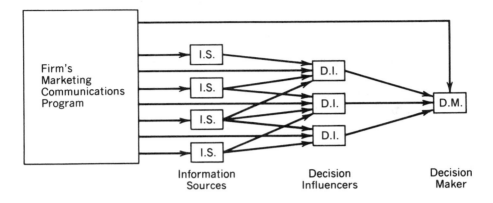

Figure 12-2: Direction of a Marketing Communications Program

things, of course, it is something they should know about if they are to serve as effective professional advisors. And, if they are to gain an image of being knowledgeable advisors, the potential problem and its implications are things that real estate agents should want to call to their clients' and prospects' attention. They may also want to inform the lenders, insurers, and others with whom they regularly deal. In other words, being perceived as a source of knowledgeable assistance is as important for many real estate agents as it is for many design professionals. If they obtain valuable intelligence, they should be eager to inform others and thus advance their own credibility and image.

Mechanical contractors may also wish to advance their credibility, especially so because they may be first in line to perform any of the mechanical work that might become necessary. They may call attention to the potential problem by discussing it with operation and maintenance staffs, who might then relate the information to property managers and/or owners.

To develop an effective marketing communications plan, then, it is essential to determine why any given public should be interested in your message and why it should want to communicate what you want it to communicate. In essence, put yourself in a public's shoes and ask, "What's in it for me?" Answering this question indicates how communications should be framed. In this case, Able Geo might contact the executive director of the local Building Owners and Managers Association (BOMA) chapter to offer the services of the firm's CEO as a guest speaker. The importance of the issue would be delineated. The firm could also prepare an article for the organization's newsletter and a brief fact sheet that interested parties could send for. By obtaining

322 PROFESSIONAL BUSINESS DEVELOPMENT

the group's mailing list, the firm could also issue a few direct mail pieces. By becoming an active, participating member of the group, it could provide more information on an "eyeball-to-eyeball" basis. Many of the same strategies could be used to reach groups of property managers, as through a local chapter of the Institute of Real Estate Managers (IREM). Real estate agents could be reached by explaining what their liability possibly might be if they fail to recommend a check of oil storage tanks before a transfer is consummated. Oil suppliers might also have a liability exposure, and Able Geo personnel could inform them about simple checking methods. Mechanical contractors would be told about their business potential and how they could go about realizing it.

Assessing Publics' Importance

The amount of time and money allocated for marketing communications depends upon the importance of the MUs involved. Recognize, however, that one public may be of value for a number of MUs, depending on the firm's services. As such, by joining an organization that caters to that public, one could probably kill a number of birds with one stone. Given this potential, it is appropriate to evaluate each public's importance to the firm. This can be done by using a form similar to the one shown in Figure 12-3. MUs considered in the firm's business plan are listed on the vertical axis, with the relative importance of each being rated on a 1-to-5 scale. Each public is listed along the horizontal axis, and, at points of intersection, the public's importance (rated on a 0-to-5 scale) is determined for the MU. The MU weighting factor is then multiplied by the public's importance factor, so the total at the bottom of each vertical column would indicate the importance of a given public to the firm. Scanning

MU No.	MU Weight (1–5)	Public 1 (0–5)		Public 2 (0–5)		Public 3 (0–5)		Public 4 (0–5)		Public 5 (0–5)	
1	2	3	6	4	8	2	4	0	0	1	2
2	3	3	9	2	6	1	3	4	12	2	6
3	4										
4	5										
5	5										
6	4										
7	1										
8	1										
9	3										
10	2										
Totals											

Figure 12-3: Form to Evaluate Relative Importance of Various Publics (Partially Complete)

the table horizontally, one can see which publics are involved with a given MU, giving the form additional value as a reference that can be useful in deciding which communications techniques should be employed.

Marketing Communications Techniques

In almost all cases, an effort should be made to achieve the serendipity factor through a variety of programs. The effectiveness of each program should be evaluated in terms of the image it presents to the public being reached, as well as the appeal and impact of the message being conveyed. Cost is an important factor, of course, and the firm should always try to obtain maximum "bang for the buck."

General Programs General programs are those that affect virtually all publics in all MUs. They comprise certain passive but nonetheless important image-forming factors, such as the graphics used in printed materials, the appearance of the firm's offices and personnel, and the manner in which the telephone is handled.

Printed Materials Printed materials often have one commonality: a firm's logo or corporate symbol. Some firms spend thousands of dollars to develop a logo that is evocative, but, unless it appears next to the firm's name, it is meaningless. Others use a distinctive typeface to spell out the name of the firm and this becomes a logo. Some believe the evocative symbol is a costly nicety; others think it indicates a firm has "arrived" and thus is important. Suffice it to say that good graphics are important, because they suggest concern about image and professionalism.

For purposes of image management, everything representing a firm in print should be consistent: letterheads, envelopes, business cards, brochures, and so on. If there is a need to establish a different image for a specific public or group of publics, then there usually is a need to engage in niche marketing by establishing some type of subsidiary firm with an image of its own.

Work with a professional commercial artist in establishing graphics. The individual selected should be informed of the general type of image that the firm wishes to establish for itself, since graphic materials will convey general messages—for example, contemporary or conservative.

Unless you want your firm to be recognized as cheap, produce high-quality materials, that is, well-designed materials that are printed on high-quality paper. Being cost-effective does not mean being cheap. If you want to associate your firm with quality, and assuming the association is justified, obtain quality in print. Remember that reality is what it is perceived to be.

Office Appearance Office appearance is a concern relative to "outside publics," such as clients and prospective clients. It also is an important consideration for internal publics, such as the employee public, since they see it every day. However, employees are part of office appearance from outside publics' point of view, a concern that could be reflected in terms of dress codes and conduct requirements. Some firms may say, "We are designing our offices as we prefer, for our own reasons. If some clients don't like it, then we're not the firm for them." Such an attitude is completely acceptable for firms that can afford it. Otherwise, decisions should consider what will be most effective, given the predispositions of the publics likely to visit the space, as well as those who use it day to day.

Telephone Techniques There is a great deal of truth in the old adage, "First impressions often are lasting." And, undeniably, a firm has only one chance to make a first impression. Since a firm's first impression often is made by telephone, effective telephone technique can be extremely important. It is also important for maintaining image. When people use the telephone, only their sense of hearing is at work. Since people use whatever cues they can obtain to establish an image, what is said and how it is said can have a significant effect in determining how imagination "fills in the blanks." If calls are handled in a professional manner and if call-back promises are kept, a positive image will be created. A firm that is not very professional is not likely to create a favorable impression by telephone. Regrettably, some highly professional firms create a disastrous image for themselves, simply because people do not use the phone correctly.

Note that proper use of the telephone involves more than just good manners. Personnel must be attuned to the need for documentation of calls, including reference to attitudes of callers, when appropriate. Personnel must also be aware of telephone technique's importance with regard to liability loss prevention. Someone with a problem is going to be far angrier and will perceive the problem as being far more serious if the firm's personnel put the caller on hold for an indefinite period or forget to give the message to the proper party.

Specific Programs Specific programs are those pursued by a firm to reach specific publics and MUs. Some of the most important of these are discussed below.

Association Participation Many trade and professional associations catering to a given public offer associate membership that permits involvement of "trade allies." For this reason, many firms that are actively involved in the single-family residential development MU also are associate members of the local National Association of Home Builders (NAHB) chapter.

Identifying groups that merit consideration by your firm is not a difficult task, once you know which publics are important. Simply contact representatives of these publics to identify the associations in which they participate. You could also check with the local Chamber of Commerce to determine the associations it represents through membership; determine if the local United Way campaign office keeps a list of associations, and contact national associations to identify local chapters. Most national associations are listed in a book called the *Encyclopedia of Associations*, something that many local libraries keep on hand.

Joining an association is not enough. If all you intend to do is pay dues, do not bother joining. To obtain positive results, *become active.* Attend most meetings. Volunteer for committee service and attend committee meetings. When volunteers are sought, raise your hand. In other words, become a vital member of the association, someone who is known and trusted to do a good job. The image you create for yourself will be the image you create for your firm. Obtain business by *not* looking for it. Wait for people to ask you what you do. You will sell best by not selling at all.

In most instances, it is advisable to have one representative for each such association, permitting effective in-depth involvement. This usually is worthwhile, because such participation can become the single most effective marketing communications strategy of all. It gives you the chance to make eyeball-to-eyeball contact with a number of important individuals, and it can also result in your becoming part of the "network."

Public Speaking The importance of public speaking cannot be overemphasized. It gives you an opportunity to be recognized as a knowledgeable source of information about a specific situation or problem. It also permits you to display yourself and your capabilities (as well as your firm and its capabilities) before a large number of people who are important to you and your firm. In most instances, your appearance before a group will be made known in advance through articles in its newsletter, as well as through news releases. (Your office can offer assistance if the association involved does not know how to handle these public relations techniques.) During your appearance, you can distribute an outline of your talk, perhaps with some back-up information, all on your company letterhead and/or with a card attached. Alternatively, you can obtain a list of attendees and send information to them, perhaps keeping them on a mailing list so that you can send them additional materials. In some cases, it is a good idea to speak before a group of which you are a member. In other cases, it may not be a good idea, and you should offer to obtain someone else from your firm as the speaker.

To obtain a speaking engagement before a group in which you or your firm does not hold membership, contact a member to determine who the program chairperson is or whom to contact to find out who the program chairperson is. Speak with the individual directly about programs the organization puts on. Offer your services about subjects that would be of interest. The best subjects tend to be those that relate to pocketbook issues, for example, the impact of new regulations governing underground storage tank leakage and steps that can be taken to minimize liability exposure and cost.

Seminars Serving as a seminar leader is similar to public speaking. Also consider the possibility of your firm sponsoring a seminar on an important subject. In many cases, firms have found that this is not only effective in establishing the firm as a knowledgeable source of assistance but also can be a money-making activity, since the seminar should not be free. Imposing a registration fee can also permit you to give discounts or free passes to certain individuals, as a gesture of goodwill.

Any seminars sponsored by your firm should be promoted in a professional manner. News releases should be issued. Professionally prepared flyers or brochures should be sent to clients and prospects, using internal mailing lists, as well as those obtained from area associations. Note that even those who do not attend will be "reached" through receipt of descriptive material.

News Releases News releases are a fundamental aspect of marketing communications programs. Specific techniques should be followed in putting them together, and some excellent instructional materials are readily available.

News items based on news releases are sometimes called "free advertising." Understand that news releases are not free. One must consider the cost of printing, postage, and preparation labor, in addition to the time or fee required to write a release. The space in which the news items appears does not cost anything, of course, but the news item is most definitely *not advertising.* It is a news item (usually) and sometimes may be treated as a feature article, if it is long enough. In any event, it has far more credibility than advertising and thus can have far more impact.

News releases comprise an excellent vehicle for keeping a firm's name before key audiences, but not just through "pick-ups" in publications read by these publics. It is strongly suggested that firms send news releases to clients and prospective clients, as well as those who provide information to clients or help influence their decisions. News releases often get read because they are relatively unusual forms of communication for those not in journalism and because they often are short and "punchy."

Topics for coverage via news releases include new projects obtained (obtain the client's permission firm), new members of staff, the opening of a new office, papers published by members of the staff, and awards won. Even small firms should be in a position to issue at least six news releases each year.

Magazine Articles Magazine articles are extremely effective for enhancing credentials. Many design professionals tend to concentrate on professional journals, however. While these publications are eminently worthy, they are not extensively read by anyone except competitors. As such, it usually is more effective to prepare articles for publications that are read by key publics. The article itself may stimulate some inquiries; reprints of the article can be distributed (with a news release) to the firm's mailing list and can be used to embellish proposals or to back up information distributed after public speaking engagements or seminars.

It usually is easier to get published in a national periodical than a local one, because there are more national periodicals. Local ones tend to be smaller, and often a great many people are trying to get into them. To be published, first develop a concept. Pocketbook issues generally are most effective, so direct your energies to finding a topic that can be of bottom-line value to the publication's readership: how to do something faster, or with fewer problems, or at less cost, or with less trial and error. Use case histories of your own projects as examples.

Identify publications that may be interested by referring to a work such as Gayle Publishing's *Directory of Periodicals* or a similar publication that should be available in most local libraries. Call a magazine's editor and ask if there is interest in an article. If not, the editor may suggest alternative articles that would be of value or another publication that may be interested.

When an article can be based on one case history, it may be effective to have it by-lined by a client. This usually makes the client happy, and the article can be far more flattering toward a firm than it would be were a representative of the firm authoring the piece.

It may be necessary to have a free-lance writer prepare the article or other written marketing communications instruments. A number of sources may be available. Contact local newspapers to determine if any staffers free-lance or if they are aware of persons who do. Also contact English and/or journalism departments at nearby colleges and universities. In addition, speak with association staffers, some of whom may be eager to do some moonlighting in this regard, providing it is acceptable to the association's leadership.

Brochures Some engineers and architects think of their brochures as "be-alls" and "end-alls" when it comes to marketing communications. In reality, however, a brochure is little more than a glorified business card. It indicates that a firm is established, performs a given set of services, and has experience with certain types of projects. A brochure is intended to put a firm's best foot forward and to get that foot wedged in the door far enough so that the firm is contacted for additional information.

Effective brochures are brief. They should be able to communicate their message in five minutes or, preferably, less. Photographs can be used to convey information about the firm's experience. A simple listing can serve to inform readers about services the firm provides. Brochures that attempt to tell prospects everything that can be known about a firm usually are counterproductive. Let prospective clients know you understand how busy they are by keeping the brochure short and to the point.

Professional assistance is strongly recommended in brochure development. A brochure that is not well prepared can create a negative image every time it is seen and can be worse than no brochure at all.

It is essential to review the language used in a brochure (as well as that used in proposals and most other forms of marketing communications). Many honest statements can be made to reinforce a firm's image for technical superiority, professionalism, and so on. Stick to the honest ones *only*. Do *not* get carried away by promising more than a firm can deliver. Can a firm really "subscribe to the highest professional standards" when it is impossible to define what the highest standards are? Does a firm "deliver projects on time and within budget," or does it just have a reputation for doing so? In other words, realize that anything you say can be held against you later on. Use words carefully.

Newsletters Many firms have employee newsletters. Some of these are also distributed to clients. Some firms have client newsletters and employee newsletters, each written and otherwise prepared for the specific audience involved.

Newsletters can be effective for purposes of marketing communications. The trick to obtaining effectiveness is making them "want read" periodicals or even "must read." What do the readers want to know about? New regulations and how they may be affected by them? Techniques for performing certain tasks more effectively? New equipment or facilities that the firm has invested in? For the most part, a potpourri of articles is best, helping to assure each issue contains at least one item of interest for each reader. Some firms do this by including a few jokes, brain teasers, and similar light material in each issue.

Awards Competitions Participating in awards competitions creates excellent marketing communications opportunities. Merely entering can create strong positive feelings among clients and staff. Winning on local or higher levels creates excellent potential for positive publicity and can significantly enhance a firm's credentials. To help elevate chances of winning, obtain professional assistance in developing the entry material.

Advertising Many design professionals have advertised for many years by placing business card ads in journals. This is known as "tombstone advertising," not without reason. Its impact is not particularly significant, although, on occasion, it can draw an inquiry or two.

Some firms are now resorting to display advertisements. Most of these ads are institutional in nature, basically indicating that a firm is involved in a certain field. Relatively few of these ads say a firm can do work for less, but some do give this impression. Over time, it is likely that more of this latter type of advertising will be used. It will be counterproductive in the case of readers who find it unseemly for a professional firm to suggest it should be selected because it offers low fees. How many readers are in that position is anybody's guess.

For the most part, any number of promotional techniques can be far more effective and far less costly than advertising. Consider the alternatives strongly before opting for paid space.

Other Numerous other tools are available for marketing communications purposes. These include open houses, obtaining credit for work on various projects through site signs, and appearing on radio and TV talk shows. Always keep efforts geared toward the business plan, however, to help assure maximum effectiveness. Avoid promotional activities that are designed more to flatter a firm than to reach its important publics.

Sources of Assistance

Professional assistance is needed to develop and implement a marketing communications plan. If a high level of expertise is not available in-house, a firm clearly does not have the wherewithal to mount a professional effort. This does not mean that the entire function should be turned over to consultants, however. Consultants cannot possibly have as intimate an understanding of the publics involved as the firm's key personnel, nor can consultants be as familiar with the firm.

Whoever is selected must understand the special needs of a professional services organization. The methods employed for image management must complement the image being established. It also is important to have a grasp of liability issues.

For the most part, a team approach is required, because a number of disciplines are needed: different writing skills, different graphic skills, and so on. The various marketing communications tools that will be applied should be pursued in accordance with the business plan, so that efforts can be made in a coordinated fashion. One cannot hope to "bring home" the first project in an MU that is new to the firm by launching marketing communications programs one month beforehand. Overall coordination is essential. It can be gained by assigning internal responsibility to one person and then by working with a consulting firm to provide needed services. This is nothing more than effective project management, except, in this case, you are the client.

Given the importance and cost of marketing communications, outside providers should be selected with care. While a free-lancer may be able to do a good job, it would be foolish to obtain one simply based on fee. Good work is essential. Anything paid for low-quality work is money wasted.

Although many advertising agencies also claim to have capabilities in public relations and marketing, few really do. As such, if you do consider using an ad agency, be certain to evaluate what it has done with respect to the types of work you need performed. Be certain to check results and client satisfaction. A full-service public relations organization generally is in a far better position to assist you, but, again, verification of the organization's accomplishments is essential.

It may be possible to rely on a number of independent providers, such as free-lancers, each of whom you have selected for aptitude with regard to the task involved. Moonlighting journalists may be most adept at writing news releases. (Mailing lists can usually be entered into your firm's computer system, with a number of subcategories being established based on the nature of interests.) Free-lance magazine writers can probably handle articles you want prepared, but language must be carefully reviewed.

Especially when it comes to major issues, such as a logo, brochure, or magazine article, it usually is advisable to obtain input from a knowledgeable outside source. For example, if you are preparing an article for a publication read by property managers and owners, it would be wise to have a property manager or owner read the draft and comment on it. Many people will be flattered that they have been asked, and they can provide valuable insights. Note, too, that this is one of the reasons why you may want to have a consultant on hand, at least for occasional meetings. In-house personnel who prepare materials can lose their objectivity and/or may be reluctant to criticize brochure wording or magazine articles prepared by "the boss." Particularly when it comes to marketing communications, it is worthwhile for you to have someone serve as a devil's advocate. An outside party usually is in the best position to fill this role.

As firms increase in size, they generally find it worthwhile to have at least one capable writer on staff, to prepare proposals, to review and edit major items of correspondence, and to develop newsletters. Some very large firms have entire staffs—generally referred to as "marketing personnel"—to help handle this function, including graphic designers supported by typesetting equipment, binding equipment, and so on. In many instances, some of the personnel are also used for purposes of direct selling. More information on this subject is covered in the following section.

DIRECT SELLING

Direct selling refers to the person-to-person efforts made in support of the business development effort. Some firms have personnel who do nothing but this, including "bird-dogging," that is, pursuing activities designed to learn who has new projects under consideration, what they involve, and how the firm can get a "leg up" on the competition.

Problems can occur when the direct selling effort is left in the hands of specialists. They can become so enamored of selling that they lose sight of other important issues, and they can bring in business that is counterproductive. It generally is not worthwhile to pursue a project that will not contribute to attainment of the business plan, unless it is determined that the project represents an MU that was overlooked or a type of work that will fit well with the firm's competence. Those involved in the selling effort must therefore be cautioned to concentrate their efforts where they will do the most good and not to pursue work that will lead to problems, that is, that requires expertise the firm does not have, that requires working for a client with a notorious reputation for filing liability suits, or that will necessitate a fee that cannot possibly permit the rendering of a quality-oriented service.

All design professionals in a firm should be involved in the selling effort. The work required is not difficult. In fact, much of it is pleasurable.

Networking

Networking comprises the establishment of a network of relationships. Many of these relationships result in business and personal friendships. When you learn of something that should be of interest to others you know, you call or jot them a note. Ideally, this type of thoughtfulness will be reciprocated.

Organizational Involvement As already noted, involvement in client-oriented organizations can comprise an important source of networking. Consider the property management public, for example. In most cases, any given property manager is responsible for several large multifamily residential and/or commercial buildings. Establishing relationships with property managers thus gives you excellent means for entering or increasing your penetration of the existing buildings market in your GMA. Many property managers' clients are developers, however, and these developers often stay active with new projects. If you do a particularly good job for a property manager, chances are this person will recommend your firm highly for new construction. As a consequence, mechanical and electrical firms that work as interprofessionals with architects will at times be "wired" into a project; that is, the architect will be told to include a certain M&E as part of the team being put together.

Note that association involvement should not be confined to organizations that represent potential clients or to individuals who are in a position to recommend your firm to potential clients. Involvement in organizations that represent your profession from a technical or management point of view also is of vital concern. On the one hand, all design professionals should support the organizations that support them. On the other hand, peers and colleagues are frequently the source of excellent business opportunities. As such, for both professional and practical reasons, key staff members should be active in groups such as ACEC, AIA, NSPE, ASCE, American Society of Heating, Refrigeration, and Airconditioning Engineers (ASHRAE), and others. Recognize, too, that achieving a leadership position in one of these groups can be of great value insofar as elevating your own and the firm's credentials.

It also is important to consider involvement in organizations such as the Chamber of Commerce, Rotary, and others that comprise businesspeople who are in a position to use your firm's services or to recommend them to others. These types of relationships can also be established through involvement in fraternal groups, church groups, and country clubs, although the impetus for such involvement generally is not business oriented.

Person-to-Person Activities Networking involves more than simply being part of an association. The relationships developed should be built upon through person-to-person activities. This involves an occasional lunch, going to the theater together, or having people over to your home. This does not mean that you should try to act as the "bosom buddy" of someone whom you do not particularly like; that can be a grating experience. However, when you do in fact like the people you are dealing with, establishing personal relationships can be rewarding on both a personal and a business level.

Note, too, that person-to-person activities can be used to kill several birds with one stone. For example, you may wish to take a lender to lunch to discuss your firm's banking needs, as well as to remind the individual about your firm's services and their availability to the bank and its patrons. The same types of communication can occur when you take a contractor, insurers, or other providers to lunch. As the saying goes, "One hand washes the other." The most effective networking relationships are built through reciprocity. Remember that you are part of others' networks.

Maintaining Client Relations

Existing clients are the most important of all, because they are the best source of new commissions and referrals. When they have new work, time requirements can be minimized because you already are familiar with how they work and the "marketing costs" associated with obtaining the work will be minimal. In some cases, there are virtually no costs at all, as when a client calls to say, "Come on over to the office to discuss this new project I have in mind."

Client relations are maintained through marketing communications activities, by issuing news releases and so on, but far more is needed. The nature of the need is fulfilled in exactly the same way that person-to-person networking objectives are achieved. On occasion, once every two or three months or so, the client or appropriate client personnel should be taken to lunch or dinner, to a sports event, and so on. (Certain restrictions may apply when public-sector clients are involved. Be sure to follow all applicable guidelines to the letter.) This all comes down to the Golden Rule: Do unto others as you would have them do unto you. If you are an important source of business to another party, you would expect that other party to let you know that you are highly regarded. Do the same, too, for your clients.

Bird-Dogging

Bird-dogging refers to the "flushing out" of new business by pursuit of projects that have been made known to you by others or by attempting to determine what projects may be on the horizon.

Following up on Project Leads When projects are made known to you and you are interested in learning more, ask the person who has given you the information about follow-up. Who should be contacted? Can the referral source's name be used? Also, try to find out more about the prospective client, if you do not know about the individual or organization already. Is the person quality-oriented, cost-oriented, or value-oriented? Does the prospect prefer to rely on one provider or several? Do the individuals involved understand that design professionals cannot promise to provide perfection?

Assuming that the prospect is worth pursuing, make contact as soon as possible, but *not* before checking with others in the firm or using files if they have already been established, as discussed below. It is possible that other personnel know the prospect or already have relationships with senior members of the prospect's staff. Perhaps the prospect is someone another party went to school with or served on a committee with. The prospect might even be someone's brother-in-law! Especially when you are attempting to make an initial presentation, you want to lead with as much "firepower" as you can. It will be far easier for someone known to the prospect to arrange an initial meeting.

The purpose of an initial meeting is to make the prospect aware of your firm's capabilities and experience with the type of project reportedly involved, in order to be considered for it or the next one. If you are not in the position to serve as prime, then perhaps you can at least obtain a referral to the prime.

Follow-up is essential. Touch all the bases you have been told to touch. Send thank-you notes to people who have given you assistance or time. Put these people on your mailing list so that they will regularly receive newsletters, news releases, and other materials, to help keep your firm's name before them.

Cold Calls It also can be important to engage in "cold calls," that is, calls to appropriate personnel in prospective client organizations with which you have no contact. Many design professionals are averse to this type of activity, and, when other efforts are pursued well, it may be unnecessary. Nonetheless, some excellent results have been obtained from cold calls. In the case of a large prospect organization, in either the public or private sector, a call to the public information officer (PIO) or the person who fills that role by some other name can yield excellent information. An effective approach is one that is completely candid, as in the following example.

> Hello. My name is John Doe. I'm vice president of XYZ Architects and Engineers here in town, and we would like to be considered for some of the design services your firm requires. We work in both new building design and modernization of existing buildings. What I need to know is whom I should speak with in order to let them know about us.

Chances are this type of open and honest approach will result in your being given several names and numbers, and you would then call each for follow-up. If they are not the correct parties, they will probably give you the names of those who are. When you contact the correct people, you should try to set up an appointment. Otherwise, you probably will be asked to submit information. If this is the case, find out what type of information you will be asked to submit. What types of projects are the people most interested in? What types of services?

If you are asked to submit materials, deliver them in person and inquire if the person you spoke with is in. If so, ask if you could have a few moments of that person's time. Ask about any future projects that may be pursued, as well as any that may be ongoing at this time, for which your firm's services may be applicable. If there is an immediate need, you will be able to prepare an even more specific package.

If for some reason you are unable to see the person to whom you are delivering materials, call within a few days to assure the materials were received and to set up an appointment for a business lunch. You do not have to beat around the bush as to the purpose of your activities, because both you and the person with whom you are meeting know exactly what your purpose is. There is nothing wrong with saying words to the effect, "We regard you as an important potential client. We'd like to be associated with you and are confident we can do a good job for you. What can we do to help make that come to pass?"

A more specific form of cold calling occurs when you are trying to obtain general intelligence. For example, you may hear that a developer is planning to build a new mall in the area or that a firm is relocating within your GMA. By making several calls, you will ultimately find the right person to speak with, to determine if your services might be of value.

Although you may not immediately gain business from cold calling, you will at least have made your firm known to those who are in a position to request proposals from you. In some cases, those with whom you speak may not have anything coming up, but they may know of others who do. By adding these people to your mailing list, you will be able to maintain a relationship. As in other cases, however, a direct call should be placed from time to time, and you should attempt to establish opportunities for personal get-togethers, to help assure that relationships are people-to-people, not paper-to-people.

Keeping Track

If the direct selling effort is to be effective, it must be organized, monitored, and recorded. It is of vital concern in developing the business plan, since it is from direct selling activities such as networking that you can obtain vitally important marketing intelligence. It also is a concern when you are identifying new MUs or project types for pursuit and in "bringing home the bacon," that is, obtaining the types of projects needed to fulfill the plan.

One way of organizing the effort is to create a separate file for each organization contacted. Each file should indicate the address of the organization and telephone number, as well as the names of individuals with whom firm representatives have communicated, the titles of these people, the date and nature of the communication, and a synopsis of what was said. These are important facts, since firm personnel or their assignments may change, just as prospective client personnel may change. By having these files—either in hard file form or in a database—it will be easy for you to refer to the record before you make follow-up contacts to see who has previously contacted the organization or individuals within it and what was discussed. When additional contacts are made, additional information is added to the file.

On occasion, perhaps annually, it may be of value to prepare a list of organizations and individuals on file and circulate it to your staff, to determine if anyone who works for your firm knows any of the individuals listed or is somehow familiar with the organizations. If so, the situation should be studied to determine if any opportunities exist.

In fact, an up-to-date prospect file can be among the most important of a firm's assets, something that would be referred to at least annually in developing the business plan. To trust the information to memory is not wise, especially so because reliance on a computer can make the work involved simple.

KEEP THE PRODUCT IN MIND

It is worth noting that, in direct selling as in marketing communications, the single most important attribute a firm has is client satisfaction. As such, while all major pursuits of a firm are important, all should be geared first and foremost to turning out high-quality work on time and within budget. In some cases, this is the only business development activity that is required. In all cases, it makes business development far easier and far more successful.

13

Communicating as a Professional

Hey, reader. Supposing this text had been wrote different—more colloquially and less concerned about proper use of language. Wow! It really would've blown you away. You probably would've thought to yourself, "Hey, what's this? I've never read no guidance book like this before." Maybe you wouldn't even have gotten this far. We all have preconceived notions about communications. When the style used is outside our range of expectation, it is immediately noticeable. It also is less credible, and chances are we will look at it with a jaundiced eye. For professionals, the point is clear: You are expected to communicate in a manner that falls within your publics' range of expectation.

As the prior chapter explained, people communicate through more than words. We also communicate through, among other things, the clothes we wear. If you want people to recognize you as a professional, you should not wear the clothing you would don for touch football or for cleaning up the yard on a Sunday afternoon.

Clothing, of course, is easy to adjust. How you communicate verbally (in writing or orally) is not, especially if your communications skills are weak. For some professionals, this means avoiding situations in which such weaknesses might become apparent. As a consequence, many design professionals—perhaps most—refrain from writing articles or speaking in public, two of the most important marketing communications skills of all. Magazine articles can be ghostwritten, but few design professionals rely on "ghostwriters." Public speaking is something else again. No one can take your place, and, in fact, few marketing communications skills are as valuable or as important as public speaking.

This chapter concentrates on techniques you can apply to improve your verbal communications skills. The emphasis is on writing of all types, including the preparation of talks.

A FEW OBSERVATIONS ABOUT PROFESSIONAL COMMUNICATION

Before getting into some of the specific "how-to's" of professional communication, it is important to note a few general observations.

First, the leaders of any profession tend to be the most effective communicators. While performing research is important, being able to communicate the results of that research is also important. Likewise, while developing new concepts for a profession is important, people can be made aware of those concepts only through communication. In fact, the power of effective communication is underscored by the fact that some of the biggest charlatans in any profession also tend to be effective communicators. For example, many "hired gun" experts are sought out because they sound so convincing in communicating their perjured opinions. Accept as a fact of life that reality is what it appears to be. If you want people to someday think of you as a particularly learned, trustworthy, and credible professional, you must communicate the information to them.

Second, believing that improvement of your verbal skills, having articles published, and making public speaking appearances are activities that can be postponed is self-deception. Start now, because, if you do not, chances are you never will. The sooner you start, the more experience you will gain. The more experience you gain, the better your skills will become. Effective communication is a learned skill. Especially when you have sources of constructive criticism available to you, practice will help you improve.

Third, many books and seminars are available to provide more in-depth guidance. Avail yourself of these, because the effort to improve is worthwhile. Professionals are expected to communicate effectively.

Fourth, shed any and all preconceptions you may have about certain rules of writing. There are very few rules, and most of those that do exist often can be bent. The rules we often hear about are no more than those established decades ago when the enunciators of the rules heard them from their teachers: "Never end a sentence with a preposition." Why not? If it sounds acceptable and is easily understood, do it. If it seems awkward, change it.

On a more general note, do not go overboard following the dictum, "Say it positively." Following this "rule" too rigidly can result in your not frightening people who should be frightened. Do not be positive for the sake of being positive. For example:

> Following their development of structural, mechanical, and electrical plans, the design professionals involved should evaluate their work with respect to one another, to help assure there is no interference among the plans.

versus

> Failure to check structural, mechanical, and electrical plans for compatibility with one another is an omission that can lead to problems on the job. Significant unanticipated costs and delays may result from having to correct these problems,

as when ductwork or electrical conduit must be removed and then relocated. Accordingly, the workscope should always include....

The English language is an institution that verges on anarchy. Certain basics tend to hold it together; these basics must be observed. But many of the other rules are not rules at all. They are, more than anything else, prejudices of long standing that no longer apply. Following the "rules" does not make writing better. Not following the basics can result in chaos.

ORGANIZING

No matter what it is you are planning to write—be it a covering letter, a proposal, a report, a magazine article, or a speech—it is essential to begin by organizing yourself.

Gather Information

Everyone has a preferred style of getting organized. One that seems to work is clearing a desk of everything except the materials you will be working from and with. If the research data is all in your head, a clear desk helps minimize distractions.

You should read all your research sources thoroughly, so that you can gain a "feel" of how various issues interrelate with one another. This permits you to review a source while keeping related concepts in mind.

Identify Your Audience

Your audience comprises your public(s): those you will be writing or speaking to. Who are they? What do they want to know? These fundamental questions all too often go ignored, resulting in papers that are sound but of no interest or value to their audiences. What are the important points from your readers' or listeners' point of view? By keeping them in mind, it will give more direction to your work. Some of the facts you develop may be almost irrelevant; others may be partially relevant and deserving of brief mention. Still others, which may not be all that vital to you, might be key issues for your audience. If you are preparing a magazine article, speak with the editor to determine what the relevant points may be, at least in a general sense. If you will be making a public address, speak with someone who is likely to attend or whose points of view are similar. *Do not guess. Do not assume.* Understand what people want to know about and go directly after those points.

Establish Direction

Once you have gone through your research sources, you should be able to develop a general direction for the finished piece to follow. In some cases, there is little choice. For example, proposals frequently are developed along the lines of tasks. Each task describes a service or a group of services, and, for the most part, they are applied one after the other. When they are not, that is, when one task will be ongoing while another begins, sequencing can be based on start times, and you can provide a separate section on scheduling, using a bar chart or something similar to indicate overall project scheduling.

In the case of a report, many different presentation styles are used. Typically, however, they begin with an executive summary, in essence, a synopsis of major findings and conclusions. This section usually is prepared last, since it should reflect only what is presented elsewhere in the report. Following the summary, it is typical to provide an introduction indicating the purpose and scope of the report. The body of a report commonly includes a description of the research methodology employed and the key findings. Appendixes are commonly provided to present any questionnaires used, tabulations of responses, and so on. In some cases, sections may also be provided for definitions and a glossary of terms.

In preparing a magazine article or speech, the direction of the presentation will be determined by the points you are trying to make. Frequently, the topic will be instructional in nature and, in such a case, it is essential to identify the specific lessons. The manner of presentation can vary significantly. For example, a topic such as the "Five Most Important Steps in Retaining a Design Professional" would probably have to unfold in chronological order, since step two would not make sense unless the audience is familiar with step one. Chances are, however, that many of the same points could be made in an article or talk entitled "The Five Biggest Mistakes Made When Retaining a Design Professional" or "Making Design Professional Services More Cost-Effective." The objective would be to describe each of the five issues, providing, as appropriate, anecdotes from "real life." In such presentations, it is typical to explain not only how something should be done but also why it should be done. The "why" usually is based on avoidance of negative outcomes. By switching the order of presentation, that is, by describing the negative outcomes first, the focus shifts to mistakes and then to ways of avoiding them, which would probably be more interesting than explaining how to do it right and how to avoid mistakes. Where appropriate, dollars and cents should be discussed, along with time factors, because this is the type of information that client-type publics are looking for.

One of the simplest means of developing an approach is to review materials produced by others. For example, if you want to develop an article for a client-oriented magazine, obtain back copies of the periodical. This can be done when you speak with the editor about an idea you have. By having a good feel for the approach you are going to use, you can also get a feel for the types of information you should relate, and this sets the stage for the next step.

Put Thoughts into Writing

Some people can assimilate background information and then sit down and write. For those who lack the skills or confidence to do this, a different approach can be used. Relying on 3-by-5-inch cards, or similar "tools," write one thought, point, or observation on each card. If for some reason in writing something down in this manner you feel inspired to write more, do so, using a separate piece of paper. Do not waste time trying to write well at this point. And do not become wedded to using what you have written. If it works later, fine. If not, so be it. You will simply have begun the process of writing, and what you have written will help you write something more effective. Do not even consider writing the introduction at this point, unless you feel inspired to do so. Keep completing the cards.

Create the Outline

Once you have your thoughts committed to writing on cards, put them in the order that will best implement your approach. It may be a chronological presentation, like this chapter. It may be to point out the five or ten basics of something, and you may wish to present these in order of the most (or least) important first. If you feel your initial approach won't work, go through your cards several times, and, based on what you have noted, establish a new direction. This may require you to divide thoughts on each card, so that one can be used in one place and another can be used somewhere else. Make up new cards as necessary and, once they are in what appears to be an effective order, prepare an outline based on what the cards show. Commit the outline to writing and then analyze it. Do the thoughts flow in the order you deem appropriate to make your case for the audience involved? Is the order hard hitting? Does the approach reinforce itself, in that each thought builds on prior thoughts? Make adjustments if you think they are necessary.

PREPARING THE FIRST DRAFT

The purpose of the first draft is to say everything you want to say. By having everything on paper, you will be able to consolidate, eliminate, rearrange, and otherwise manipulate the manuscript to get it into a condition that is ready for the second draft. In other words, do not be overly concerned about proper word selection, the smooth flow of ideas, or any of the other considerations applied to evaluate a piece of writing. You will have time for that later.

If it will be helpful for you to begin with the introduction, begin there. The introduction often is difficult to write at this point, however, because you are not certain of what will follow. One approach that may be helpful, therefore, is to begin at that point where you feel most comfortable and are sure of what you are saying. If the proper words at times elude you, do not spend time trying to think of better ones. If need be, just leave blanks or put in the words you think are close to those you are thinking of. The goal is to get your thoughts on paper. The more you get "hung up" on proper wording or phraseology, the more distractions you create for yourself.

The same guidance applies when you feel overwhelmed by the size, complexity, or short deadline of a report. Don't panic! Start writing somewhere, anywhere. Remember the question, "How do you eat an elephant?" The answer is: "One bite at a time."

Once you have completed a section, go on either to the one before or the one after, or simply go to another—no matter where—that you are comfortable with. If the latter approach is taken, you will be able to concentrate on the points that must be made in order to connect the two sections. In some cases, as when you may be discussing "the five" or "the ten" key considerations, intermediary passages may be unnecessary. But chances are that discussion relative to any key consideration will refer to discussion under others. This will suggest the proper order and will also indicate specific issues that should be discussed under key considerations that have not yet been written or that, if they have been written, may need modification.

Once you have prepared the "guts" of your draft, come back to the introduction, if you have not written it already. A brief summary may be in order, too. In this case, in preparing these final elements, consider the classic three-part outline of a piece of writing:

1. Tell them what you are going to tell them,
2. Tell them, and
3. Tell them what you told them.

Do not worry about style, and do not be concered if the introduction and summary (if used) sound stilted, unimaginative, or in any other way unacceptable. The goal in developing the first draft is simply to have everything in place, permitting you to polish it later.

Go over the first draft lightly before considering it done. Have you said everything you wanted to say? Have all thoughts pertinent to your subject and your audience been stated? Does one thought somehow lead to another, given the organizational method you have used? If so, you are ready to have the first draft prepared in a clean version. If not, make adjustments. If you are unsure whether a given thought should be mentioned in one area as opposed to another, put it in both places.

In most instances, your first draft will consist of a number of typewritten pages (from a typewriter or printer) that are marked up heavily or handwritten pages similarly marked up. Have the material typed or do it yourself, so you have "clean" pages to work with for the second draft. In most instances, it is best to have someone else do the typing or word processing, if only because, in retyping it yourself, you will be considering making changes, resulting in redrafting while keyboarding. This can slow the overall process more than it might accelerate it, simply because the material will likely interrelate. A change made in one place might necessitate a change in another. Three pages after making a major revision it may appear that the original concept should have been left alone or that, assuming it was effective, major elements following it now require modification.

If the first-draft manuscript is in particularly "rough" shape—so rough that it may be difficult for a word processor or typist to follow—consider dictating what you have prepared. This approach generally allows some leeway for minor modifications, particularly in the area of word selection. It may be effective, even if the first draft you have prepared is easy to follow.

In having the first draft typed, employ wide margins and triple- or quadruple-spacing. This gives you room to make the modifications that usually are necessary in preparing the second draft.

PREPARING THE SECOND DRAFT

Preparation of the second draft begins with analyzing the first draft for overall direction. Have you implemented the approach you established? If not, is it because the approach was inappropriate or because you developed a better one? If the alternative approach is one that you believe will work better, have you implemented it well, at least insofar as organization is concerned? Does it lend itself to reinforcing the points you want to make? The same questions apply if you have attempted to implement the approach you decided upon initially. Does it work? If not, now would be the time to change it, through reorganization.

In essence, the approach you select, as indicated by your outline, is the superstructure of your writing. If there are flaws in it, adding words to it will usually make the flaws more evident. Too often at that point an attempt will be made to use still more words to mitigate the flaws. While the flaws might become less noticeable, the overall piece will probably lose a great deal of its strength, since progression from thought to thought will of necessity rely on circumlocutions and strained transitions. If these problems exist in your first draft, restructuring is required. Some believe this is done easiest with a word processor. Others might be more inclined to use the traditional approach of cutting and pasting, since it can be done faster, and more alternatives can be tried. "What will happen if I put this thought here instead of there?" Using a cut-and-paste approach, possibly with a photocopy of the original as opposed to the original itself, answers the question. The necessary words can be added to create a transition into the new material, and new words likewise will be needed to close the hole created by moving a thought. Before you change words at either location, however, it may be easier just to speak the words out loud, to see if there is a "fit" between the thoughts.

Fine Tuning

Once you have what you believe to be the final order, the next steps comprise fine tuning. Note that most of the fine tuning suggested assumes a written document is being prepared. It is suggested that you *always* take this approach, unless your talk will be principally off-the-cuff. There are two basic reasons for this suggestion.

First, you generally will derive a better result by preparing your talk in writing, since this forces discipline. A talk should be more casual, but preparing it in a casual manner can result in a presentation that wanders. By preparing it as a written piece and then becoming familiar with it, you can help assure your presentation is to the point and well organized, while being delivered in a relaxed manner. Do not make an oral presentation that comprises a reading of something you have written. There is a major difference between written English and spoken English. Reading a written piece aloud results in a stiff, lifeless presentation.

Second, almost anything worth talking about before an audience is also worth writing about. The text prepared for a talk often can be used as an article or technical paper submitted for publication. As a result, you get a double impact, and both can be highly beneficial.

The fine-tuning technique discussed begins by evaluating word selections and then going into the structure of sentences. For making effective word choices, a dictionary and thesaurus are essential. In using these, however, recognize that simpler is better. Do not use a ten-dollar word where a fifty-cent selection will do just as well.

Identifying the Role of the Reader   You have already identified who your readers or listeners are. But where do they stand relative to your piece? Are they behind you, learning from your observations, or are they in the audience, taking direction? Consider the following:

> All too many professionals allow clients to have their way without first informing them of the alternatives. Effective professionals will at least caution their clients and advise them of optional approaches and their respective consequences.

This passage places readers behind the writer. They are looking through the writer's eyes and see two types of professionals or two approaches: one performed by one type of professional and the other, by a different, better type. Now consider the following:

> All too many professionals allow their clients to have their way without first informing them of the alternatives. Do not let yourself get caught in this trap. As a professional, it is your responsibility to advise clients, and sometimes this may mean saving them from themselves or at least trying to. If you believe clients are not making effective decisions, let them know. Do it in a diplomatic manner, of course, so as not to offend them and to avoid appearing arrogant.

The reader's position has been changed. Instead of observing a wrong way and a right way, they are now receiving direct instruction, because the writer refers to "you," whether or not the word is stated. The "you" approach is generally more effective, because it permits shorter, tighter, more hard-hitting sentences. However, in some formal pieces, or in pieces where the audience is mixed, it may not be effective and could feasibly lead to some confusion.

Neither way is right or wrong. The principal concern is selecting the approach that best suits what it is you are saying and the context of the piece itself. Another concern is consistency. When you start something one way, it usually is advisable to stay with the approach. It is possible to switch out of one mode into the other, however, providing you do it clearly. Consider the following:

> All too many professionals allow their clients to have their own way without first informing them of the alternatives. Effective professionals will at least caution their client, and advise them of optional approaches and their respective consequences. Assume you are in the position of discussing structural design. The client says to you, "The approach we used last time will work well." How should you respond?...

By carefully moving from one point of view to the other, the reader can move easily at your command. In the opening of the passage, the reader is behind your back, surveying two approaches: the good and the bad. Then the reader is carefully placed in front of you, to receive specific guidance: a demonstration of the point through an everyday experience. In many cases, this shift in position can make your writing more interesting and effective, but it requires more skill to keep it under control.

Review what you have written. Is the role of the reader clear? Have you used a consistent approach? Can you make the piece more effective by switching back and forth?

Note that references to yourself and to the audience you are writing for can sometimes get bogged down in certain conventions. One of these is referring to "one," as in, "One should not do this." As soon as you use this word, you tend to lose the audience or your reader, because it lacks "punch." It is far more compelling to say, "You should not do this," or, "Clients should not do this."

In referring to yourself, use "I." Assuming you are preparing a piece in which you have been identified and are writing from a position of authority, "I" will lead to direct insights to your perceptions or actions. By all means, avoid "the author" when referring to yourself in articles and "the undersigned" when writing letters. For example:

> When asked to prepare a proposal, the author is tempted to examine the client's request and....

versus

> When I am sent a client-prepared workscope and request for a proposal, I will sometimes examine the workscope looking for....

Note that the latter seems more alive than the former, and both have more life than a passage created by turning yourself into a group; for example,

When some architects are sent a client-prepared workscope, they will. . . .

Eliminate Useless Words and Phrases There are certain useless words and phrases in this sentence, in that some of those selected are not necessary to create awareness of meaning. For example, "awareness of meaning" is redundant. If one is not aware of meaning, there is no meaning. Therefore, the end of the sentence could be changed to ". . . create meaning."

Now look at the beginning of the previous paragraph: "There are. . . ." Try to eliminate such phrases, because they comprise slipshod uses that diminish the power of your writing. Eliminating them forces you to restructure the sentence. In this instance, it should be seen that deleting "There are" leads to the following construction:

Certain useless words in this sentence are not necessary to create meaning.

Note, however, that "useless" itself becomes useless, in that words not necessary to create meaning have no use. Accordingly:

Certain words in this sentence are not necessary to create meaning.

Since we only uses words to create meaning, the last three words also may be unnecessary:

Certain words in this sentence are not necessary.

The sentence is still true, in that "unnecessary" could be substituted for "not necessary," but then we have a sentence that, because it illustrates a point, makes no sense at all!

Figure 13-1 lists a number of typical unnecessary words and phrases you should learn to eliminate. Figure 13-2 lists circumlocutions and, in parentheses, briefer alternatives.

Eliminate Improper Words Your text may contain improper words. This includes *absolutes*, such as *all* and *every*. In a proposal, contract, or report, where you are bound or may be bound by law to fulfill your words, reliance on an absolute can have serious repercussions. *All* does *not* mean "almost all." It means just what it says: all—*without exception. Every* means basically the same. Similarly, *assure* can mean "guarantee." There is nothing wrong with using such words per se, providing they convey the meaning intended. Figure 13-3 lists some commonly used absolutes and alternatives.

Another form of improper word usage is the "slipshod synonym," that is, a

what had been done *in this direction*
set an example *from this point*
consideration was given *to the factor of*
connected *together*
exploratory *in nature*
this, and other *available* information
a *small* 20-ml flask
numerals *are used to* identify
preservation by *use of* salt
the device *in question*
the marks *provided* on the chart
overspeed is *a* critical *factor*
will not *properly* align
stop any leaks *that are discovered*
turn the handle *when required* for adjustment
horizontally level
aluminum *metal*
retract *upward*
uniformly consistent
properly tuned
two cubic feet *in volume*
read *through*
engage *with each other*
ac *current*
maximum *possible*
identical *in all ways*
value *and importance of*
assemble, fuse, attach *together*
time interval
innumerable *numbers of*
a potentiometer is *defined as*
the *factor of* radiation is important
formed by *virtue of*
the adjustments *provided* on the device
after installation of the rheostat *has been completed*
it is quite possible that the device may
gasket *located* inside the
after *completion of an* assembly
convert *over* to
by a *device known as a* chronograph
by *means of* integration
read and record
must be lowered *in order* to get a proper reading
there are some products *that*
recorded history
future planning
red *in color*
open, closed *position*

Figure 13-1: Useless Words and Phrases

hexagonal *in shape*
close proximity
recoil *back*
range *all the way* from
coil *winding*
actual experience
between *the limits of* A and B
final outcome
physical size
balance *against each other*
cancel *out*
close *off*
open *up*
contains *within*
past experience
small-*sized* sample

Figure 13-1: *(Continued)*

word that is used to mean something it does not really mean. One of these is *mute* for *moot*, as in, "That is a moot point." *Moot* in this context means subject to argument or debate. Many people say, "That is a mute point," but *mute* means "silent." Similarly, *appraise* is often used for *apprise*, as in, "John appraised Bill of the latest developments." *Appraise* means to estimate the quality or value of something. One could say, "John appraised the latest developments," meaning John evaluated the meaning or impact of the latest developments. But this was not the intention. Instead, the sentence should have read, "John apprised Bill of the latest developments," meaning, "John *informed* Bill of the latest developments."

in the nature of (like, as)
in connection with (concerning)
in view of the fact that (because)
pursuant to agreement (as agreed)
in lieu of (in place of)
hold in abeyance (wait, postpone)
be cognizant of (know, notice)
it is necessary that (must)
is equipped with (has)
pertaining to (about)
in order to (to)
with respect to (for, about)
through the medium of (by)
in the course of (while, during)
on the order of (about, roughly)
from the point of view of (for)

inasmuch as (because)
with a view to (to)
in the event that (if)
prior to (before)
with reference to (about, concerning)
for the purpose of (for)
along the line of (like, as)
for this reason (so)
caution must be observed (be careful)
it is imperative that (be sure that)
taken into consideration (considered)
for the reason that (because)
large numbers of (many)
is provided with (has)
at the present time (now)
enclosed please find (enclosed is)

Figure 13-2: Circumlocutions (and Alternatives)

stress, strain
normal, perpendicular
calibrate, adjust
construction, restriction
transitory, transient
anhydrous, hygroscopic
replace, reinstall
resistor, rheostat
maximum, optimum
refraction, diffraction
affect, effect
adhere, cohere, stick
absorb, adsorb
preclude, prevent
capacity, capacitance
transparent, translucent
round, spherical, circular
emf, voltage, potential
ambient, atmospheric
adapt, adopt
corrode, erode
homogeneous, homogenous
altitude, elevation

militate, mitigate
develop, discover
develop, design
evaluate, express opinion on
apprise, appraise
varying, various
between, among
alternate, alternative
allow, permit
imply, infer
predominate, predominant
principle, principal
resumé, resume
sometime, some time
contact, write *or* be in touch with
couple, two
most, almost
economic, economically
while, because
since, because
moot, mute
effect, impact

Figure 13-3: Selected Slipshod Synonyms

Two other common misapplications are worth special attention. They involve *that* versus *which* and *e.g.* versus *i.e.* "Someone vandalized the blue car *that* was parked over there." "Someone vandalized the blue car *which* was parked over there." Which is correct? Realistically, both are, since the two words—*that* and *which*—are increasingly being used interchangeably. Purists may insist that *which* is the correct choice, because something definite was referred to: the blue car. It is their contention that *that* should be used only when something indefinite is involved; for example, "Some may vandalize cars that are parked over there." Because "definite" is judgmental, however, it could also be argued that, even in the last sentence, *which* should have been used. In many, if not most, cases, it is up to you.

Certain Latin abbreviations are commonly used, often because they comprise another way of saying something that otherwise would appear too frequently. For example, you will see *e.g.* (foreign words and abbreviations generally should be underlined or italicized, although some have become accepted English) commonly used as a synonym for "for example." *E.g.* is an abbreviation of *exempli gratia*: translation, "for example." *I.e.* is an abbreviation of *id est*, or "that is, Writers frequently confuse the two and thus say, "Many design professions happen to be interprofessional in nature, i.e., mechanical, electrical, and structural engineering." In this context, *i.e.* is incorrect, because the three types of engineering exemplify the statement. For *i.e.* to be correct, what follows it would have to be explanatory, for example, "Many design professions happen to be interprofessional in nature; i.e., they are commonly pursued in support of other design work, such as architecture."

Figure 13-4 lists a number of commonly misapplied words. Refer to a dictionary if you are uncertain why some words do not mean the same thing. Note, however, that "common usages" now make some of these substitutions acceptable. "Common usage" means that a certain word or phrase has been used incorrectly for so long that it is now more or less acceptable. Perhaps it would not do to use these words in a text prepared for grammarians, but for most of the writing you are likely to do, such uses would be acceptable.

Be Careful with Jargon   Certain words or phrases may be well known to you, or to other engineers or architects, but they may be unknown to the public you are writing for. To the extent that it facilitates your forming of the first draft, rely as much on jargon as you want. Once your initial thoughts are in place, however, you should examine the jargon. If it is necessary to keep it, be sure it has been defined. For example:

> One should first contact a consulting mechanical and electrical engineer (M&E) to obtain an opinion. Then, working with the M&E. . . .

It may even be appropriate to explain what it is a mechanical and electrical consultant does or to focus on one discipline or the other:

> One should first contact a mechanical engineer, that is, a consulting engineer who specializes in the design and investigation of a building's mechanical systems, such as its heating, cooling, and ventilation systems. The mechanical engineer, or mechanical, can then. . . .

Is it acceptable to use "mechanical" for "mechanical engineer," as shown? Perhaps, but why? It would not take a great deal of time and effort to refer to the individual consistently as a "mechanical engineer." While it may be uncomfortable for you, realize that the reader is your guest. Make your guest comfortable.

You can use jargon somewhat more in writing than in a talk. When comments are in writing, the reader can refer to what has gone before in order to clarify whatever needs clarification. This is not the case for the audience listening to a talk. If the speaker defines something and listeners do not follow along, they may remain lost for the rest of the talk.

Avoid Imprecision   Design professionals commonly are taught to avoid absolute words, something that has already been discussed. In response, however, some tend to "weasel word" much of what they write. There is no real benefit to doing this. It is relatively easy to express limitations and nonetheless retain precision. Here are two examples.

Our observations suggested the thermostat perhaps was not properly calibrated.

Our observations verified employee complaints of overheating. This situation could be due to improper thermostat calibration, improper location of the thermostat, failure to close the blinds to reduce solar heat gains, or a combination of these.

The danger in being more precise, from a legal standpoint, is the possibility of creating a list, such as the one given above, but omitting something from it. Accordingly, it may be wise to add something such as the following,

> ... or a combination of these. We recommend that the area be monitored to identify the problem's actual cause. It could include factors not mentioned above and of which we are unaware.

Remember, professionals are not engaged to be imprecise. People expect professionals to be knowledgeable about certain issues, and professionals should indicate this knowledge in their writing. This does not mean that a professional must at all times be definite. If there is not enough information available to provide an answer or form an opinion, this situation can be related ... precisely.

Warnings and Limitations In reports, proposals, and other forms of writing for which you may be legally liable, it is essential to make any and all warnings or notices of limitations crystal clear. Do not attempt to downplay a condition or in any other way minimize a potential problem. Consider the following two versions of the same report of a home inspection:

> On Monday, September 7, 1991, I inspected a single-family house located at 123 Elm Street, in Franklin, Massachusetts. There are several signs of some differential settlement, in that windows on the west wall of the house are out of plumb. This settlement appears to be somewhat long standing in nature. In that the house is eighty years old, it is reasonable to conclude its settlement has stopped or, if it is continuing, it is extremely slight.

versus

> The following comprises a report of visual observations made at a single-family residence located at 123 Elm Street, Franklin, Massachusetts. Observations were made on Monday, September 7, 1991. The purpose of the observations was to note any apparent signs of structural weakness. Other weaknesses may exist, such as those related to the heating system. Any such conditions, while they may or may not have been observed, are not herein related because they fall outside the purpose for which the consultant was engaged.
>
> *Note:* Other conditions may have an effect upon the overall condition of the house, its value, or saleability. The prospective purchaser is advised that

additional observation may be necessary. In addition, the prospective purchaser is advised that visual observation is not effective in determining all conditions that affect structural soundness. (An uneven floor does not necessarily connote that a building is not structurally sound. Likewise, a level floor does not necessarily indicate that a building is structurally sound. Certain hidden conditions can be identified only through different types of testing, services that the client specifically excluded from this engagement.)

The only visual indication of potential structural weakness relates to windows on the west side of the house. These are out of alignment. One possible explanation of this is differential settlement along the western side. If that is the case, the problem may or may not be cause for concern. The house is eighty years old. If settlement occurred during the house's first few years of existence, it is reasonable to assume the foundation currently is stable. However, if settlement has been ongoing or is of recent vintage, there is far more potential for it to continue. The only techniques available to establish actual conditions involve subsurface testing and settlement testing. *Note*: The prospective purchaser is advised to authorize this work before making an investment.

Please note that this report comprises a statement of professional opinion only and that this statement of opinion is based on visual observation only. The report does not constitute a guarantee or warranty of any type, and none should be inferred.

If the design professional preparing the report was working for the homeseller or real estate agent handling the sale, chances are the client would be disappointed. In fact, even a prospective home purchaser might be disappointed, due to lack of definitive guidance, except the suggestion for testing and monitoring. These attitudes are unfortunate, but under no circumstances should the design professional make warnings or limitations less apparent.

The first example given is approximately what a design professional actually did state and, because of it, wound up spending some $30,000 in defense of a claim, in part because "structural soundness" was left undefined, along with the purpose of the observations. Particularly when dealing with homeowners and those in a similar position, be extremely cautious in expressing yourself. Those who do not customarily deal with design professionals often have unrealistic expectations. The courts leave it to design professionals to help prevent misunderstandings. That is part of your responsibility, too. Note that very sophisticated builders may also allege that a report or warning was not clear enough. Again, you want your position to be unmistakable. For example:

> An experienced construction monitor should be on hand to evaluate the condition of the concrete before a stripping order is implemented.

versus

> *Warning:* Premature stripping of forms could result in serious damage to the structural integrity of the building and possible injury to workers. Formwork

should be stripped only after the condition of the concrete has been appropriately observed and tested, and findings indicate that it is safe to remove forms. *All monitoring* should be performed by individuals who are properly trained and otherwise qualified. All testing should conform with. . . .

On the one hand, by making such warnings and limitations clear, it cannot be alleged later that the design professional was somehow negligent in transmitting guidance. More important, this approach helps assure that warnings and limitations are seen, read, and followed, thus reducing the potential for problems that result in claims.

Gender Reference It has, for many years, been common practice to refer to an indefinite gender in the masculine, for example, "The competent civil engineer will not take this approach. Instead, *he* will. . ." Women have protested this convention, and it does not take a great deal of insight to understand why. Several techniques are used for dealing with it. One of the most common is an advisory or apology, along the lines of the following:

> Masculine pronouns have been used solely in keeping with convention. No specific gender is implied, and none should be inferred.

Some women react very negatively to this approach. To them it says, "I understand the problem, but I don't care enough to do anything about it."

Alternatives are available. One of these is continually to say "he or she," "him or her," and so on, throughout a text. While this may be more gender-neutral, it is awkward and can get in the way of good writing. Another alternative, which is far superior to the first two, is pluralization. By using it, references can be made to "they" and "them," with an occasional "he or she" if necessary. Still another good approach is to refer to the category of individual involved; for example, "The competent civil engineer will not take this approach. Instead, the civil engineer. . . ." Taken to an extreme, however, the latter approach creates undue wordiness.

Overall, the best approach is to rely mostly on pluralization and category references. This approach has been used throughout this text.

Active versus Passive When you are writing in the active tense, the subject of the sentence does something. When you are writing in the passive tense, something is done by someone.

When the passive tense is used, its purpose is usually to call more attention to the object of the action than to the person or thing that makes the action occur. Sometimes it is used to eliminate the subject from consideration, or from obvious consideration. For example:

> Subsequent to completion of the investigatory phase, raw findings will be reviewed and categorized. A draft report then will be prepared and submitted for

comments. Comments received will be considered in development of a second draft report.

This approach makes all action appear to occur without human intervention. While this may be acceptable in a proposal, use of the active tense brings far better results. Consider the same project description in which the proposal writer seeks to mention the name of the firm making the proposal—Able Associates—and the client, Big Retail:

> Once its investigations are complete, the Able project team will review and categorize data. The team will then develop a draft report and submit it to Big Retail representatives for review and comment. The Able team will review these comments and apply them in preparing its second draft of the report.

By putting people back into the description, the writing becomes far livelier. A process that could otherwise be regarded as machine-like becomes one in which people do the work.

The particular style used above resulted in the active description being longer than the passive. Usually it works out in just the opposite way, because the passive tense tends to rely heavily on nouns that in many instances could be expressed in their verb forms; for example:

> Assignment of personnel will be based upon a review of their current and prior work loads, and their experience and familiarity with the type of project in question.

versus

> Persons experienced with the type of work required will be assigned to the project.

As a general rule in writing, "The briefer, the better."

Pare Sentence Length In presenting a talk, sentence length is not too much of a concern. In written English, it is. For the most part, power is obtained from short, declarative sentences. Consider the following two versions of the same paragraph:

> The process begins with a thorough review of the structure, relying on trained individuals with the knowledge and experience to know what they are looking at, and familiar with the various testing equipment at their disposal to examine conditions they cannot see.

versus

> The process begins with a thorough review of the structure. Experienced

evaluators examine it visually, looking for indications of defects. Then they use testing equipment to check for hidden conditions.

Long clauses connected with words such as *and, also,* and *but* often can be broken into two or more sentences. The same applies to those connected by relative pronouns. While breaking thoughts apart may sometimes result in more words being needed, the power gained can make it worthwhile:

> We will assign this project to Bill Smith, who is uniquely qualified to handle it.

versus

> We will assign this project to Bill Smith. He is uniquely qualified to handle it.

By making the second thought stand on its own, it is more prominent and thus seems to be far more than an afterthought.

Analyze Dangling Modifiers Trying to develop an example of a dangling modifier, the author prepared this sentence. There is nothing wrong with dangling modifiers, except when they are used improperly, which is common, or when they are overused, which also is common.

Consider the opening sentence of the previous paragraph. A common mistake would have been:

> Trying to develop an example of a dangling modifier, this sentence was prepared.

Did the sentence prepare the dangling modifier? Of course not, yet that is what the sentence says. As a general rule, the first identity after the dangling modifier should be what is modified. Here is another example of an incorrect dangling modifier:

> Having completed the field work, the report was prepared.

This indicates that a report was evidently outdoors, working out, perhaps preparing itself for a marathon. For the most part, sentences with dangling modifiers can be restructured.

Agreement Between Subject and Verb A variety of correct phrases sounds incorrect. Sounds? Correct. The subject of the sentence is "a variety," and "a variety" is singular. "Of" is a preposition, and "phrases" is the object of the preposition, not the subject of the sentence. If the proper constructions sound awkward, change the sentence, to derive, "Some phrases that sound correct are not correct."

As you are probably aware, multiple subjects joined by *and* or other words require a pluralized verb: "Architects, as well as engineers, are affected by this situation." However, words joined by *or*, connoting separateness, do not necessarily require a pluralized verb: "An architect or an engineer *is* affected by this situation." When *or* connects a singular and a plural, the plural form often is used; for example, "An architect or the architect's engineers are affected by this situation." You could also say, "An architect is, or the architect's engineers are affected by this situation," but that construction could be deemed pedantic and makes for weak writing.

Applying English Prime For purposes of discussion, it can be assumed you now have a piece of effective writing. Is it as effective as it could be? One way of finding out is to apply the concept of English Prime, that is, writing without the verb "to be" except when it is used to indicate tense. By attempting to eliminate at least some of your reliance on "is" and "are," you will be forced to turn to verbs that establish action, lending strength to your writing. Consider the following:

> An adaptive reuse project is challenging in many respects, particularly because there are so many unknowns. Surprises—some nasty ones, too—may be behind many walls. This is why comprehensive predesign review is so necessary.

versus

> An adaptive reuse project presents many challenges, in part because of the surprises—some nasty—that designers may encounter. What an architect presumes to exist behind a wall may differ substantially from the actual composition. We can lessen the potential for such an occurrence by performing a comprehensive predesign review.

How many times have you used the verb "to be" in your text? Can some uses be eliminated? Try. The result will be writing that has more thought—and power—behind it.

Completing the Second Draft

At this point, you should have a second draft that is ready for retyping. Performing the various exercises indicated does not assure you of having an effective piece, but it should be sufficient to help you avoid some common major problems.

DEVELOPING THE THIRD DRAFT

The third draft is what you would like to be your final draft, but this is not necessarily the case. You should subject it to some testing before considering it final.

Gunning's Fog Index

Several techniques have been developed to quantify writing style and thereby indicate how difficult it is to follow. One of these is a system developed by Robert Gunning, called the Fog Index. The number developed from applying Gunning's rules indicates the grade level an individual should have obtained in order to understand the writing involved. A Fog Index of 6 suggests that your writing is understandable to someone with a sixth-grade education. A 17 indicates that only someone with postgraduate education would be capable of following your writing.

The Fog Index is based on the number of words per sentence and the percentage of words that have three or more syllables. Begin by randomly selecting one hundred consecutive words (beginning with the first word of a sentence in any passage you have prepared). Obtain a word count per sentence by dividing one hundred by the number of sentences in the passage. Note, however, that a sentence in this case is defined as a complete thought. Some longer sentences often contain two or more complete thoughts, and thus would be counted as two or more sentences. Add to the result the number of words with three or more syllables. (In calculating multisyllabic words, do *not* count: proper nouns, such as the names of places or people; verbs made into three syllables through the addition of *-ed* or *-es*; or compound words built on short ones, such as *bookkeeper* or *butterfly*.) Once you have the total, multiply it by 0.4.

By applying the Gunning Fog Index to this section, beginning with "Several techniques . . ." two paragraphs earlier, it could be said that this text has a Fog Index of about 12, meaning that someone with a high school education should be able to follow it.

Note that the Fog Index does not address the quality of writing, or its style or effectiveness. It is merely a rough assessment device designed to establish an objective means for evaluating certain aspects of your writing. The key word is *objective*. Writers generally are not in a position to evaluate their own writing.

The Acid Test

The acid test of writing is obtained from the candid evaluations of those to whom you are writing. In most instances, you know several people who fall into this group. In the case of reports, proposals, and other types of business writing, your peers within the firm should be capable of providing comment. You may even wish to follow on a local level a program developed many years ago by ASFE. The program is still active and is geared principally to proposals and the engineering reports resulting from the execution of work proposed. Any specific form of writing could be targeted, however. Assume it is proposals. A local chapter of ACEC, AIA, NSPE/PEPP, or others could easily establish an effective program. Each member would submit two copies of a typical proposal, in which the names of people and projects were made anonymous, to the chapter office. The chapter office would then send two different proposals to each submittor. Each submittor then would comment on both proposals and return them to the office. Then the office would redistribute materials to the original submittors. In that way, each submittor would receive two separate sets of comments.

In terms of magazine articles prepared for a specific group of some type, you could easily contact a few appropriate persons and ask for assistance. Merely explain your intent and ask for their advice and comment. Advise them to "be tough." Most will be happy to lend a hand, and some will probably be flattered to be asked for help.

In reviewing comments, steel yourself. Not everyone is capable of phrasing criticism constructively, for example, saying, "Wouldn't it be better to substitute infer for imply," rather than, as opposed to, "Infer is wrong." Regard any criticism as constructive, realizing that comments received are made with the desire to help, no matter how they may be phrased. If a person has difficulty understanding something you have written, you might say, "Well, he's just stupid," because it makes complete sense to you. That would be an unfortunate attitude. Look at English as a number of languages, not just one. If someone has difficulty understanding your point, you are using the wrong language, and your reviewer is forthright enough to say so. It is up to you to speak the language of those to whom your piece is addressed.

In the case of articles prepared for lay audiences, you may even want to have your writing reviewed by junior or senior high school students, since that is the grade level to which most writing should be geared. Ask questions of the students. Have them explain your thoughts to you, to help assure your intent is clear.

Yet another approach you may want to take is to work with a high school (or lower grade) English teacher. Pay the person for review and explanations. You may feel embarrassed. *Don't*. Many technical professionals have never learned the basics well to begin with, which is one of the reasons why so few write.

Examine any areas where difficulties are noted and rewrite as necessary. Go over your rewritten versions with those who commented on them. In the case of a magazine article, send a copy of your submission to any reviewers, to keep them apprised, and—as appropriate—give them credit in the by-line, for example:

> John Doe is a senior designer with Jones and Smith, AIA, a 25-year-old architectural firm located in Franklin, Massachusetts. Mr. Doe wishes to express his thanks to Arthur Green for his assistance in preparation of this piece.

The editor may cut the credit line, but that would not be your fault. Your reviewer will appreciate your courtesy.

FROM WRITTEN TO ORAL

To translate your written presentation into an oral one, simply take your final version and make notes of the key points made. This can be done on 3-by-5-inch or 5-by-7-inch cards. Then talk your way through. You should be familiar with your subject matter, and you will thus be able to deliver it using "spoken English." Some people believe that professional delivery means virtually reading a written piece. Do not fall victim to this stereotype. A professional is just as human as anyone else but does need to convey a sense of authority and intimate understanding of the subject matter. This does not at all mean that a stilted style is required. You can leave people impressed with your knowledge and understanding by demonstrating it even within a context laced with humor and witty insights. This makes you seem far more "level-headed" and approachable, which is often one of the goals of public speaking to begin with.

INTERPERSONAL COMMUNICATIONS

All of the above discussion relates to one-way communication. The vast majority of your communicating will be two-way, when you are engaged in conversation with others. By increasing the amount of writing you do and by analyzing that writing closely, you can help assure better language even in casual conversation. But using proper language is not nearly as important as saying the right thing. All too often people say the wrong thing because they respond too quickly.

In all of your conversations, think of the person with whom you are communicating. What is that person's point of view and attitude? Ideally, you should want whomever you are speaking with to take the action you suggest. For this to occur, you must be able to see things from the other person's point of view. In most discussion situations, a number of arguments can be raised as to why something should be done one way or another. To argue from your position, you often will be pushing against others' direction. By arguing from the others' viewpoint, you are pushing them along in their direction, to where you want them to go. For example, if you are dealing with clients who want to minimize expense, you should consider promotion of additional services as a means to save money. After all, a more complete design service can lower construction costs by reducing the need for contingencies in a bid, lower the likelihood of unanticipated conditions, and so on. Similarly, when directing a subordinate, you want to select words that express thoughts that will best complement the subordinate's viewpoint. By speaking without thinking first, you can say something you will regret later, especially so because of the image formation process. One sentence in one hundred may be used as the "telltale clue" which in the receptor's mind reveals what you are really thinking.

Most people do not imagine a professional as someone who "shoots from the lip." Accordingly, until such time as speaking in a diplomatic manner becomes second nature, try pausing a few moments before replying to something someone else says. If the comment is inflammatory, use the pause not only to think about what your response should be but also to allow your emotions to dissipate, so that you do not get caught up in a shouting match. One of the colloquial definitions of a diplomat is "someone who can tell someone else to go to hell... and the other guy can't wait to pack his bags." Developing such a skill will help assure you convey a professional image, even in casual conversation. Perhaps more important, the image will be an accurate reflection of what really exists.

Index

Absolute words, 111, 142
Accounting, 218-234
   chart of accounts, 220-233
   classifying accounts, 218-233
   methods, 233-234
Actual cause, 72-73
Actuaries, 153
Adversarial procedures, *see* Dispute resolution
Advertising:
   by design professionals, *see* Marketing communications
   by lawyers, 20-21, 68
   for personnel, *see* Human resources management, advertising
Advocacy, 68, 71, 72, 112
Affirmative defense, 189
Aggregate coverage, 159
Agnew, Spiro T., 28
Allied Properties v. John A. Blume & Associates, 80
Alternative dispute resolution, *see* Dispute resolution, alternative
American Arbitration Association, 201-202, 203, 208
American Consulting Engineers Council, 36, 39, 97, 220, 331, 356
American Institute of Architects, The, 3, 9, 36, 39, 97, 331, 356
American Institute of Certified Public
Accountants, 219
American Society of Civil Engineers, 3, 9, 18, 39, 97, 331
American Society of Heating, Refrigerating, and Airconditioning Engineers, 331
Appeal process, 194
Apprenticeship, 2, 7
Arbitration, *see* Dispute resolution
Asbestos, *see* Hazardous materials
As-built plans or specifications, *see* Record documents
ASFE/The Association of Engineering Firms Practicing in the Geosciences, 17, 39, 58, 72, 136, 139, 162, 170, 184, 211, 212, 213, 216, 273, 356
Assets, *see* Economics, of professional practice
Association involvement, *see* Marketing communications, association involvement
Assumption of risk defense, 75-76
Assumption(s), 17, 28, 37, 48, 52, 53, 72, 92, 120, 122, 174, 179-180, 184, 199, 302
Attorneys, *see* Lawyers
Attraction, of clients, *see* Client attraction; Competition, for clients
Audio-visual aids, 49, 55, 59
Awards, *see* Marketing communications, awards competitions

362 INDEX

Barratry, 150
Basic ordering agreement, 101
Bidding, for contractor services, 57–59
Bird-dogging, see Direct selling, bird-dogging
Birthday greetings, 283
Boards of registration, see Registration, boards of
Body of theory, 4–5
Bonuses, see Human resources management, compensation
Brochures, see Marketing communications, brochures
Brooks, Jack, 27
Brooks Law, 27–28
Building Owners and Managers Association International, 38, 321
Buildings as products, 77
Business cards, see Marketing communications, graphics
Business communications, see Marketing communications
Business development, 316–335
    direct selling, see Direct selling
    marketing communications, see Marketing communications
Business planning, 286–315, 319, 328–329, 330, 334
Business practices, 21
Buying in, see Low-balling

Capacity, of insurance industry, 155–157
Career ladders, see Human resources management, career ladders
Case histories, see Marketing communications, case histories
Census Bureau, 301
Certification, of nonprofessionals, 7
Certification(s), 80, 83, 105, 106, 115–118. See also Warranty
Chamber of Commerce, 38, 324, 331
Changed conditions, 127, 199
Change orders:
    contractors', 57, 74
    design professionals', 29, 257
    field, 122
Civil justice system, 13, 20–21, 64–91, 100, 105, 131, 187–194, 201
Claims made, see Insurance, claims made
Claims managers, 154
Class-action suits, 69, 74
Clergymen, 3
Client attitudes, 103, 110–111, 118, 121, 122–123, 126, 137, 138, 141, 145, 175, 257

Client attraction, 8, 319, 330. See also Marketing communications
Client education, 37–39, 42, 56, 63, 103, 104–105, 106, 107, 109–111, 114, 119–120, 122–123, 127–128, 131, 136, 137, 142, 148, 159–160, 234–235, 351
Client maintenance, see Client relations
Client-professional relationships, see Professional-client relationships
Client relations, 327, 332
Client responsibilities, 103
Clients, new, 246
Client satisfaction, 217–218, 316, 335
Client selection, 21, 110, 170–171, 173–175, 330, 332–333, 351
Codes, 8
Code(s) of ethics, see Ethics, code(s) of
Cold calls, see Direct selling, cold calls
Colleague-to-colleague relations, 8–9
Collections, see Payment
Commercial disparagement, 88–91
Communications:
    business, see Marketing communications
    interpersonal, 358–359
    professional, see Professional communication
    supervisor-to-employee, 278–279, 358
Community sanction, see Sanction(s) of the community
Comparative negligence, doctrine of, see Negligence, comparative
Compensation, see Human resources management, compensation
Competition:
    for clients, 8–9, 35, 306, 312, 332–333
    fee-based, 248–249
    intraprofessional, 8–9, 29, 300, 330
    unfair, 88–91, 267
Complaint, 188
Complete professional service, 35, 62, 63
Computer-aided design and drafting, 16
Condominiums, 69–77, 137, 166
Confidentiality, 129, 130
    dispute resolution, see Dispute resolution, confidentiality
    personnel policies, 276
Consequential damages, 90, 102, 116, 118–119, 146
Construction claims, 41
Construction management, 62
Construction monitoring, see Field observation
Construction site safety, see Site safety

INDEX **363**

Construction Specifications Institute, 97
Construction superintendence, 62
Consulting Engineers Council of
    Metropolitan Washington, 29, 31
Consumer protection, 7
Contingencies, 245
Contingency fees, 67
Contingency planning, 32
Contract formation, 50–53, 78, 96, 101, 113, 122–123, 175, 280, 308
Contract law, 65, 104–105, 109, 265
Contractor coordination, 62
Contractor selection, 57–58, 84–88, 129
Contract provisions, 80–81, 87–88, 97, 102–103, 106–147, 187
Contract review, 47, 52–53
Contracts:
    acceptance, 94
    addendum, 96, 104
    of adhesion, see Disparate bargaining power
    agreement, 94
    amendments, 101, 191
    attestation, 95
    authentication, 94, 95
    bilateral, 93
    binding, 93, 94–95
    breach of, 93, 104, 125–126, 140, 157
    competent parties, 94, 95
    consideration, 94, 109
    enforceable, 93
    express, 93
    extralegal conditions, 103
    formats, 95–101
    implied, 93, 94
    inclusion by reference, 104
    interpretation of, 109
    joinder, 195
    legal form, 94
    legal purpose, 94, 95
    model(s), 97
    multiple, 97–98, 165, 308
    negotiation of, 112–113
    offer, 94
    one-sided, 98
    oral, 93–94, 100–101
    payment, 246
    for professional services, 92–147, 170–171, 184
    quasicontracts, 93–94
    renegotiation of, 126–127
    separate, 98, 308
    severability, 94
    termination of, 126–127, 140–141
    unenforceable, 93
    unilateral, 93
    verbal, 93–94
    void, 93
    voidable, 93, 119
    written, 93–94, 345
Contractual liability, see Liability, contractual
Contractual relations, interference with, 88–91
Contributory negligence, doctrine of, see Negligence, contributory
Copyright, 142
Corporate structure, see Economics, of professional practice
Cost estimates, 119–121, 183, 246
Counter-claim, 189
Cover-ups, 182
Craft worker guilds, 2
Criminal law, see Statutory law
Critical path method, 257

Damages, types of, 86, 89, 91. See also Consequential damages; Punitive damages
Deceit, doctrine of, 81–84, 118
Deep Foundations Construction Industry Round-table, 212
Defamation, 84–88, 89, 129, 182
Defense, cost of, 159
Definitions, 111, 117, 118, 121, 145
Delays, 181, 183
    of construction, 58
Department of Justice, see Justice, Department of
Depositions, 83, 169. See also Discovery
Depreciation, see Economics, of professional practice, depreciation
Design concepts, formation of, 47–49
Design intent, 60, 61
Design-phase services, 54–60
Design team:
    assembly of, 43–45, 52
    composition, 47
    morale, 44
    timing of assembly, 44
Detail report, 258
Deutch, James, 60
Directed verdict, 192–193
Direct selling, 330–335
    bird-dogging, 330, 332–334
    client relations, see Client relations

Direct selling *(Continued)*
  cold calls, 333-334
  networking, *see* Networking
  tracking, 334-335
Discovery, 68, 83, 159, 169, 190, 197, 200, 202, 205, 207, 209. *See also* Dispute resolution, litigation
Disparate bargaining power, 105-106, 118
Dispute resolution, 103, 186-213
  ADR by covenant, 199, 213
  adversarial, 194-195, 197, 211
  advisory arbitration, 204
  alternative, 101-102, 103, 186-213
  appeals, 194, 196, 203, 204-205, 209
  arbitration, 202-203, 204
  benefits of ADR, 196-197, 208
  binding, 196
  binding mediation, 211-212
  conciliatory, 194-195, 200, 211
  confidentiality, 196, 206, 209
  contract requirements, 197
  cost of, 196, 200, 206, 213
  court-annexed arbitration, 204-205
  court-appointed masters, 205-206
  court-ordered arbitration, 204-205
  discovery, *see* Discovery
  expedited binding arbitration, 203-204
  informal spontaneous negotiation, 199-200, 211
  joinder, 195, 203
  litigation, 101-102, 112, 186-194, 208
  mandatory, 195
  mandatory binding arbitration, 201
  mandatory nonbinding arbitration, 204
  mandatory pretrial negotiation, 201
  mediation, 210
  mediation/arbitration, 211-212
  mediation/arbitration II, 212
  mediation-then-arbitration, 212
  Michigan Mediation, 207
  mini-trials, 205, 208, 209
  neutral, 195, 196, 197, 204, 207, 208, 211, 212
  neutral preselection, 202-203, 204, 211, 212
  non-binding, 196
  100-day document, 212
  private litigation, 209, 212
  rent-a-judge, 209, 212-213
  resolution through experts, 213
  settlement masters, 205-206
  specialized arbitration, 202
  specialized binding arbitration, 203
  summary jury trial, 208-209
  trial on order of reference, 209
  voluntary, 195
  voluntary nonbinding arbitration, 204
  voluntary prehearing negotiation, 200-201
Disputes, 42, 53, 54, 58, 75, 103
Documentation, 129, 178-179
  importance of, 74, 183, 184
  meetings, 59
Dodge reports, 301
Double-envelope system, 31-32. *See also* Professional, engagement
Downside, discussion of, 17-18
DPIC Companies, 167, 210
Duty of care, 19, 56, 66, 68, 69, 70-71, 75, 100, 107, 133, 137, 180-181

Economics, of professional practice, 214-262
  accounting, *see* Accounting
  accounts receivable report, 254
  assets:
    current, 218, 250
    fixed, 215, 218-219
    liquid, 215, 218
  balance sheet, 250
  bonuses, 242
  borrowing, 215
  cash flow, 254
  charts of accounts, 220-233
  contingencies, 245
  contingency reserves, 215
  corporate structure, 216
  depreciation, 219-220
  direct labor, 235
  dividends, 216
  equity, 219
  expenses, 219, 233
  federal regulations, 243-244
  fee establishment, 9, 18, 26, 33, 34-35, 149, 175, 176, 244-249
  financial reporting, 249-254
  forecast of assets and liabilities, 254
  general and administrative expenses, 235
  indirect labor, 235
  liabilities:
    current, 219, 250
    long-term, 219
  line of credit, 254
  multiplier, 239, 240
  net worth, 219
  overhead, 35, 37, 135, 234-244
  payroll burden, 235
  profit, *see* Profit motive
  revenue, 219, 233
  taxes, 219-220, 243

value-based pricing, 247-248
Employee piracy, *see* Human resources management, employee piracy
Employee stock ownership plans, 282
Employment agencies, *see* Human resources management, personnel recruiters
Engagement of professionals, *see* Professional, engagement
Engineer, title of, 6
Engineers' Joint Contract Documents Committee, 97
Equity, interest in firm, *see* Human resources management, ownership opportunities
Errors and omissions, 15-16, 17, 19, 29, 50, 56, 59, 60, 61, 68, 70, 74, 78, 82, 105, 122, 131-132, 145, 169, 217-218. *See also* Negligence
Establishment, the, 12-13
Ethics, code(s) of, 7-9, 14, 90, 91
Excluded services, 121, 127-128
Exculpatory wording, 111-112
Execution of professional commissions, *see* Professional commissions, execution of
Executive search firms, *see* Human resources management, personnel recruiters
Exit interviews, *see* Human resources management, exit interviews
Experience, 44
Expert services, 18, 20, 71-72, 75, 150, 187, 190, 191-192, 201, 207, 213, 249, 337
Express negligence, doctrine of, 109, 130
Express warranty, *see* Warranty, express
Extralegal conditions, 103

Facsimile machines, 101
Fact sheets, 321
Fast-track construction, 134-135, 166
Federal government as client, 243-244
Fee-based professional procurement, *see* Professional, engagement
Fee-bidding, *see* Professional, engagement
Feedback network, 178-179
Fee inflation, 175-176
Fee premium, for risk funding, *see* Risk funding
Fee-quality relationships, 9, 16, 18, 23, 29, 33, 35, 149, 175-176, 308, 330
Fees, *see* Economics, of professional practice, fee establishment
Felonies, 65
Field observation, 23, 61-62, 77, 80, 81, 105-106, 121-125, 141, 146-147, 175, 182
Field personnel, 59, 182

Financial Accounting Standards Board, 219
Financial management, 214, 249-254. *See also* Economics, of professional practice
Foreseeability, 71, 73
Freedom to report, 128-130, 182
Frivolous suits, 12, 68, 75, 150
Functional specificity, 6

General Accounting Office, 33
General conditions, 52-53, 58, 61, 95, 96, 99, 103, 113
Goals, of business, *see* Marketing, business goals
Golden Rule, 22, 24, 90, 273, 332
Graphics, *see* Marketing communications, graphics
Greed, 20
Greenwood, Ernest, 4-12, 17, 22, 48, 316
Guild system, 1-3
 craftworker guilds, 2, 3
 decline of, 3
 extinction of, 3
 merchant guilds, 2-3
Gut reactions, 184

Hazardous materials:
 contract provisions, 126-127, 134
 definition of, 111
 insurance coverage, 151-152, 158
 risks of, 103, 124, 303
 services market, 173, 303
 site remediation, *see* Site remediation
 unanticipated discovery of, 126-127
Head hunters, *see* Human resources management, personnel recruiters
Health and safety, protection of, 8, 42, 56, 76, 118, 127, 134, 149, 180
Hired guns, *see* Expert services
Human resources management, 21, 262, 263-285
 advertising, 270
 career ladders, 279-280
 compensation, 270, 275, 281, 282
 culling, 270-272
 drug screening, 272, 276
 employee piracy, 267
 exit interviews, 273, 283
 former employees, relations with, 283-284
 hiring, 264-273
 interviews, 271-272
 job descriptions, 265-266, 273
 liability, 284-285
 mentors, 274

366 INDEX

Human resources management *(Continued)*
  moonlighting, 277
  motivation, 279, 328
  orientation and training, 264, 273–274
  ownership opportunities, 282–283
  performance reviews, 278–279, 284
  personality factors, 265
  personnel files, 284–285
  personnel policies, 274–278, 284
  personnel recruiters, 269–270
  promotions, 266
  recruitment, 265–270
  reference checking, 272–273
  sexual harassment, 276
  supervisor-to-employee communications, 278–279
  team building, 54, 56, 280–282
  termination of employment, 283–284
  turnover, 263–264, 267–269
  unjust dismissal, 275

Image, 263, 264. *See also* Professions
  development of, 317–318, 323, 331
  management of, 318–319, 323, 324, 329, 336, 337, 359
Implied warranty, *see* Warranty, implied
Incorporation by reference, 100
Indemnifications, 66, 106–111, 130–132
  client-proposed, 103, 106–109, 113, 114, 130–132
  design professional-proposed, 103, 109–111, 117, 124, 128, 161, 171–172, 175
  implied, 108
  partial, 57, 103, 158, 171, 175. *See also* Risk allocation
Information furnished by others, 144–145
Innovation, 30, 32, 123–124, 308
Institute of Chartered Accountants, 3
Institute of Real Estate Management, 332
Institution of Civil Engineers, 3
Institution of Mechanical Engineers, 3
Instruments of service, 77–78, 80. *See also* Plans and specifications
  ownership of, 141–144
Insurance:
  agents, 152–153, 154, 158, 160, 163, 164, 165, 183
  claims made, 158–160
  commercial general liability, 153, 157, 158
  contractor's general liability, 133–134
  contributory coverage, 163
  endorsements, 160
  establishing premiums, 163–168

  occurrence-based, 158, 160
  owner's protective policy, 108
  professional liability, *see* Professional liability insurance
  project, 160–161
  role of, 122
  self, *see* Self-insurance
  workers' compensation, 132–135, 281
Insureds, 152
Insurers, 153–155
  dispute resolution, 198
  formation of, 162–163
Interference with prospective economic advantage, 91
Interrogatories, 83, 169. *See also* Discovery
Interviews, of firm representatives, 26, 49–50

Job descriptions, *see* Human resources management, job descriptions
Jobs, *vs.* commissions, 41
Johnson, Lyndon B., 12
Joinder, *see* Dispute resolution, joinder
Joint and several liability, *see* Liability, joint and several
Journeymen, 2
Judge, *see* Trier of fact
Judicial image, 212
Jury, *see* Trier of fact
Justice, Department of, 9, 14

Kennedy, John F., 12
King, Dr. Martin Luther, 12

Labor unions, 2
Lawsuits against professionals, 20–21, 58, 75, 168–169, 187. *See also* Negligence
Lawyers, 3, 20–21, 67–68, 71, 92, 108, 112, 114, 169, 186, 187, 191, 192, 196, 197–198, 200–201, 202, 204, 206, 207, 208
Letterhead, *see* Marketing communications, graphics
Liabilities, financial, *see* Economics, of professional practice, liabilities
Liability:
  assumption of, 104–105, 116
  contractual, 104, 116, 131, 151, 157, 327–328
  employee, *see* Human resources management, liability
  joint and several, 19, 67–68, 108, 131, 150
  limitation of, *see* Risk allocation
  personal, 19, 152
  professional, 11, 14, 23, 37, 54, 62, 68–77, 82–83. *See also* Negligence; Professional

liability loss prevention
  strict, 77–78, 80–81, 115–118, 142
  tort, 104
  vicarious, 98, 150, 167
Life-cycle costs, 33
Limitation of liability, *see* Risk allocation
Litigation, *see* Dispute resolution
Litigious mania, 112, 156
Logos, *see* Marketing communications, graphics
Loss reserves, *see* Risk funding
Low-balling, 29, 34, 58

Magazine articles, *see* Marketing communications, magazine articles
Maintenance of service, 140–141
Marketing, 16, 42–43, 149, 179, 286–315
  annual analysis, 293–296
  assistance, sources of, 300–302
  business goals, 286–287, 288, 307, 309
  communications, *see* Marketing communications
  contingency planning, 315
  decision makers, 303
  forecasting, 305–307
  geographic marketing area, 290, 302, 311
  growth, 286–287
  market conditions, 288
  marketing research:
    external, 300–305
    internal, 290–300
  marketing units, 287–315, 319, 322–323, 330
  market penetration, 288
  market segmentation, 287–288, 289–290
  market share, 288
  niche marketing, 310, 323
  organizational research, *see* Marketing, marketing research, internal
  plan finalization, 311–315
  plan updating, 315
  project analysis, 291–293
  selecting future options, 309–311
  third-party analysis, 304–305
  trend analysis, 296–300
    external, 297–300
    internal, 296–300
Marketing communications, 269, 316–330. See also Professional communication
  advertising, 326, 328
  assistance, 327, 329–330
  association involvement, 321–322, 324–325, 331
  awards competitions, 326, 328
  brochures, 323, 325, 327–328, 330
  case histories, 327
  decision influencers, 320–322, 326
  decision makers, 319–322, 326
  direct selling, *see* Direct selling
  graphics, 323, 325, 329–330
  image, role of, *see* Image
  magazine articles, 326–328, 329, 330, 336, 337, 338, 357
  mailing lists, 325, 326, 329, 333, 334
  newsletters, *see* Newsletters, in-house
  news releases, 325, 326, 332, 333
  office appearance, 323–324
  plan, 319–322, 329
  promotion, 8, 286
  publics, 319–323, 326, 338
  public speaking, 38, 323, 325, 326, 336, 337, 338, 343, 358
  seminars, 325–326
  techniques of, 321–322, 323, 329
  telephone techniques, 324
  thank-you notes, 333
Maryland, 17, 28–29, 30, 31, 36, 45
Masters, 2
Mediation, *see* Dispute resolution
Meeting documentation, *see* Documentation, meetings
Memoranda, 178, 180, 182, 241
Merchant guilds, 2–3
Mergers, 220
Meritless claims, *see* Frivolous suits
Microfilm, 143
Misdemeanors, 65
Missing elements defense, 75
Monitoring construction, *see* Field observation
Moonlighting, *see* Human resources management, moonlighting
Multiple contracts, *see* Contracts, multiple
Mutual plan review, 16, 56–57, 58–59, 73
Mutual workscope development, 16, 23, 24, 25–28, 30

National Association of Home Builders, 324
National Society of Professional Engineers, 9, 16, 39, 97, 331, 357
National Union of Teachers, 3
Negligence, 66, 67, 68–77, 78, 81–82. *See also* Errors and omissions; Professional liability loss prevention
  claims of, 168–169
  comparative, 75, 131
  contributory, 75, 189

Negligence *(Continued)*
express, doctrine of, *see* Express negligence, doctrine of
professional, doctrine of, 68–76, 104, 116, 142
Negotiation, of claims, 15
Networking, 331–332, 334, 335. *See also* Marketing communications, association involvement
Neutral, *see* Dispute resolution, neutral
Newsletters, in-house, 283, 321, 328, 330
News media, 20–21, 55
News releases, *see* Marketing communications, news releases
Niche marketing, *see* Marketing, niche marketing
Norms of behavior, 65–66. *See also* Professional, norms

Opinion(s), 82, 84, 117, 120–121, 129
Opinion of probable construction cost, *see* Cost estimates
Organizational research, *see* Marketing, marketing research, internal
Orientation, of new employees, *see* Human resources management, orientation and training
Overhead, *see* Economics, of professional practice
Overtime, 177
Ownership transition, 282

Parole evidence rule, 191
Payment, 151, 187, 246–247
Payment schedules, 53
Peculiar risk, doctrine of, 151–152
Peer review, 11, 23
internal, 56, 175
Peremptory challenge, 191
Performance bonds, 100
Performance schedule, 54, 96, 113, 176–177, 181
Personal diplomacy, 199, 359
Personnel assignments, 176–177
Personnel management, *see* Human resources management
Plans and specifications, *see also* Instruments of service
conformance to, 61
copyright of, 142, 144
details, 56–58
development of, 56
implementation of, 62, 80, 121–125

interpretation of, 124–125
as products, 77–78, 80, 141–144
retention of, 142–143
review of, *see* Mutual plan review; Peer review
unauthorized reuse, 142, 143
Pleadings stage, 188–189. *See also* Dispute resolution, litigation
Pollution, 13. *See also* Hazardous materials
Pollution exclusion, 153, 164
Postconstruction-phase activities, 62–63, 73–74
Preacquisition site assessments, *see* Preliminary site assessments
Prebidding conferences:
for contractors, 58–59, 61
for design professionals, 30, 50
Preconstruction conference, 59–60, 61
Predesign conference, 54–55, 59
Predesign planning, 54–55
Preliminary site assessments, 320, 322
Premiums, *see* Insurance, establishing premiums
Press releases, *see* Marketing communications, news releases
Price competition, 8–9
Prime-interprofessional relationships, 42–45, 50–57, 97–98, 138, 140–141, 175, 180
Prior acts coverage, 158–159
Privilege, 87–88
Privity, lack of, 19, 70. *See also* Duty of care
Procurement of professionals, *see* Professional, engagement
Product liability, *see* Liability, strict
Professional:
arrogance, 6
associations, 10
authority, 5–6, 8, 17
behavior, *see* Professional, norms
culture, 9–10
definition of, 21–24
demeanor, 184
distinction, 11–12
engagement, 16–17, 18, 23, 25–39, 46–53, 54, 137, 256–257
image, 8, 11, 120, 316–317
judgement, 5–6
licensure, 6–7
monopoly, 9
norms, 9–10, 11, 316
regulation, 6–7, 9, 14
self-protection, 14
social values, 9–10

stereotypes, 10, 22
symbols, 9–10
Professional–client relationships, 14–15, 17–18, 21–24, 25, 27, 29, 31, 32, 36, 37, 38, 40, 49, 51–53, 76, 92, 101–102, 136, 182, 184, 257. *See also* Client relations
Professional commissions, execution of, 40–63
Professional communication, 336–359
  absolutes, 345, 349
  active *vs.* passive tense, 352–353
  circumlocutions, 342, 345, 347
  dangling modifiers, 354
  English prime, 355
  evaluation, 356–357
  first draft preparation, 340–342
  gender of references, 352
  Gunning's Fog Index, 356
  interpersonal, *see* Communications, interpersonal
  jargon, 349
  organizing, 338–340, 342
  precision, 349–350
  second draft preparation, 342–355
  sentence length, 353–354
  slipshod synonyms, 345, 347, 348
  subject–verb agreement, 354–355
  third draft development, 355–357
  useless words and phrases, 345, 346–347
  warnings and limitations, *see* Warnings
  word selection, *see* Word selection
  written-to-oral conversion, 358
Professional liability insurance, 20, 37, 53, 80, 105, 108–109, 117–118, 135, 148–169, 171
  captive insurers, 162–163
  deductibles, 161, 163–168
  endorsements, 117–118, 160
  establishing premiums, 163–168
  insurer formation, 162–163
  limits, 163–168
  premium credits, 167
  prior acts, 158–159, 164
  project coverage, 160–161, 164, 166
  risk retention groups, 162–163
  underwriting, 167
Professional liability loss prevention, 44, 46, 51, 53, 54, 56, 62, 76, 79, 90, 92, 96, 100, 170–185, 273–274, 307–309, 324, 329, 349, 350
Professional-to-professional relations, 90
Professionals, suits against, *see* Professional liability loss prevention

Professional service, complete, 123–125
Professional service delivery systems, 14
Professional services, composition of, 40–63
Professions:
  attributes of, 4–10
  duty to, 11
  evolution of, 1–24
  formation of, 3–4
  policing of, 14
  regulation, 6–7, 9, 14
  rise of, 3
Profit motive, 8, 19–20, 214–218, 245
Project budget, 257–262
Project management reports, 254–262
Project managers, 49, 54, 55, 56–57, 245, 246, 292–293
Project notebook, 55
Project reports, 257–262
Project selection, 173–175
Project summary report, 258–259
Promotion, 310. *See also* Marketing communications
Pronouns, 111
Property managers, 38, 320, 330, 331
Proposal meeting, 44, 46–47
Proposal-phase services, 42–53
Proposals, 40, 46–47, 326, 327, 330, 339, 345, 356–357. *See also* Professional, engagement; Professional communication
  competitively priced, 99
  contract format, 95–96
  cost of, 25, 47
  development of, 265, 350, 353
  fee, 95–96
Proximate cause, 72–74
Public:
  health and safety, *see* Health and safety, protection of
  recognition, 3
  scrutiny, 11
  sentiment, 11
  support, 3, 11
  trust, 3, 9, 11, 13, 20, 21
Public relations, *see* Marketing communications
Publics, *see* Marketing communications, publics
Public speaking, *see* Marketing communications, public speaking
Punitive damages, 66–67, 86, 116, 131, 132
Purchase order(s), 99

Qualifications, statement(s) of, 26, 30
Qualifications-based selection, *see* Professional, engagement
Quality control, 17, 23, 35, 51, 54, 56, 62, 176, 185, 241
Quality-fee relationships, *see* Fee-quality relationships
Quality management, *see* Quality control

Rationality, 4–5, 48
Real estate agents, 320–321, 322
Real pro, 21–24
Reasonable person doctrine, 66
Recommended Practices for Design Professionals Engaged as Experts for the Resolution of Construction Industry Disputes, 72
Record documents, 144–146
Records retention, 142–143
Recruitment, of personnel, *see* Human resources management, recruitment
Referrals, 8, 174, 293, 316, 331, 332
Registration, boards of, 9, 10, 14, 18
Rehabilitation, 124, 126–127
Reinsurance, 107, 149, 154–155, 162
Remediation monitoring, *see* Field observation
Report preparation, 341, 345, 350
Reports, 179–180, 241
Request(s) for proposal (RFPs), 99, 300
Request(s) for quotation (RFQs), 30, 38, 99
Restraint of trade, 9, 10, 14, 16
Retirement planning, 216
Risk allocation, 17–18, 52–53, 95, 102, 103, 119, 125, 135–140, 158, 161, 171–172
Risk funding, 107–109, 113, 118, 137–140, 172–173, 247
Risk management, 27, 43, 51, 53, 76, 83, 96, 102–113, 127, 134–135, 136, 142, 157, 161–162, 167, 170–185. *See also* Professional liability loss prevention
Risk retention, *see* Risk funding
Risk retention groups, 162–163
Risk transfer, 107–112, 118, 136–137, 158, 161, 171–172
Rotary International, 38, 331
Royal College of Physicians, 3, 7
Royal College of Veterinary Surgeons, 3
Royal Institute of British Architects, 3

Safety, protection of, *see* Health and safety, protection of
Salary, *see* Human resources management, compensation

Sampling, 82
Sanction(s) of the community, 1–2, 6–7, 9, 14, 21, 37
Schedule, *see* Performance schedule
Scope(s) of service(s), 20, 23, 40–63, 66, 83, 96, 113, 128, 245, 308. *See also* Mutual workscope development; Unilateral workscope(s)
Self-insurance, 118, 148–149, 159, 161–162
Seminars, *see* Marketing communications, seminars
Separate contracts, *see* Contracts, multiple
Serendipity factor, 320, 333
Services, scope of, *see* Scope(s) of service(s)
Service sector, of economy, 3–4
Settlement hearing, 190
Sexual harassment, *see* Human resources management, sexual harassment
Shop drawing review, 17, 23, 56, 60–61, 175
Short list, 48
Shotgun suits, 68, 134
Sick building syndrome, 69–77
Single-envelope system, 31. *See also* Professional, engagement
Site remediation, 61
Site safety, 62, 106, 132–135, 146
Slander, *see* Defamation
Slow pay, *see* Payment
Social values, of a profession, 9
Society:
    needs of, 2–3, 11, 76
    sanction of, *see* Sanction(s) of the community
Specialization, 5, 14, 41
Spirit of rationality, *see* Rationality
Staff education, 177–179
Staff maintenance, 45
Standard forms, 46, 254, 255
Standard(s), 8, 30, 73
    of care, *see* Standard(s), of practice
    of practice, 11, 18–19, 41, 54, 57, 69–70, 71–72, 74, 75, 81, 82–83, 116
Statements of qualifications, *see* Qualifications, statement(s) of
Stationery, *see* Marketing communications, graphics
Statutes of limitations and repose, 19, 76–77, 93, 102, 103, 137, 142–143
Statutory law, 64–65
Stipulation, 188
Stock option plan, 282
Stop-work authority, 133, 146–147
Subpoenas duces tacum, *see* Discovery
Surveyors Institute, 3

Systematic body of theory, *see* Body of theory

Taxes, 6
Tax laws, 301
Team building, *see* Human resources management, team building
Team orientation, 41, 54
Telephone, proper usage of, 178, 278, 324
Television, 12-13
Terra Insurance Company, 162
Third-party beneficiaries, 150-151
Third-party claims, 19, 112, 139-140, 160, 180, 181-182
Tort law, *see* Civil justice system
Tort liability, *see* Liability, tort
Totally clean rule, the, 75
Trend analysis, *see* Marketing, trend analysis
Trier of fact, 64, 71, 187-194, 195, 201, 205, 207, 209, 211
Turnkey projects, 77, 81
Turnover, *see* Human resources management, turnover

Unanticipated conditions, 181
Unanticipated expense, 74-75, 119, 181
Underground storage tanks, 320-322, 325

Underwriters, 153-154
Unilateral workscopes, 16, 28-32, 37, 46-53, 99. *See also* Professional, engagement
U.S. Supreme Court, 12, 16
Unjust enrichment, 74-75, 83

Vicarious liability, *see* Liability, vicarious
Vietnam War, 12-13
Voir dire, 191, 192

Warnings, 76, 84, 337, 350-352
Warranty, 78-81, 123
   breach of, 78-81, 118
   express, 79-80, 81
   implied, 80-81
Warren, Earl, 12
Wealth, accumulation of, 3
White collar workers, 3-4
Word selection, 111, 144-145, 327-328, 329, 340, 342-353, 355. *See also* Professional communication
Work rejection, *see* Stop-work authority
Workers' compensation insurance, *see* Insurance, workers' compensation
Workscope(s), *see* Scope(s) of service(s)
Writing, *see* Professional communication